Lecture Notes in Earth Sciences 119

Huilin Xing

# Advances in Geocomputing

Springer

Dr. Huilin Xing
The University of Queensland
Earth Systems Science Computational Centre (ESSCC)
St. Lucia, Brisbane, QLD 4072
Australia
h.xing@uq.edu.au

Additional material to this book can be downloaded from http://extras.springer.com
For all Lecture Notes in Earth Sciences published till now please see final pages of the book

ISBN: 978-3-662-51867-0 ISBN: 978-3-540-85879-9 (eBook)

DOI 10.1007/978-3-540-85879-9

Lecture Notes in Earth Sciences ISSN: 0930-0317


Softcover reprint of the hardcover 1st edition 2009



*Cover design:* deblik, Berlin

Printed on acid-free paper

9 8 7 6 5 4 3 2 1

springer.com

# Preface

Numerical modelling is being an advanced tool in geoscience and geoengineering. Idealized experiments and field observations have been the main pillars of geoscience for decades, while the rapid development of supercomputers leads to a paradigm shift towards geocomputing. High-performance computing based simulations offer outstanding opportunities to get insights into increasingly complex geoscience and geoengineering problems.

Several new institutes and initiatives with special emphasis on high-performance geocomputing have been established around the world, such as ACcESS MNRF (Australian Computational Earth System Simulator, Major National Research Facility, http://www.access.edu.au) and AuScope (an organisation for a National Earth Science Infrastructure Program, http://www.auscope.org.au) in Australia; GEON (GEOscience Network, http://www.geongrid.org/), PRAGMA (Pacific Rim Applications and Grid Middleware Assembly, http://www.pragma-grid.net/ ), SERVO (Solid Earth Virtual Research Observatory, http://www.servogrid.org/), PetaSHA (Petascale Cyberfacility for Physics-based Seismic Hazard Analysis from Southern California Earthquake Center, including TeraShake etc. platforms, http://scecdata.usc.edu/petasha/) and CIG (Computational Infrastructure for Geodynamics, http://www.geodynamics.org) in the United States; GeoFEM (http://geofem.tokyo.rist.or.jp) and CHIKAKU system (http://www.riken.go.jp/lab-www/CHIKAKU/index-e.html), and the Earth Simulator Center (www.es.jamstec.go.jp/) in Japan; the Laboratory of Computational Geodynamics of Chinese Academy of Sciences in China; and the iSERVO seed project (iSERVO-international Solid Earth Virtual Research Observatory, http://www.iservo.edu.au) aims to foster ongoing international cooperation on simulation of solid earth phenomena. iSERVO is the natural follow-on to ACES (APEC Cooperation for Earthquake Simulation, http://www.aces.org.au).

This book provides a concise overview of the recent developments in geocomputing, covering model construction, advanced computational theory, visualization of the results, and high-performance software development on supercomputers. We present applications spanning the different temporal and spatial scales of geoscience. Those exemplary simulations focus on topics from geodynamics, crustal dynamics, earthquakes, tsunami and rock physics. The book is composed of 8 chapters written by 35 authors from 6 countries – Australia, China, Germany, Japan, Switzerland and the Unites States, which reflected the current state-of-the-art achievements and the future research direction in the field. All mention of colour in the legends can be seen only in the enclosed DVD-ROM and in the online version in addition to the animation files of the amazing simulation results.

The collection of topics aims to reflect the diversity of recent advances in geocomputing. Such a broad perspective may be useful for scientists as well as for graduate students in geophysics, geology, geochemistry, computational science, environmental and mining engineering, and software engineering. I hope this book will be relevant and valuable to the whole geoscience community and serve to both define and advance the state of geocomputing.

The last but not the least, I would like to express my deep appreciation to all the authors and reviewers for their outstanding contribution to this book, and to Professor Dave Yuen of University of Minnesota, Dr. Chris Bendall and Janet Sterritt-Brunner of Springer for their kind encouragement and help to have such diverse topics on geocomputing published as a book.

*Huilin Xing*

The University of Queensland, Australia

# Contents

# Contributors

John R. Baumgardner
University of California, Department of Earth & Planetary Science, 307 McCone Hall, Berkeley, CA 94720-4767, USA, jrbaumgardner@cox.net

Amit Chourasia
San Diego Supercomputer Center, 9500 Gilman Drive, MC0505, La Jolla, CA 92093, USA, amit@sdsc.edu

Yifeng Cui
San Diego Supercomputer Center, 9500 Gilman Drive, MC0505, La Jolla, CA 92093, USA, yifengcui@gmail.com

Florian Fusseis
Earth & Geographical Sciences, University of Western Australia, WA 6009, Australia, fusseis@cyllene.uwa.edu.au

Oliver Gaede
Earth & Geographical Sciences, University of Western Australia, WA 6009, Australia, gaede@cyllene.uwa.edu.au

Klaus Gessner
CSIRO Exploration and Mining, Bentley, WA 6102, Australia; Earth & Geographical Sciences, University of Western Australia, WA 6009, Australia, kgessner@cyllene.uwa.edu.au

Klaus-D. Gottschaldt
University of Queensland, ESSCC, PO Box 6067, St Lucia, QLD 4067, Australia; now at: Deutsches Zentrum für Luft- und Raumfahrt, Institut für Physik der Atmosphäre, 82234 Oberpfaffenhofen, Germany, klausgottschaldt@web.de

Bruce Hobbs
CSIRO Exploration and Mining, Bentley, WA 6102, Australia; Earth & Geographical Sciences, University of Western Australia, WA 6009, Australia, bruce.hobbs@csiro.au

Thomas Jordan
University of Southern California, 3651 Trousdale Parkway, Los Angeles, CA 90089, USA, tjordan@usc.edu

Jie Liu
CSIRO Exploration and Mining, Bentley, WA 6102, Australia, Jie.liu@csiro.au

Mian Liu
University of Missouri, Columbia, MO 65211, USA, lium@missouri.edu

Yingchun Liu
Graduate University of Chinese Academy of Sciences, Beijing, China; Department of Geology & Geophysics and Minnesota Supercomputing Institute, University of Minnesota at Twin Cities, Minneapolis, MN 55455, USA, spring.yingch@gmail.com

Philip Maechling
University of Southern California, 3651 Trousdale Parkway, Los Angeles, CA 90089, USA, maechlin@usc.edu

Reagan Moore
San Diego Supercomputer Center, 9500 Gilman Drive, MC0505, La Jolla, CA 92093, USA, moore@sdsc.edu

Peter Mora
ESSCC, The University of Queensland, St. Lucia, QLD 4072, Brisbane, Australia

Gabriele Morra
ETH Zürich, Institute of Geophysics, 8093, Hönggerberg, Switzerland, gabrielemorra@gmail.com

Hans B. Mühlhaus
University of Queensland, ESSCC, PO Box 6067, St Lucia QLD 4067, Australia, CSIRO Exploration and Mining, Bentley, WA 6102, Australia, h.muhlhaus@uq.edu.au

Kengo Nakajima
Information Technology Center, The University of Tokyo, 2-11-16 Yayoi, Bunkyo-ku, Tokyo 113-8658, Japan, nakajima@eps.s.u-tokyo.ac.jp

Kim Olsen
San Diego State University, 5500 Campanile Drive, San Diego, CA 92182, USA, kbolsen@sciences.sdsu.edu

Alison Ord
CSIRO Exploration and Mining, Bentley, WA 6102, Australia; Earth & Geographical Sciences, University of Western Australia, WA 6009, Australia, Alison.ord@csiro.au

Thomas Poulet
CSIRO Exploration and Mining, Bentley, WA 6102, Australia; Earth & Geographical Sciences, University of Western Australia, WA 6009, Australia, Thomas.poulet@csiro.au

Klaus Regenauer-Lieb
CSIRO Exploration and Mining, Bentley, WA 6102, Australia; Earth & Geographical Sciences, University of Western Australia, WA 6009, Australia, Klaus@cyllene.uwa.edu.au

Gideon Rosenbaum
School of Physical Sciences, University of Queensland, Brisbane, QLD 4072, Australia, g.rosenbaum@uq.edu.au

Erik O. D. Sevre
Department of Geology & Geophysics and Minnesota Supercomputing Institute, University of Minnesota at Twin Cities, Minneapolis, USA, esevre@gmail.com

Yaolin Shi
Graduate University of Chinese Academy of Sciences, Beijing, China, shiyl@gucas.ac.cn

Delphine Siret
CSIRO Exploration and Mining, Bentley, WA 6102, Australia, delphine.siret@free.fr

Dave R. Stegman
Earth Sciences, The University of Melbourne, Victoria 3010, Australia, dstegman@unimelb.edu.au

Uwe Walzer
Friedrich-Schiller-Universität Jena, Institut für Geowissenschaften, Burgweg 11 07749 Jena, Germany, u.walzer@uni-jena.de

Yucang Wang
ESSCC, The University of Queensland, St. Lucia, QLD 4072, Brisbane, Australia, wangyc@esscc.uq.edu.au, yucang_wang@hotmail.com

Roberto Weinberg
School of Geoscience, Monash University, Clayton, VIC 3800, Australia, Roberto.Weinberg@sci.monash.edu.au

Huilin Xing
The University of Queensland, Earth Systems Science Computational Centre, QLD 4072, Australia, xing@esscc.uq.edu.au, h.xing@uq.edu.au

Youqing Yang
University of Missouri, Columbia, MO 65211, USA, yangyo@missouri.edu

Wenhui Yu
The University of Queensland, Earth Systems Science Computational Centre, QLD 4072, Australia & Department of Engineering Mechanics, Dalian University of Technology, Dalian, China, yuwenhui1981@163.com

David A. Yuen
Department of Geology & Geophysics and Minnesota Supercomputing Institute, University of Minnesota at Twin Cities, Minneapolis, MN 55455, USA, daveyuen@gmail.com

Ji Zhang
The University of Queensland, Earth Systems Science Computational Centre, QLD 4072, Australia, ji.zhang@uq.edu.au

# I. First Steps Towards Modeling a Multi-Scale Earth System

Klaus Regenauer-Lieb,[1,2] Thomas Poulet,[1,2] Delphine Siret,[1] Florian Fusseis,[2] Jie Liu,[1] Klaus Gessner,[1,2] Oliver Gaede,[2] Gabriele Morra,[3] Bruce Hobbs,[1,2] Alison Ord,[1,2] Hans Muhlhaus,[1,4] David A. Yuen,[5] Roberto Weinberg[6] and Gideon Rosenbaum[7]

[1]CSIRO Exploration and Mining, Bentley, WA 6102, Australia
[2]Earth & Geographical Sciences, University of Western Australia, WA 6009, Australia
[3]ETH Zürich, Institute of Geophysics, 8093, Hönggerberg, Switzerland
[4]ESSCC, The University of Queensland, St. Lucia, QLD 4072, Australia
[5] Department of Geology & Geophysics and Minnesota Supercomputing Institute, University of Minnesota at Twin Cities, Minneapolis, MN 55455, USA
[6]School of Geoscience, Monash University, Clayton, Victoria 3800, Australia
[7]School of Physical Sciences, University of Queensland, Brisbane, QD 4072, Australia

Recent advances in computational geodynamics are applied to explore the link between Earth's heat, its chemistry and its mechanical behavior. Computational thermal-mechanical solutions are now allowing us to understand Earth patterns by solving the basic physics of heat transfer. This approach is currently used to solve basic convection patterns of terrestrial planets. Applying the same methodology to smaller scales delivers promising similarities between observed and predicted structures which are often the site of mineral deposits. The new approach involves a fully coupled solution to the energy, momentum and continuity equations of the system at all scales, allowing the prediction of fractures, shear zones and other typical geological patterns out of a randomly perturbed initial state. The results of this approach are linking a global geodynamic mechanical framework over regional-scale mineral deposits down to the underlying micro-scale processes. Ongoing work includes the challenge of incorporating chemistry into the formulation.

For this, we use computational experiments on micro-scale processes and build a Preliminary Reference Earth Material Database PreMDB. Gibbs

H. Xing, *Advances in Geocomputing*, Lecture Notes in Earth Sciences 119,
DOI 10.1007/978-3-540-85879-9_1

energy minimization techniques are used to solve equilibrium compositions in chemistry and to predict the basic mechanical properties from chemistry. Physical properties that cannot be extracted directly from the thermodynamic potential functions are complemented by empirical data. The next level of models concerns itself with a homogenization of the mechanical properties to larger scale reproducing micro-chemical and microstructural observations. The predictive power of these models is currently tested based on field data from mineral deposits at micro-decameter scale. The next steps will be to up-scale the approach to meter and kilometer scale. The global scale modelling will provide better forward simulations for the genesis of giant ore/mineral deposits and other processes of global interest. The approach presented here is intended as a first step for such future cross-scale simulations in geology. Advanced multi-scale formalisms are beyond the scope of this paper.

## 1 Introduction

When applying laboratory data directly to Earth System modeling it is impossible to reproduce key observations and investigate a number of apparent paradoxes, such as:

(1) The subduction initiation paradox; the generation of weak trans-lithospheric faults requires special pleading in classical models (McKenzie, 1977).
(2) The Brace-Goetze (Christmas tree) crustal strength paradox; the continental lithosphere is found to be too strong. Cold continental breakup (for surface heat flow < 60 $mW/m^2$) is not possible under normal geodynamic forcing (Kusznir and Park, 1984a).
(3) The mid-crustal detachment paradox (Axen and Selverstone, 1994); weak crustal detachments are observed exactly where classical strength envelopes predict a strength maximum.
(4) The jelly sandwich paradox (Jackson, 2002); the upper mantle fails to present significant strength and does not deform in a seismogenic manner.
(5) The upper plate paradox (Kusznir, 1991; Weinberg et al., 2007); the brittle crust is deforming much less than the ductile lower crust and mantle.

These paradoxes prompt us to rethink about simple extrapolations of the laboratory strength estimates. Clearly we must improve in our way of modeling the lithosphere. We postulate here, that most, if not all these paradoxes are derived from the fact that numerical models do not take into account feedback effects within a fully coupled momentum-energy-continuity system. We will show that the feedback between deformation, heat production and the mechanical response of the system can resolve these paradoxes. Additional feedback (e.g. melt generation, fluid release) might be important, but are not required to resolve the paradoxes and are therefore considered to be future refinements. Here we will formulate only a simple formulation showing that this necessarily leads to a multi-scale approach.

## 2 Multiscale Non-Equilibrium Thermodynamics

### 2.1 The Equilibrium Yardstick

Predicting the way the Earth works from a fundamental physics based approach is a present challenge in computational geophysics. At conceptual level, many approaches have been suggested (Ben-Zion and Sammis, 2003; Fleitout and Froidevaux, 1980; Ord and Hobbs, 1989; Regenauer-Lieb and Yuen, 2003; Yuen et al., 1978), but there has been no development of a clear roadmap for the practical implementation of this approach.

The key to coupling length and time scales is the identification of specific scales relevant for Earth dynamic processes. Candidate for the large scale is the thermal diffusion length scale, which potentially provides a minimum equilibrium yardstick equivalent to a quantum energy state for earthquakes. Microstructure evolution on the other hand relies on a much smaller scale which is of the order of the chemical diffusion length scale. This suggests that thermal-mechanical modelling must explore the equilibrium of these chemical gradients. The chemical diffusion length scale may become the dominant equilibrium yardstick. Although considerable work has been devoted to exploring these approaches in the past, the science of multi-scaling in thermodynamics has not yet made a breakthrough in geology. This is chiefly because coupled thermodynamic modelling is computationally demanding and has not yet become state of the art in the geoscience.

The coupled energy approach has its natural antecedents in planetary scale convection simulations. Figure 1 shows that the same approach applied to smaller scales is promising for exploring pattern formation at these scales. At the large planetary scale, pattern formation is calculated by solving the problem of how a planet looses heat. It is well known that such a planet can reach a critical energy state where convection transfers heat more efficiently than conduction. This occurs when the positive feedback given by the product of buoyancy forces and heat advection overcomes the negative feedback defined by the product of viscous forces and heat conduction. A fully coupled momentum-energy-continuity equation calculation can resolve this instability. Nonlinear feedbacks lead to the onset of convection, meaning that convection emerges self-consistently. The important and underlying assumption for calculating this phenomenon is that any scale(s) below the dominant wavelength for growth of Rayleigh-Taylor instabilities are unimportant for the evolution of these instabilities. Processes at this scale are approximated by an effective viscous rheology. The dominant wavelength for Rayleigh-Taylor becomes the equilibrium yardstick for calculating planetary convection. Note that this distinction of scales is arbitrary; other scales may just not be resolved and therefore parameterized in the model and they could be still important. However, this approach is now well established in computational geodynamics.

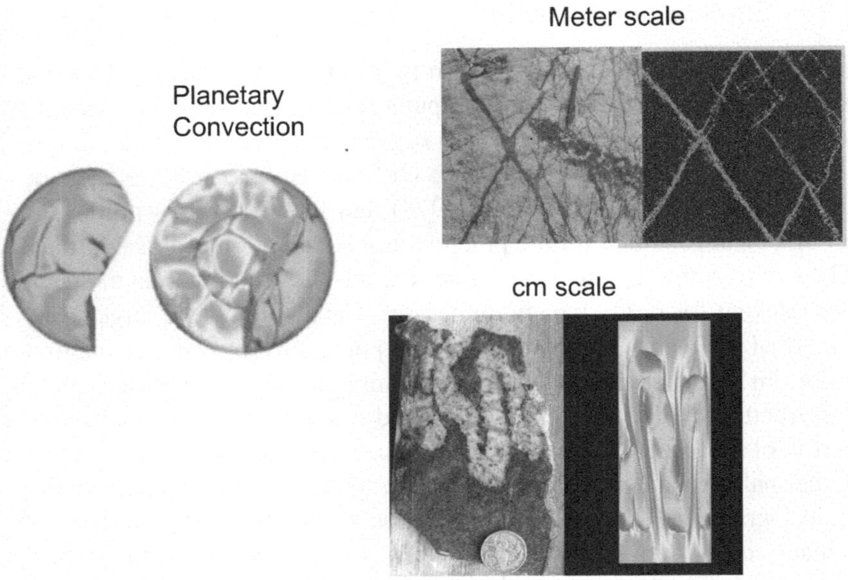


**Fig. 1** Far from equilibrium processes and Earth patterns. Computational thermal-mechanical solutions are now allowing us to understand pattern formation from consideration of basic physics and chemistry. However, a common framework is still lacking. The planetary convection simulation shows the temperature field inside the Earth as resulting from a spherical geodynamical mantle convection model (Bunge et al., 1997). We are proposing here to derive similar Earth patterns at smaller scale emerging out of random perturbations of the basic energy fluxes. Our, first such thermal-mechanical self-consistent results are compared with structures observed in nature. Photographs show conjugated Cu-veins and a cm scale fold (courtesy of Yanhua Zhang and Andy Tomkins, respectively). The attached movie material (Gosford.avi + fold.avi, available on accompanying DVD) show simulation of meter scale fracturing of Gosford sandstone and cm-scale folding, respectively. The Gosford movie shows a particle simulation where particles cracks either in tension (*red*) or shear (*yellow*). With increasing vertical loading, the primary shear zones emerge through the centre of the specimen, with further damage zones, of similar orientation but of lesser scale, emerging off-centre, and reflecting off the boundaries (Movie courtesy of Yanhua Zhang). The cm-fold shown in the other movie emerge out of thermal-mechanical simulation for an initially perfectly layered feldspar-quartz composite (contours show strain). If thermal-mechanical feedback is switched off, the same simulation shortens homogeneously by pure shear

Patterns at meter scale can also be obtained by fully coupling the mechanics to the energy equation. The particle simulation in Fig. 1 shows a surprising similarity to fracture patterns observed in nature. Here the equilibrium yardstick is the grain size and it is assumed that there is no property change below this scale. A continuum thermodynamic approach applied to cm scale folding, also leads to convincing results. Here, the equilibrium yardstick is the diffusion of some chemical species. The continuum mechanical approach needs to resolve this length scale and the physics of chemical diffusion in order to calculate the patterns emerging out of shortening a randomly perturbed stratified layer. While these results may be qualitatively appealing they do not provide a unified framework.

## 2.2 Non Equilibrium Thermodynamics and Multiscaling

Most natural processes are non-equilibrium processes. A local equilibrium assumption can be used as a first step towards a multi-scale computation. The assumption on the local equilibrium is an approximation but it holds for many systems. This assumption is valid if it is possible to distinguish two characteristic time scales, that is, the time required to reach the equilibrium in the entire system and the time required to reach the equilibrium in some volume, which is small compared to the size of a system under study.

A brute force computational non-equilibrium method is a better choice, but it will rely on efficient numerical schemes which allow an adaptive multi-scale resolution. Such methods are under development but are not available yet. The local equilibrium assumption works particularly well on a geological time scale because of the logarithmic relationship between relevant processes and their time scales indicated in Fig. 2. The separation of scales relies on the "multi-physics" nature of their underlying processes, spanning from molecular dynamics at the microscale to continuum mechanics at planetary scales. A detailed description of the underlying processes including a summary of numerical approaches can be found in Tables 1, 2 in (Regenauer-Lieb and Yuen, 2003). The large separation of scales has led to a separation of scientific disciplines which are: nanochemistry dealing with atomic scales; structural geology and laboratory physics for analysing microstructures at grain size scale; field geology, structural geology and seismology for analyzing fold and fault length scales; geodynamics for plate tectonic scale and planetary physics for planetary convection scales.

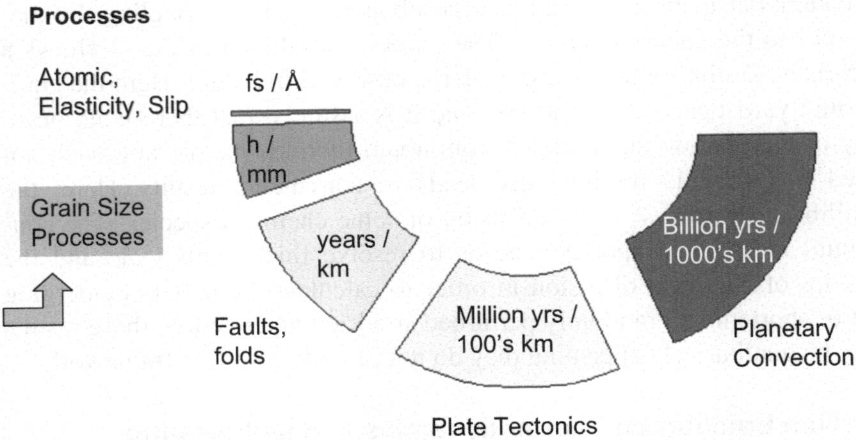


**Fig. 2** Time and spatial processes are coupled. For the equilibrium calculation of elastic constants, time steps of molecular dynamics simulations are on the order of half a femtosecond and the spatial scale is on the order of Angstrom resolution. At the grain size scale with hours of deformation, these elastic properties may be assumed to have reached local equilibrium but faults are out of equilibrium. The same principle may be applied in a staggered sense to the larger plate tectonics and planetary convection scales. An example is highlighted in the figure, namely that of grain size elasticity observed in the laboratory over time scales of seconds to hours, where it can safely be assumed to be in equilibrium. As a simple approach we suggest to derive equilibrium properties for the next larger scale by relaxation to equilibrium calculations at the smaller scale.

The staggered solution method presented here is conceived as the most basic approach towards crossing the traditional scale separations. In doing so the mathematical approach to "multi-physics" interactions is boiled down to a continuum mechanics framework extended by chemistry introducing heterogeneity at grain size scale. We wish to emphasize that such a framework is at best applicable to the "real world" by "nudging" of solutions to observations at various scales. This complication arises because the relaxation to equilibrium at small scale is a function of the large scale hierarchical driver, i.e. different local equilibrium states exist for different large scale out of equilibrium boundary conditions. This important macro-microscale feedback is not yet implemented. The approach is, intended as the first step towards a "heterogeneous multi-scale method" (Enquist and Huang, 2003) which would overcome the deficiency.

## 2.3 Coupling Mechanics and Chemistry

The simplified local equilibrium approach is a simple avenue for coupling chemistry with mechanics. In order to do so we propose to use local equilibrium for chemical processes and derive material properties from Gibbs free energy minimization method. Reversible material properties such as thermal expansion coefficient, specific heat, elastic shear modulus, bulk modulus and density, can be directly calculated from this method. These material properties are thus derived self-consistently from thermodynamics.

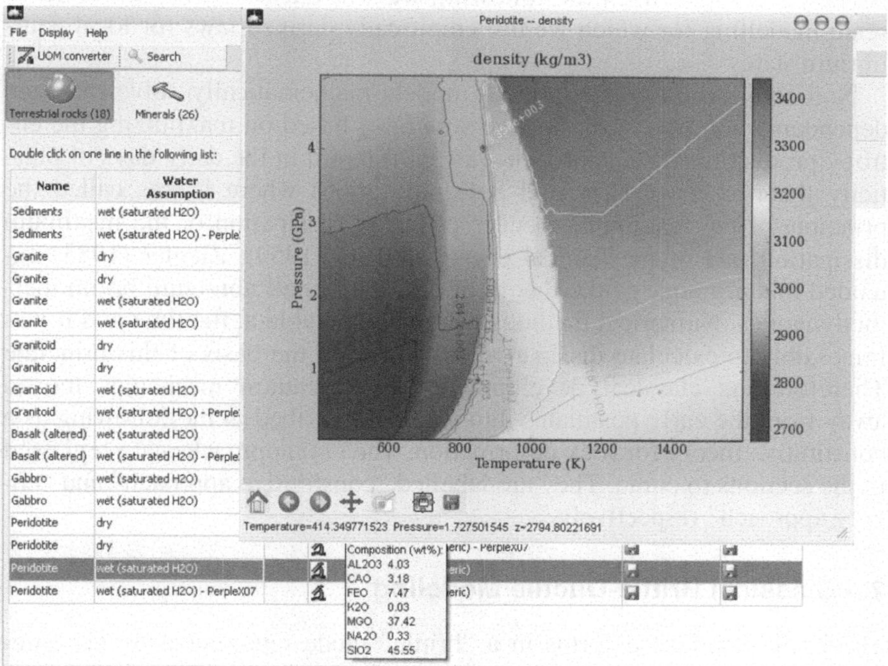


**Fig. 3** An example of the density at local equilibrium given as a function of pressure and temperature from PreMDB (Siret et al., 2008). The equilibrium density is predicted from Gibbs free energy minimization for the given chemical composition (Peridotite, wet). In addition to density, thermal expansion coefficient, specific heat, elastic shear modulus and bulk modulus can also be derived from the chemical equilibrium

Currently the compositions of 48 major rock forming dry and wet minerals and nine terrestrial rocks have been incorporated into a reference database Preliminary Reference Earth Material Database PreMDB, representing a standard for the sedimentary part of the crust, the upper and lower continental crust, oceanic crust and mantle (pyrolite and peridotite). A total of 20 material properties are obtained and prepared for coupling with finite element models to run non-equilibrium geological, geotechnical and geodynamical models driven by chemical and thermal gradients. Work on an implementation of this chemical solver for larger scales is still ongoing. Of particular concern is revisiting the behaviour of polyphase rocks as predicted from homogenization calculations of its mineral constituents (Tullis et al., 1991) and its chemical and thermal gradients. The goal of this sub-millimetre (microstructural) – decimetre (meso-) scale work is the derivation of better flow laws for the metre (macro-) scale (Fig. 2). This work is aimed at allowing more realistic coupling of chemistry and mechanics but is not yet completed. In the following we will only discuss plate tectonic scale modelling for which we use empirically derived laws for local equilibrium states.

Non-equilibrium computational models mathematically solve the time dependent evolution of dissipative structures based on maximizing the entropy production. This approach was put forward in the early days of plasticity theory (Martyushev and Seleznev, 2006) where it was called the principle of the maximum specific power of dissipation or the maximum dissipation rate of mechanical energy (Prager, 1959). Ziegler (1983) extended this principle of the theory of plasticity to all non-equilibrium thermodynamics. Numerical methods were not available at the time and it was impossible to calculate dissipative structures on the basis of this principle Therefore the classical developments in continuum mechanics moved away from the early postulates into what is described in the following as a constitutive theory for rock deformation. The two approaches are reviewed in the sections to come. They are labelled "constitutive approach" and "energy approach" respectively.

## 2.4 Classical Brittle-Ductile Modeling

The Earth's surface deforms in a "brittle" mode represented by pressure-sensitive, temperature insensitive elasto-plastic behavior. The rocks below the brittle-ductile transition (BDT) deform by "ductile" creep represented by pressure insensitive, temperature sensitive visco-elastic behavior. Advanced modeling approaches of the brittle-ductile transition (Albert and Phillips, 2002) employ a combined elasto-visco-plastic approach in which all three behaviors are allowed to occur simultaneously in series. The BDT then emerges self consistently as a narrow transition zone between the plastic (brittle) and the viscous (ductile) regimes above and below, respectively.

While such classical models can reproduce key features observed in large scale geodynamics these numerical predictions fail to reproduce observations on the brittle-ductile transition from microstructural and laboratory analyses. These call for the existence of a broad transition zone named "semi brittle" (Kohlstedt et al., 1995). The semi-brittle regime is the region where brittle and ductile processes overlap. It is thought to be around 10 km wide (Kohlstedt et al., 1995). It is therefore inferred to be considerably wider than the sub-kilometer scale inferred by numerical models. A second important problem is that in the classical numerical models localized shear zones are mainly driven by brittle deformation while in geology the importance of mylonitic ductile shear zones is very well documented (Christiansen and Pollard, 1997).

Finally, the traditional approach to modeling crustal deformation is based on the balance of momentum with no attempt to consider energy dissipation within the system as a key to solving the "mechanical" evolution of state. Dynamic, time dependent processes, driven by the energy fluxes occurring during an earthquake, for example, are never considered in these approaches. These incompatibilities of observations, nomenclature and fundamental theories form well recognized communication gaps between geodynamicists, rock mechanicians, field geologists, and seismologists (Handy et al., 2001), and a unified approach is needed. This approach is a non-equilibrium thermodynamics formulation (Regenauer-Lieb and Yuen, 2003) which is based on stored and dissipated energy potentials and the principle of maximum entropy production. The mathematical formulation is relatively compact in terms of non-equilibrium thermodynamics and it allows for the incorporation of chemistry. However, since we have not yet routinely implemented the chemical potentials we choose to present the thermo-mechanical formulation in order to relate the two approaches and show the similarities and the differences between them.

We differentiate between two basic modeling strategies for localization phenomena: (1) The classical "constitutive approach", basically a mathematical/engineering approach to failure which is ideal for the lifetime of engineering structures and is used as a first order approximation to geological time scale (2) the new "energy approach" which uses basic physics to calculate failure and may be more appropriate for long geological timescales, since diffusion processes control the degree of potential weakening self-consistently. The energy approach features, in principle, a simplified thermodynamic method. However, we refrain from calling it a "thermodynamic model" since the consideration of energy feedback provides necessary but not sufficient conditions for a unified thermodynamic theory which should in the future include chemistry. In the present version of our model many additional feedback mechanisms with their associated energy states are not yet considered (Lyakhovsky et al., 1997). We describe a very basic setup, where only the entropy change through competition between two simple feedback loops; isentropic thermal expansion (Benallal and Bigoni, 2004) and dissipative shear heating (Braeck and Podladchikov, 2007; Gruntfest, 1963; Kaus and Podladchikov, 2006; Ogawa, 1987; Regenauer-Lieb and Yuen, 1998) differentiate the "energy approach" from the classical "constitutive approach". In the discussion, we will show that this simple difference can resolve all of the above mentioned discrepancies and paradoxes.

# 3 Mathematical Formulation

## 3.1 Classical Constitutive Approaches for the Lithosphere

The classical constitutive approach assumes scale invariant material behavior and neglects the principle of maximum dissipation which implies neglect of the conversion of mechanical work into heat. In this approach the uppermost part of the crust is commonly represented by Byerlee's law which is essentially a plastic failure criterion (Albert and Phillips, 2002). The lower part of the crust is then represented by a viscous law (commonly power-law creep). Elasticity is sometimes incorporated into these representations; however, for large scale modeling its effect is often neglected. In the most basic approaches, the switch from one kind of behavior to the other is somewhat arbitrarily defined for a given strain-rate and geothermal gradient, where the stress in the viscous material matches that in the plastic material resulting in a "Christmas Tree" distribution of strength downwards through the crust.

Within the context of the Mohr-Coulomb constitutive relation, Byerlee's law implies a zero value for the cohesion and a friction angle of approximately 50°; the dilation angle is never explicitly stated but, assuming that this dilation angle is not equal to the friction angle, non-associative behavior is implied with the consequence that localization according to the Hill postulate (Rice, 1977) is possible. However for this to happen, Byerlee's law needs to be expressed in a formulation that allows non-associative behavior in a continuum and not as a failure criterion on a single discontinuity. This allows a prediction of brittle faults without the consideration of the energy dissipation.

One problem of this approach is an arbitrary switch between rheologies at a certain predefined strain rate. This can be avoided through the use of an elasto-visco-plastic rheology, where the brittle-ductile transition can adjust to different boundary conditions. However, when applying only this elasto-visco-plastic strain rate addition, without further couplings, invariably narrow transition zones at mesh resolution are obtained.

In contrast to these constitutive models, our energy models employ a single constitutive relation for the entire crust, namely, that of a modified elasto-plastic von Mises potential (Albert and Phillips, 2002; Regenauer-Lieb et al., 2004) with an added power-law creep term. More precisely, in our approach the yield function at low pressure is a Drucker-Prager function growing in a linear manner with hydrostatic pressure to the classical von Mises yield surface. The elasto-plastic rheology is then commonly extended by adding a non-linear viscous creep constitutive law assuming power-law creep.

$$\dot{\varepsilon}_{ij}^{tot} = \dot{\varepsilon}_{ij}^{el} + \dot{\varepsilon}_{ij}^{pl} + \dot{\varepsilon}_{ij}^{cr}$$

$$\dot{\varepsilon}_{ij}^{el} = \frac{1+\nu}{E}\frac{D\tilde{\sigma}'_{ij}}{Dt} + \frac{\nu}{E}\frac{Dp}{Dt} + \alpha\frac{DT}{Dt}\delta_{ij}$$

$$\dot{\varepsilon}_{ij} = \dot{\varepsilon}_{ij}^{el} + \left(\dot{\varepsilon}^{pl}\frac{\sigma'_{ij}}{2\tau}\right)_{plastic} + \left(A\sigma'_{ij}J_2^{n-1}\exp\left(-\frac{Q}{RT}\right)\right)_{creep}$$

where $E$ is Young's modulus, $\nu$ is Poisson ratio and $\alpha$ is the coefficient of thermal expansion. $\dot{\tilde{\sigma}}_{ij}$ is the objective co-rotational stress rate and $\delta_{ij}$ is the Kronecker delta. The plastic yield stress is $\tau$; $A$ and $n$ are power-law material constants. $Q$ is the activation enthalpy and $R$ is the universal gas constant. $J_2$ is defined as the second invariant of the deviatoric stress tensor:

$$J_2 \equiv \sqrt{\frac{3}{2}\sigma'_{ij}\sigma'_{ij}}$$

$\sigma'_{ij}$ is the deviatoric stress, which is defined by:

$$\sigma'_{ij} \equiv \sigma_{ij} + p\delta_{ij}$$

where

$$p = -\frac{1}{3}trace(\sigma_{ij})$$

is the trace of the Cauchy stress tensor, or the pressure.

We thus use the same formulation as in the classical theory, the only difference being that we do not hardwire shear zone formation into a Mohr-Coulomb approach and we consider the conversion of mechanical work into heat. In our approach shear zones do not appear without this energy coupling.

## 3.2 Energy Approach

Processes that operate in deforming rocks are commonly strongly coupled, in that one process has a first order feedback influence on other processes. Such feedback is neglected in the classical *constitutive theory* for rock deformation. This is done through choice of a special rheology, namely non-associated Mohr-Coulomb (labeled constitutive theory in Fig. 4).

Figure 4 illustrates the mathematical framework underlying the feedback of the thermal diffusion process, which is fundamentally important for the emergence of planetary convection in Fig. 1. The same approach can also predict planetary scale shear zones, if shear heating is considered in addition. Therefore, in a simple model at planetary spatial and temporal scales, the thermal diffusion length scale is sufficient. Thermodynamic models gain complexity when considering smaller scale processes.

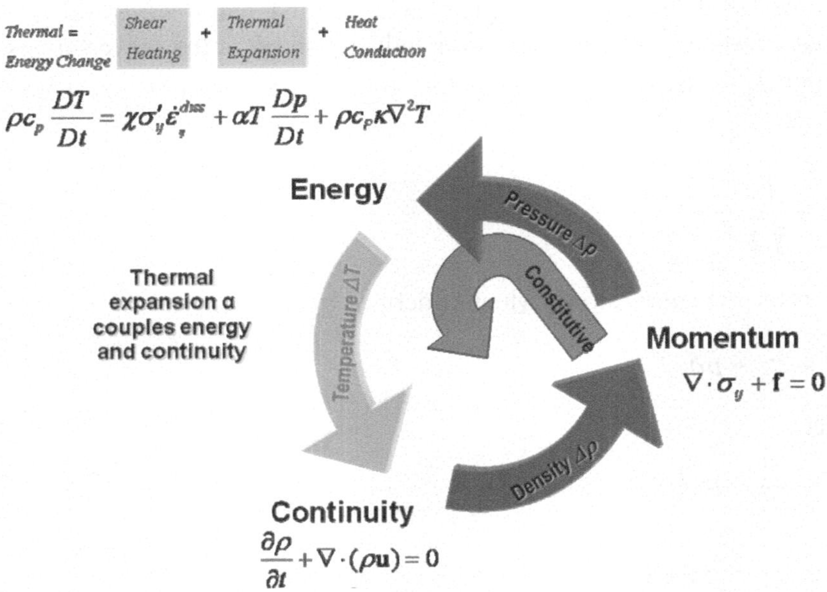


**Fig. 4** Processes in a rock involve small changes in the feedback variables density, pressure and temperature which have a large effect on the deformation of rocks; in particular their localization phenomena. Two key physical processes are described by the two opposing mechanical terms in the energy equation, the dissipated shear heating term (*first grey box* see text for explanation of parameters) and the recoverable, isentropic thermal expansion (*second grey box*). The classical *constitutive approach* bypasses the energy equation and hardwires pattern formation into a constitutive equation

All constitutive equations in our formulation are non-localizing at the outset. However, the rheology must be chosen such that the maximum entropy production applies. Localization behavior can only arise locally through energy feedback. These energy feedbacks close the balance laws via the energy equation. When investigating the feedback from heat/mechanical power generated through mechanical dissipation two important processes have been identified: (a) shear heating feedback by ductile (viscous) processes assessed through linear stability analyses presented two decades ago (Hobbs et al., 1986); (b) thermal-plastic instabilities arising from elastic thermal-expansion couplings in brittle processes (Benallal and Bigoni, 2004). The latter mechanism is one of a family of mechanisms that incorporate volumetric deformation and density change. Thermal expansion feedback has been found to be a critical phenomenon in experimental studies of the semi-brittle regime (Lu and Jackson, 1998). However, it has only recently been discussed in a thorough theoretical analysis (Benallal and Bigoni, 2004). Now we have the numerical tools (ABAQUS/Standard, 2000; Regenauer-Lieb and Yuen, 2004) available to go beyond the quasistatic linear stability framework with the new approach.

We only use thermo-mechanical coupling through thermo-elasticity and shear heating which leads to flow localization. Thermo-elastic feedback relies on the importance of thermal expansion which in our approach is the only feedback/localization mechanism considered for the brittle field. Shear heating feedback relies on the Arrhenius-temperature dependence of power law creep and is the only feedback loop considered for shear zone formation at greater depth. Thermal expansion feedback introduces weak pressure dependence to the constitutive relation which is equivalent to isentropic dilation. This has the effect that the material develops localization on a length scale that complies with heterogeneities in the spatial distribution of thermal expansion. Thus localization is permitted in both the brittle and ductile regimes arising solely from thermal-elastic dilation effects. We interpret the development of localization through this mechanism in the brittle regime as brittle fracturing, whilst localization in the ductile regime corresponds to ductile shear zone development.

It should be noted that this simple formulation can be extended, but does not, as yet, describe brittle fracture near the surface. At the surface, the plastic dilatancy is several orders of magnitude larger than thermal expansion. However, we suggest that our simple approach adequately captures the elementary physics at higher temperature and pressures corresponding to more than about 3 km overburden.

In our approach the brittle-ductile transition (BDT) emerges self-consistently as a region where both feedback mechanisms overlap at approximately the same magnitude. The principal diagnostic difference to the above described constitutive non-associated approaches for the BDT is that there are two, and not just one, mechanisms for flow localization available, a brittle localization and a ductile localization process. The latter mechanism is missing in the classical constitutive approach and is the reason for its failure to predict the BDT as a wide transition zone.

We use these two processes to address key issues in geodynamics and earthquake mechanics. While other mechanisms clearly exist, the two different principal physical mechanisms, shear heating and thermal expansion instabilities, are found to be sufficient to explain localization phenomena in all materials, other mechanisms clearly exist, but are not considered for simplicity.

Our new approach is summarized in Fig. 4. This figure highlights the difference between our formulation and previous approaches to flow localization. We employ the feedback between the fully coupled continuity, momentum and energy equations which are:

$$\frac{\partial \rho}{\partial t} + \nabla \cdot (\rho \mathbf{u}) = 0 \qquad \textbf{Continuity equation}$$

where $\rho$ is the density and $\mathbf{u}$ is the local material velocity vector of the volume under consideration; the second term describes the divergence of the velocity field. The continuity equation incorporates time as a derivative, which is implicitly derived from the evolution of isentropic (thermal expansion) work in the energy equation.

The momentum equation describes equilibrium of forces

$$\nabla \cdot \sigma_{ij} + \mathbf{f} = 0 \qquad \textbf{Momentum equation}$$

where $\nabla \cdot \sigma_{ij}$ is the divergence of the Cauchy stress tensor and $\mathbf{f}$ is the body force.

The energy equation describes the energy fluxes which in our case are

$$\rho c_P \frac{DT}{Dt} = \chi \sigma'_{ij} \dot{\varepsilon}_{ij}^{diss} + \alpha T \frac{Dp}{Dt} + \rho c_P \kappa \nabla^2 T \qquad \textbf{Energy equation}$$

where $c_P$ is the specific heat and $\frac{DT}{Dt}$ is the material derivative of the temperature. The first term on the right side describes shear heating through mechanical dissipative processes where $\chi$ is the efficiency of converting mechanical work into heat ($\chi \leq 1$). The shear heating efficiency of most materials is commonly 85% and 95% for large strain (Chrysochoos et al., 1989). The second term on the right describes the temperature change through isentropic/recoverable work arising from thermal expansion and the last term describes the temperature change arising from the phonon part of heat conduction with thermal diffusivity $\kappa$.

### 3.3 Scale Dependence of Ductile Shear Zones

Elasto-visco-plastic modelling with feedback includes all ingredients necessary for the investigation of the transient phenomena leading to the self-consistent nucleation of shear and fault zones. The drawback of this approach is that a proper implementation of the multi-level feedback is computationally expensive due to its inherent multi-scale nature. The computational cost relies on the high degree of spatial and hence temporal resolution that is required to resolve feedback. Different homogenization scales apply to different physical processes. These scales may be decoupled if they are sufficiently wide apart and the smaller scale may be assumed to have reached thermodynamic equilibrium within the time scale relevant for the large scale process. If this assumption is true the calculations can indeed be performed independently. We are here proposing such an approach for the plate tectonic scale.

The spatial scale of resolution can be derived from one-dimensional calculations and from theoretical considerations (Regenauer-Lieb et al., 2006a). Up to now the basic progress in this field has been mostly made in metallurgy. However, for metals the intrinsic material length scales of plasticity and thermal feedback (Lemonds and Needleman, 1986) collapses into the micron-scale. This makes the above mentioned separation of scales impossible and the calculations somewhat more complicated. In geology thermal feedback and meso-scale plasticity spreads out owing to the slow deformation and the low diffusivity of rocks. On the question regarding nucleation of shear zones, a separation of the length scales for shear zone formation is a fundamental issue.

The intrinsic material length scale of deformation by dislocations can be shown to govern the width of shear bands in metals (see Aifantis (1987)). The fundamental physics of this length scale hinges on a breakdown of the classical continuum mechanics below a homogenization scale where dislocation can be referred to as "statistically stored dislocations". Below 10 microns the discrete nature of dislocations is felt and there appear so called "geometrically necessary dislocations" which are related to the gradients of plastic strain in a material. Recently, nano-indentation and micro-torsion experiments have given support to this theoretically postulated limit. It was found that it is 200–300% harder to indent at nano-scale than at large scale (see Gao et al. (1999) for a review). The immediate outcome of this is that in plasticity there appears an intrinsic material length parameter characterizing the energy of defects. This defect energy governs the strain gradient of plasticity at mesoscale. This strain-gradient plasticity theory recovers at large homogenization scale the power law hardening relationship when a macroscopic population of statistically distributed dislocations is achieved. While this length scale relies on the shear gradients it has been suggested to expand the theory to include a second length scale for stretch gradients (Fleck and Hutchinson, 2001). All of these length scales are below tens of micrometer scale and are below the length scale of interest for geodynamic processes. For the purpose of our calculations we assume that dislocations can be described in a homogenized way by the power law and calculate feedback processes at larger scale.

Each feedback process has its own diffusional length scale in the quasi-steady state limit, if such a limit can be achieved during the deformation. In this respect, the *energy approach* can be expanded naturally for other weakening/strengthening effects such as those arising from chemical reactions. The limit is defined by a characteristic diffusive length scale $l_i$:

$$l_i = 2\sqrt{\frac{\kappa_i}{\dot{\varepsilon}_{\max}}}$$

where *i* refers to the *i*th weakening mechanism associated with an effective diffusivity $\kappa_i$.

Considering, for instance, mechanical weakening through the effect of a chemical diffusion process, the mechanical shear zone width, $l_i$, would be expected to be of the order of centimeters or less, because this is a typical diffusion length scale for chemical species in crustal environments. We assume here that there is no porous flow across the BDT, hence temperature must be the fastest diffusion process trailed by chemistry. Therefore, for plate tectonic length scales of kilometers and time scales of millions of years, the main diffusional scaling length is that of thermal diffusion. In this approach a number of heat source terms become very important (radiogenic heat, heat of reaction/solution, latent heat effects of mineral transformations etc.). In the numerical approach discussed above the main large scale mechanism that is capable of supplying a planar heat source important for a plate bounding fault is shear heating. Its associated feedback in creep deformation and thermal expansion is hence the main factor for plate tectonic localization. This thermal-visco-elastic feedback ensures that energy dissipated by the deformation is capable of weakening the material if temperature dependent mechanisms dominate. We wish to emphasize that this mechanism of shear localization in ductile rocks is intended as the most basic approach. A future comprehensive multi-scale approach should additionally include ductile damage and other micro-scale processes.

It is hence obvious that in this approach the ductile shear localization and weakening is limited by thermal diffusion. The advantage of the new approach is that brittle deformation is also linked to temperature through the isentropic thermal dilation effect. In our formulation brittle shear zones are therefore also governed by a length scale that is controlled by the thermal diffusivity which, in the quasi-steady state limit, describes equilibrium between shear heating and cooling by conduction. Assuming representative values of $\kappa_{thermal}$ of the order of $10^{-6}$ m$^2$ s$^{-1}$ and $\dot{\varepsilon}_{\max}$ of the order of $10^{-12}$ s$^{-1}$, $l_{therm}$ for the shear zone is of the order of 1 km for quasi-steady state, which is a reasonable value for a plate bounding mylonitic shear zone.

This length scale is a new aspect of the physics introduced thermal diffusion. It is the fundamental quantity controlling the final post-localization equilibration width of shear heating controlled shear zones (Shawki, 1994; Shawki and Clifton, 1989) when heat conduction and shear heating are in approximate thermal-mechanical equilibrium. It is also the quantity that governs the resolution criteria for numerical thermal-mechanical modeling of shear zones (Regenauer-Lieb and Yuen, 2003). In order to be able to see thermal feedback in a numerical simulation, we need to resolve below the thermal length scale.

Summing up these length scales, we would want to have a maximum element size of the order of tens to hundreds of metre in order to be able to resolve all feedback mechanism within a single numerical analysis. Now, a typical 2-D geodynamic calculation would comprise an area of 1000 km × 100 km. This would imply a minimum of ten million nodes in the calculation, although in practice not all of the model needs to be resolved at such high resolution, if it does not localize. However, it becomes apparent why ductile shear zones are hard to capture in geodynamical calculations. Ductile shear zones are however not beyond the reach of current computers.

## 3.4 Intrinsic Length Scales for Brittle Faults

There are important differences in the intrinsic length and time scales of elastic, plastic and viscous deformation. Visco-plastic deformation is transmitted by the motion of line defects, so it is a rate limited process controlled by atomic relaxation times. Elastic strain relies on electromagnetic waves, so it is determined by electronic relaxation times. The length scale of plastic deformation relies on the size of line defects and magnitude of lattice vibrations. The length scale of viscous deformation relies on thermal and chemical diffusion through lattice and crystal sizes, while elastic deformation relies on electronic (ionic or covalent) bonding only. In the classical brittle elastic theory the electronic relaxation time can be neglected, thus a time-invariant formulation may be adopted. One may come to the conclusion that the development of a multi-scale brittle theory is simpler than the ductile theory.

This is erroneous because the concept of time enters through the back door, by the evolution of damage during brittle-elastic deformation, which of course, is again linked to the energy/temperature evolution inside a brittle shear zone. For the evolution of brittle damage inside the fault zone it is also important to know how much of the plastic work is dissipated as heat because both together describe the total state evolution inside the shear zone. It is, hence, logical to start with a solution that can describe crack evolution and crystalline slip by dislocation (largely appearing as heat) together. Such theories are emerging now and they are briefly described.

Multiscale computer simulations are now coming up with a totally new dimension and insight into the dynamics and micro-physics of shear zones. The calculations are conventionally first done at atomistic level with say about 100 million atoms displaying the dislocation evolution around a crack tip (Bulatov et al., 1998). In such dislocation-dynamics simulations, computational efficiency is achieved through a less detailed description of dislocations in which atomic degrees of freedom are replaced by piecewise straight lines, and a mesh spacing (a few nanometres) is used that is larger than the crystal lattice parameter. This means dislocation mobility and close-range interactions are not determined as atomic-level processes, but are specified by external parameters known as "local rules". For this approach to be predictive, atomistic behaviour of dislocation cores has to be integrated into meso-scale dislocation-dynamics simulations. This can be done by a simple step up in scale (Bulatov et al., 1998). A micro-to-mesoscale connection is proposed in which the local rules are derived from the physically occurring dislocation core processes in an atomistic simulation.

The most complete method for spanning the length scales of dynamic simulations is to embed molecular dynamic calculations into larger scale finite element continuum calculations. This has been done for a crack in a silicon slab where five nested computational dynamic regions have been used (Abraham et al., 1998): the largest continuum finite-element (FE) region; the atomistic molecular-dynamics (MD) region; in between the quantum tight-binding (TB) region; the Finite Element-Tight-Binding (FE-TB) "handshaking" region; and the MD-TB "handshaking" region.

Another interesting method for multi-scaling is by going straight from discrete (molecular dynamics) to continuum style calculations without an intervening discrete element (dislocation dynamics) step. An inter-atomic-based-potential FEM has been developed, that is capable of reproducing stability criteria at both the atomistic and the continuum levels thus providing a fundamental potential framework for the physics of deformation (Li et al., 2002).

While the multi-scale approaches from atomistic to micro-scale are thus growing rapidly for the case of individual cracks, the next level up of describing the behavior of crack populations appears to be still outside the realm of fundamental atomistic based calculations. For the purpose of doing this, in a first step without the rigor underpinned by MD simulations, we introduce some basic assumptions that serve a priori as an empirical framework.

## 3.5 Scale Dependence for Brittle Faults

Thermodynamic solutions to the problem of brittle shear zone formation will provide insight into the quasi-periodicity of earthquakes while solutions to the ductile shear zones give insights into the problem of cyclic-like nature of plate tectonics. We have shown above that multi-scaling analyses in ductile shear zones are emerging as a new standard in material sciences. However, in the brittle field the multi-scale thermal-dynamic material properties are less well constrained than the ductile properties. From the last chapter it becomes apparent that we cannot resolve the physics of multiple interacting cracks at large plate tectonic scale chiefly because of lack of computer power. Experiments and field observations suggest, however, that the brittle strength of the lithosphere is probably overstated. Significant scale dependence of the brittle properties of rocks have for instance been reported in the literature on brittle rock experiments and their temperature sensitivity see e.g. (Shimada, 1993). In field observation (mine site load bearing jacks) the brittle compressive failure strength of a rock is for instance found to be at least a factor of three magnitude smaller at meter scale than at cm scale (Pinto da Cunha, 1993). Above 1 m there appears to be a statistical satisfactory number of planes of weaknesses in rocks so that the failure strength does not decrease further. Unfortunately, huge testing machines are necessary to obtain mechanical data relevant for the larger scale. The necessity for assessing the large scale has been realized only for the laboratory assessment of friction (Dieterich, 1979). An equivalent approach is lacking for the compressive failure strength of rocks.

On the weight of the above described observations and in the absence of precise data we assume that the characteristic scale of 1 $m^3$ is the homogenization scale and we assume that that rock strength is defined in the high temperature limit by Goetze's criterion (Caristan, 1982) rather than the

Byerlee law (Byerlee, 1978). Our particular emphasis lies on the high temperature rock strength since this is where the largest deviatoric stresses are expected inside the lithosphere.

## 4 Discussion

The observations of multi-scale material strength from different disciplines reported above suggest that classical scale-invariant material property extrapolations might give an upper bound of strength of the lithosphere. Conventionally, laboratory data are obtained at cm scale and may result in overestimations of lithospheric strength when applied to the hundreds of kilometers scale.

In effect, all of the plate tectonic paradoxes reported in the introduction can be assumed to result from the overestimation of the dynamic strength of the lithosphere by constitutive methods. We briefly summarize results from the "energy approach" which overcome these shortcomings.

The first observation concerns the subduction initiation paradox and the generation of weak trans-lithospheric faults (McKenzie, 1977). This observation required special pleading for classical strength models. It is conceivable that forcing convergence across fracture zones and at the same time requiring enormously high fluid pressures could initiate subduction (Hall et al., 2003). It can be shown on the other hand that the shear heating feedback becomes critical when allowing for a small amount of water inside the lithosphere (Regenauer-Lieb et al., 2001). Energy feedback thus naturally lead to the development of extremely weak km wide shear zones with an effective viscosity as low as $5 \times 10^{20}$ Pas (Regenauer-Lieb et al., 2001; Regenauer-Lieb and Yuen, 2003) and the weak shear zone need not be assumed a priori.

The Brace-Goetze (Christmas tree) crustal strength paradox (Brace and Kohlstedt, 1980) implies that significant deformation of the lithosphere would be prevented for thermal conditions below 75 mW/m$^2$ in compression and a value of 60 mW/m$^2$ in extension (Kusznir and Park, 1984a; 1984b). While such heat flows are not entirely off limits it would imply that any tectonic activity would be impossible for areas below these threshold values. We know that this is not the case. The Asian intraplate has for instance an average surface heat flow lower than 75 mW/m$^2$ (Wang, 2001). The Brace-Goetze strength profile implies that the Asian continental lithosphere is too strong to be indented by the Indian indenter. Likewise, cold continental breakup such as in Galicia margin would be impossible under normal geodynamic forcing. This paradox is solved by considering weakening through shear heating feedback (Weinberg et al.,

2007). Specifically, two key feedback effects both related to thermal energy, and natural in geological systems, have been included in large scale numerical models of continental deformation driven by plate tectonics: (a) shear heating, whereby heating around a small strength perturbation leads to decreased viscosity, which triggers increased strain rate thus increasing heating, causing the development of a ductile shear zone; (b) thermal expansion feedback, where the same temperature heterogeneities, lead to pressure fluctuations; regions of temperature decrease lead to a pressure decrease, whereas regions with a positive temperature increase lead to a pressure increase thus, triggering the onset or arrest of brittle failure, respectively. Cross scale calculations including such effects can be performed with sufficient local resolution (100 m) nowadays. The solutions show that there is considerable weakening of the strong layers in continents so that the continents are much weaker than previously estimated.

The mid-crustal detachment paradox (Axen and Selverstone, 1994) implies that weak crustal detachments are observed exactly where classical strength envelopes predict a strength maximum. This paradox is conventionally explained by either high fluid pressure or weak rheologies. However, it can be shown that mid-crustal detachment develop naturally out of energy feedback through its effect on continental strength (Regenauer-Lieb et al., 2006b).

The jelly sandwich paradox (Jackson, 2002) highlights the fact that the upper mantle fails to present significant strength and does not deform in a seismogenic manner – yet in the classical constitutive model it would be expected to be the strongest part of the lithosphere. Our models support the hypothesis that it is indeed time to abandon the jelly sandwich. The implications for upper mantle strength and seismicity can be explained by the efficiency of feedback in the mantle (Weinberg et al., 2007) and the fundamentally different seismogenic behavior of quartz compared to olivine (Regenauer-Lieb and Yuen, 2006). This comparison shows that olivine having a high activation energy for creep behaves in a fundamentally different manner to quartz which has a much lower activation energy. The main difference lies in the efficiency of the shear heating feedback. In quartz shear heating feedback is inefficient and therefore cannot easily cross the scales. This leads to unstable dynamic slip pulses with heterogeneous faults. Olivine on the other hand behaves in a much more stable manner because the efficient weakening by shear heating quickly establishes large scale shear zones. These shear zones are characterized by a quasi-equilibrium of heat production on the shear planes and thermal diffusion away from the shear zones.

The upper plate paradox (Kusznir, 1991) implies that the brittle crust is deforming much less than the ductile lower crust. This mismatch frequently observed between stretching values inferred from surface extension and bulk crustal thinning can be explained by assuming a priori an exceptionally weak middle crust (Nagel and Buck, 2007). However, the weakness of the middle crustal layer is also a natural feature of the energy feedback at the brittle-ductile transition and below (Weinberg et al., 2007).

While it is always possible to argue for special pleading to explain the individual observations we believe that the weight of the sum of the observations is compelling enough to consider the possibility that the lithosphere can become a lot weaker than previously thought through multiscale feedback processes. We have shown here how we can apply large scale planetary thermal-mechanical modeling techniques to small scale features and have presented a first cut of incorporation of chemistry for future more robust up-scaling methods.

**Acknowledgements** We wish to acknowledge support through the Pmd*CRC, CSIRO, the University of Western Australia and the Western Australian Government through the Premiers Science Fellowship Program. David Yuen acknowledges support from ITR and CSEDI programs of National Science Foundation.

## References

ABAQUS/Standard (2000), 384pp., Hibbit, Karlsson and Sorenson Inc.

Abraham, F. F., et al. (1998), Spanning the length scales in dynamic simulation, *Computers in Physics*, *12*, 538–546.

Aifantis, E. C. (1987), The physics of plastic deformation, *International Journal of Plasticity*, *3*, 211–247.

Albert, R. A., and R. J. Phillips (2002), Time-dependent effects in elastoviscoplastic models of loaded lithosphere, *Geophysical Journal International*, *151*, 612–621.

Axen, G. J., and J. Selverstone (1994), Stress state and fluid-pressure level along the whipple detachment fault, California, *Geology*, *22*, 835–838.

Ben-Zion, Y., and C. G. Sammis (2003), Characterization of fault zones, *Pure and Applied Geophysics*, *160*, 677.

Benallal, A., and D. Bigoni (2004), Effects of temperature and thermo-mechanical couplings on material instabilities and strain localization of inelastic materials, *Journal of the Mechanics and Physics of Solids*, *52*, 725.

Brace, F. W., and D. L. Kohlstedt (1980), Limits on lithospheric stress imposed by laboratory experiments, *Journal of Geophysical Research*, *50*, 6248–6252.

Braeck, S., and Y. Y. Podladchikov (2007), Spontaneous thermal runaway as an ultimate failure mechanism of materials, *Physical Review Letters*, *98*.

Bulatov, V., et al. (1998), Connecting atomistic and mesoscale simulations of crystal plasticity, *Nature*, *391*, 669–672.

Bunge, H. P., M. A. Richards, and J. R. Baumgardner (1997), A sensitivity study of three-dimensional spherical mantle convection at 108 Rayleigh number: effects of depth-dependent viscosity, heating mode, and endothermic phase change, *Journal of Geophysical Research*, *102*, 11991–12007.

Byerlee, J. D. (1978), Friction of rocks, *Pure and Applied Geophysics, 116*, 615–626.
Caristan, Y. (1982), The transition from high temperature creep to fracture in Maryland Diabase, *Journal of Geophysical Research, 87*, 6781–6790.
Christiansen, P. P., and D. D. Pollard (1997), Nucleation, growth and structural development of mylonitic shear zones in granitic rock, *Journal of Structural Geology, 19*, 1159–1172.
Chrysochoos, A., et al. (1989), Plastic and dissipated work and stored energy, *Nuclear Engineering and Design, 114*, 323–333.
Dieterich, J. H. (1979), Modeling of rock friction .1. Experimental results and constitutive equations, *Journal of Geophysical Research, 84*, 2161–2168.
Enquist, E., and Z. Huang (2003), Heterogeneous multiscale method: a general methodology for multiscale modeling, *Physical Reviews B, 67*, 092101: 092101–092104.
Fleck, N. A., and J. W. Hutchinson (2001), A reformulation of strain gradient plasticity, *Journal of the Mechanics and Physics of Solids, 49*, 2245–2271.
Fleitout, L., and C. Froidevaux (1980), Thermal and mechanical evolution of shear zones, *Journal of Structural Geology, 2*, 159–164.
Gao, H., et al. (1999), Modeling plasticity at the micrometer scale, *Naturwissenschaften, 86*, 507–515.
Gruntfest, I. J. (1963), Thermal feedback in liquid flow – plane shear at constant stress, *Transactions of the Society of Rheology, 7*, 195–207.
Hall, C. E., et al. (2003), Catastrophic initiation of subduction following forced convergence across fracture zones, *Earth and Planetary Science Letters, 212*, 15–30.
Handy, M. R., et al. (2001), Rheology and geodynamic modelling: the next step forward, *International Journal of Earth Sciences, 90*, 149–156.
Hobbs, B. E., et al. (1986), Earthquakes in the ductile regime, *Pure and Applied Geophysics, 124*, 310–336.
Jackson, J. (2002), Strength of the continental lithosphere: time to abandon the jelly sandwich? *GSA Today, 12*, 4–10.
Kaus, B., and Y. Podladchikov (2006), Initiation of localized shear zones in viscoelastoplastic rock, *Journal of Geophysical Research, 111*, B04412, doi:04410.01029/02005JB003652.
Kohlstedt, D. L., et al. (1995), Strength of the lithosphere: constraints imposed by laboratory measurements, *Journal of Geophysical Research, 100*, 17587–17602.
Kusznir, N. J. (1991), The distribution of stress with depth in the lithosphere: thermo-rheological and geodynamic constraints, *Philosophical Transaction of the Royal Society of London Series A-Mathematical Physical and Engineering Sciences, 337*, 95–110.
Kusznir, N. J., and R. G. Park (1984a), Intraplate lithosphere deformation and the strength of the lithosphere, *Geophysical Journal of the Royal Astronomical Society, 79*, 513–538.
Kusznir, N. J., and R. G. Park (1984b), The strength of intraplate lithosphere, *Physics of the Earth and Planetary Interiors, 36*, 224–235.

Lemonds, J., and A. Needleman (1986), Finite element analyses of shear localization in rate and temperature dependent solids, *Mechanics of Materials*, *5*, 339–361.

Li, J., et al. (2002), Atomistic mechanisms governing elastic limit and incipient plasticity in crystals, *Nature*, *418*, 307–310.

Lu, C., and I. Jackson (1998), Seismic-frequency laboratory measurements of shear mode viscoelasticity in crustal rocks II: thermally stressed quartzite and granite, *Pure and Applied Geophysics*, *153*, 441.

Lyakhovsky, V., et al. (1997), Distributed damage, faulting, and friction, *Journal of Geophysical Research-Solid Earth*, *102*, 27635–27649.

Martyushev, L. M., and V. D. Seleznev (2006), Maximum entropy production principle in physics, chemistry and biology, *Physics Reports-Review Section of Physics Letters*, *426*, 1–45.

McKenzie, D. P. (1977), The initiation of trenches: a finite amplitude instability, in *Island Arcs Deep Sea Trenches and Back-Arc Basins*, edited by M. Talwani and W. C. Pitman, pp. 57–61, Maurice Ewing Ser. Vol. 1.

Nagel, T. J., and W. R. Buck (2007), Control of rheological stratification on rifting geometry: a symmetric model resolving the upper plate paradox, *International Journal of Earth Sciences*, *96*, 1047–1057.

Ogawa, M. (1987), Shear instability in a viscoelastic material as the cause of deep focus earthquakes, *Journal of Geophysical Research*, *92*, 13801–13810.

Ord, A., and B. E. Hobbs (1989), The strength of the continental crust, detatchment zones and the development of plastic instabilities, *Tectonophysics*, *158*, 269–289.

Pinto da Cunha, A. (1993), *Scale Effect in Rock Masses 93*, A.A. Balkema, Rotterdam.

Prager, W. (1959), *An Introduction to Plasticity*, Addison Wesley, Reading, Massachusetts.

Regenauer-Lieb, K., and D. Yuen (1998), Rapid conversion of elastic energy into shear heating during incipient necking of the lithosphere, *Geophysical Research Letters*, *25*, 2737–2740.

Regenauer-Lieb, K., and D. A. Yuen (2003), Modeling shear zones in geological and planetary sciences: solid- and fluid- thermal- mechanical approaches, *Earth Science Reviews*, *63*, 295–349.

Regenauer-Lieb, K., and D. A. Yuen (2004), Positive feedback of interacting ductile faults from coupling of equation of state, rheology and thermal-mechanics, *Physics of Earth and Planetary Interiors*, *142*, 113–135.

Regenauer-Lieb, K., and D. Yuen (2006), Quartz rheology and short time-scale crustal instabilities, *Pure and Applied Geophysics*, *163*, 1915–1932.

Regenauer-Lieb, K., et al. (2001), The initiation of subduction: criticality by addition of water? *Science*, *294*, 578–580.

Regenauer-Lieb, K., et al. (2004), On the thermodynamics of listric faults, *Earth Planets and Space*, *56*, 1111–1120.

Regenauer-Lieb, K., et al. (2006a), From point defects to plate tectonic faults, *Philosophical Magazine*, *86*, 3371–3392.

Regenauer-Lieb, K., et al. (2006b), The effect of energy feedbacks on continental strength, *Nature*, *442*, 67–70.

Rice, J. R. (1977), The localization of plastic deformation, in *Theoretical and Applied Mechanics*, edited by W. T. Koiter, pp. 207–220, North-Holland, Amsterdam.
Shawki, T. G. (1994), An energy criterion for the onset of shear localization in thermal viscoplastic material, Part II: Applications and implications, *Journal of Applied Mechanics*, *61*, 538–547.
Shawki, T. G., and R. J. Clifton (1989), Shear band formation in thermal viscoplastic materials, *Mechanics of Materials*, *8*, 13–43.
Shimada, M. (1993), Lithosphere strength inferred from fracture strength of rocks at high confining pressures and temperatures, *Tectonophysics*, *217*, 55–64.
Siret, D., et al. (2008), PreMDB, a thermodynamically consistent material database as a key to geodynamic modelling, *Geotechnica Acta*, DOI 10.1007/s11440-008-0065-0.
Tullis, T. E., et al. (1991), Flow laws for polyphase aggregates from end member flow laws, *Journal of Geophysical Research*, *96*, 8081–8096.
Wang, Y. (2001), Heat flow pattern and lateral variations of lithosphere strength in China mainland: constraints on active deformation, *Physics of the Earth and Planetary Interiors*, *126*, 121–146.
Weinberg, R., et al. (2007), Mantle detachment faults and the break-up of cold continental lithosphere, *Geology*, *35*, 1035–1038.
Yuen, D. A., et al. (1978), Shear deformation zones along major transform faults and subducting slabs, *Geophysical Journal of the Royal Astronomical Society*, *54*, 93–119.
Ziegler, H. (1983), *An Introduction to Thermomechanics*, North Holland, Amsterdam.

# II. 3D Mesh Generation in Geocomputing

Huilin Xing,[1] Wenhui Yu[1,2] and Ji Zhang[1]

[1]The University of Queensland, Earth Systems Science Computational Centre, St. Lucia, Brisbane, QLD 4072, Australia
[2]Department of Engineering Mechanics, Dalian University of Technology, Dalian, China

**Abstract** Mesh generation has been widely used in engineering computing, but seems to be relatively "new" for geoscience community. This paper firstly lists such relevant progresses of mesh generation for the engineering computing and then discusses the possibility for applying/extending them to geoscience computing (i.e. geocomputing). For geoscience, the available input data are normally a large quantity of point data in the 3D space rather than the defined shapes and dimensions with reasonable tolerances provided by the industrial designers, thus quite different from the engineering cases. To deal with such geoscience data, this paper briefly introduces the relevant progresses on geometrical modeling, hexahedral and tetrahedral shaped mesh generation, and then focuses on the applying and/or extending the related methods to generate all hexahedral/tetrahedral shaped meshes in 3D for geoscience purposes, which is described through the different practical application examples, such as the all-hexahedral shaped mesh generation for a fracture dominated reservoir system, the South Australia interacting fault system and the entire earth model without or with the simplified/practical plate boundaries; and all-tetrahedral shaped mesh generation for a multi-layer underground geological model, and visualizing and meshing with the microseismicity data recorded during a hydraulic stimulation process in a geothermal reservoir.

H. Xing, *Advances in Geocomputing*, Lecture Notes in Earth Sciences 119,
DOI 10.1007/978-3-540-85879-9_2

## 1 Introduction

With the rapid development of supercomputers, high-performance computing based simulation is internationally recognized as paradigm shift that offers an outstanding opportunity for advancement to better understand and quantify the earth systems science, materials science and engineering etc. Especially for the geoscience, short-term idealized experiments and field site observations have helped people to understand the relevant phenomena, while the prediction and risk assessment of complex earth systems may be better to be achieved numerically, that includes the finite element method (FEM), the finite difference method (FDM). FDM normally requires the simplified regular square/cubic mesh (even with overlap), which is much easier to deal with. Therefore, the mesh generation for FEM analysis is focused here. Several commercial FEM software packages are available and widely used in the practical for industrial engineering design and analysis, such as ABAQUS, ANSYS, ADINA, MSC Software and LS-Dyna3D. Following the above successful stories in the mechanical and civil engineering computing, the finite element based numerical modeling of the geoscience offers an outstanding opportunity to gain an understanding of those dynamics and complex system behaviour, and to develop the scientific underpinning for geoscience, such as ground motion, geodynamics, interacting fault system, and earthquake and tsunami forecasting.

Mesh generation is a critical step before finite element analysis could be carried out, which is defined as a process of dividing a continuous physical domain into a grids (elements) for the further numerical solution. Mesh generation and optimization process may be achieved by numerous in-house and/or commercial software programs, and many researchers are still working on it. The Sandia National Laboratories' 16th International Meshing Roundtable (http://www.imr.sandia.gov/16imr/main.html) and the 6th Symposium on Trends in Unstructured Mesh Generation (http://www.andrew.cmu.edu/user/sowen/meshtrends6/index.html) was just held in 2007, which reflects the related research history, current outcomes and problems. Robert Schneiders maintains a website to provide information on mesh and grid generation:. people working in the field, research groups, books and conferences (http://www-users.informatik.rwth-aachen.de/~roberts/meshgeneration.html); Owen also maintains a meshing research corner (see http:// www.andrew.cmu.edu/user/sowen/mesh.html) and did a survey of unstructured mesh generation technology used in both in-house and commercial software (see http://www.andrew.cmu.edu/user/sowen/survey/index.html); Xing and Mora (2003) did a similar survey but focused on the technologies mostly relevant to mesh generation of crustal fault systems. Based on these surveys, surface domains may be subdivided into triangle or quadrilateral shaped mesh and volumes may be subdivided primarily into tetrahedral or hexahedral shaped mesh by using various of software, where the following algorithms are mostly used: Octree (e.g. Frey et al. 1994; Schneiders and Bunten 1995, 1996; Schneiders 1996, 1997; Shephard and Marcel 1991, 1992; Tu and O'Hallaron 2004; Yerry and Shephard 1984), Delaunay (e.g. Baker 1989; Borouchaki and Lo 1995; Borouchaki et al. 1996; Borouchaki and George 1997; Borouchaki and Frey1998; Borouchaki et al. 2000; Du and Wang 2003; George et al. 1991; George and Seveno 1994; George1997; Joe 1991a, b, c, 1995; Lee and Schacter 1980; Shewchuk 2002; Weatherill and Hassan 1994; Wright and Alan 1994), advanced front algorithm (e.g. Lau and Lo, 1996; Lee and Lo 1994; Lee and Hobbs 1999; Lee 2003; Lee and Lee 2002, 2003; Lo 1991a, b, 1992; Lohner 1996a, b, c; Lohner and Cebral 2000; Lohner and Eugenio 1998; Owen et al. 1999; Owen and Saigal 2000; Yamakawa and Shimada 2003), and sweeping/mapping method (e.g. Cheng and Li 1996; Knupp 1998, 1999; Lai et al. 2000; Scott et al. 2005; Staten et al. 1998, 2005). A brief summary is listed in Table 1. Automatic generation of triangle and quadrilateral shaped mesh in 2D and tetrahedral shaped mesh in 3D for a normal continuous domain is already quite mature, but for a high quality hexahedral shaped mesh, the automatic mesh generation of a general 3D physical domain is still not available, which is mainly achieved through the case-by-case trial-and-error techniques. Moreover, for a discontinuous complex domain, further research for automatic mesh generation of tetrahedron/hexahedron shaped meshes is still required.

**Table 1** Various algorithms used in mesh generation and their advantages/disadvantages

| | Delaunay | AFT | Quadtree/ Octree | Indirect/ Sweeping/ Mapped Meshing |
|---|---|---|---|---|
| Supported Element Type | Triangle/ Tetrahedron | Triangle/ Tetrahedron/ Quadrangle/ Hexahedron | Triangle/ Tetrahedron/ Quadrangle/ Hexahedron | Hexahedron |
| Algorithm Efficiency | O(Nlog(N)) | O(Nlog(N)) | O(Nlog(N)) | |
| Element Quality | Normally good | Normally good | Good except boundary | Normally good, but not always achievable |

(Continued)

**Table 1** (Continued)

| | Delaunay | AFT | Quadtree/ Octree | Indirect/ Sweeping/ Mapped Meshing |
|---|---|---|---|---|
| Adaptivity | Yes | Yes | Yes | Case by case |
| Automatic generation | Yes | Yes | Yes | Case by case |
| Known problems | Boundary recovery/ Sliver Decomposition | Convergence problem | Boundary fitting | Sometimes not achivable |
| Commecial Software/ Public Code | ANSYS[a]/ Qhull[b]/ Triangle[c] | ANSYS/GID[d]/ Hypermesh[e] | ICEM CFD[f]/ MEGA[g]/ MESH[h]/QMG[i] | ANSYS/ GID/ Hypermesh/ CUBIT[j] |

[a]see http://www.ansys.com for more information.
[b]see http://www.qhull.org for more information.
[c]see http://www.cs.cmu.edu/~quake/triangle.html for more information.
[d]see http://gid.cimne.upc.es for more information.
[e]see http://www.altair.com/software/hw_hm.htm for more information.
[f]see http://www.icemcfd.com/ for more information.
[g]see http://www.scorec.rpi.edu/ for more information.
[h]see http://www.synopsys.com/products/tcad/mesh_ds.html for more information.
[i]see http://simon.cs.cornell.edu/Info/People/vavasis/qmg-home.html for more information.
[j]see http://cubit.sandia.gov for more information.

The existing commercial and industrial strength in-house graphics software as above are, on the whole, designed for the mechanical and civil engineering industry (such as the automobile industry). In these industries, the domain normally has defined shapes and dimensions with reasonable tolerances provided by the designers. Many applications use a "bottom-up" approach to mesh generation. Vertices are firstly meshed, followed by curves, then surfaces and finally solids. The input for the subsequent meshing operation is the result of the previous lower dimension meshing operation. For an example, nodes are firstly placed at all vertices of the geometry. Nodes are then distributed along the geometric curves. The result of the curve meshing process provides input of a surface meshing algorithm, where a set of curves define a closed set of surface boundaries. Decomposing the surface into well-shaped triangles or quadrilaterals is the next stage of the meshing process. Finally, if a solid model is provided as the geometric domain, a set of meshed surfaces defining a closed volume is provided as input of a volume mesher for automatic generation of tetrahedra, hexahedra or mixed element types (see Owen's survey as indicated above).

However, difficulties have been encountered by using the existing commercial graphics packages (including the industrial strength in-house software) to construct a computational model of a certain geoscience case, such as an interacting fault system. Specifically, it is very difficult and time-consuming to use the existing engineering-focused software to construct practical 3D models with complex fault geometries, which are given by the converted fault data from digital images or a serial of point sets. This is mainly because the existing commercial graphics software are, on the whole, designed for the mechanical and civil engineering industry (such as the automobile industry) as described above. In these industries, the domain normally has defined shapes and dimensions with reasonable tolerances provided by the designers, whereas, in the cases of geoscience, these are usually specified by a large quantity of point data, such as the fault data. Therefore, special techniques are required for treatment of the different cases from the geoscience community. In addition, mesh generation seems to be quite "new" for the geoscience community despite it is widely used in the engineering computing. To follow the successful stories in the mechanical/civil engineering computing etc., a preliminary introduction/training is necessary to ensure such successful engineering-focused mesh generation procedures and relevant software be helpful and applicable to the geoscience filed.

Automatic generation of triangle and quadrilateral shaped mesh in 2D is already quite mature, thus the 3D cases are focused here. To generate meshes containing faults (discontinuity) described by a large amount of point data set (including digital images) in 3D space, tetrahedral shaped elements seem to be more easily achieved automatically by using the Delaunay algorithm (the mostly common one) and the advancing front technique (AFT). While the conventional Delaunay algorithm is not ideal since it is difficult to guarantee that nodal points are exactly located at the fault/boundary interfaces specified by the input point data. The AFT may generate a mesh with nodes located exactly at fault interfaces, although very few codes are available with this algorithm for automatic mesh generation due to a few of its limitations. Hexahedral shaped elements provide an alternative to tetrahedral shaped meshes. So far, there is no software tool available to automatically generate a high quality hexahedral mesh for a general 3D physical domain including the fault system, despite a lot of outcomes have been achieved, such as CUBIT (http://sass1693.sandia.gov/cubit/).

In summary, mesh generation is well developed and widely used in the engineering field as above, but it seems to be quite "new" for the geoscientists, thus the successful experiences and outcomes in the other field would be quite helpful and useful for the geoscience community. Here, we focus on applying and/or extending such relevant successful experiences and methods to generate all hexahedral/tetrahedral shaped mesh in 3D for the geoscience purposes through several different practical application examples.

## 2 Geometrical Modeling

With the rapid development of advanced digital image technology and its application in the earth sciences, more and more image information of the Earth is available and may be used to build computational models and even to validate numerical results. However, the available input data for geoscience are quite different from the engineering case as described above; it is a large quantity of point data (including digital images) rather than defined shapes and dimensions with reasonable tolerances provided by the industrial designers. Thus there are a number of challenges that must be overcome, such as how to obtain and use geological point data set to construct the computational model and in particular generate usable meshes (elements) for the further finite element analysis. Such a pre-processing normally includes geometrical modeling and mesh generation. Here geometrical modeling aims to construct and manage the related geological data, and it can also be applied to reconstruct the tectonic evolution of the whole Earth, or a specified region such as the western Mediterranean since the Oligocene (e.g. Rosenbaum et al. 2002). The task of geometrical modeling here concentrates on the geological model construction, which aims at editing and managing the tectonic data, and constructing 2D or 3D geological surface/solid model involving faults and/or plate boundaries. A specially-purposed tectonic CAD/Database module was developed within the CHIKAKU system (Kanai et al. 2000; Xing et al. 2001), which aims to manage the data and construct a computational model for an interacting fault system. It is composed of two subsystems: tectonic database and CAD. The main purpose is to construct 2D and/or 3D solid models including faults and plate boundaries and to output the solid model of the specified region directly for the CHIKAKU Mesh (Xing et al. 2001, 2007a) and/or in the standard IGES data format for easily interfacing with other software packages (such as I-DEAS and Patran) for further mesh generation. The developed geometrical modeling module includes the following three aspects: (1) Data input: The data may include (a) the underground structural data of stratum boundary point, stratum borderline, stratum and plate/fault, which could be available through the natural and/or man-made earthquake data inversion, drilling well and advanced digital image etc.; (b) observational data: hypocentral distribution and distribution of the distortion and (c) reference data for coastlines, the administration field, rivers, the Earth's surface and the ocean floor. (2) Data management and editing: conversion of digital images and other data to the necessary formats for constructing the fault and plate models, visualization and editing of the above data; (3) Geometrical modeling: construction and editing of the lines and curves defining faults and plate boundaries, construction of the parametric surfaces using the above curves, construction and editing of the solid models using the generated surfaces, editing and output of a solid model of a specified region. The above function may be also partly or totally achieved by other in-house or commercial software such as those specified in the above surveys with the similar procedure. Upon completion of the geometrical modeling, the mesh generation and optimization are carried out together with the specification of loading, boundary conditions and material properties and so on.

## 3 Hexahedral Mesh Generation

### 3.1 Introduction

There is the often-held position that quadrilateral and hexahedral shaped elements have superior performance over tetrahedral shaped elements when comparing an equivalent number of degrees of freedom, and also more suitable for nonlinear finite element analysis in some cases. For finite element analysis of incompressible or nearly incompressible nonlinear behavior, such as the large deformation of the Earth in 3D, it seems to be necessary to use a hexahedral mesh rather than a tetrahedral mesh to obtain sufficiently accurate results. The hexahedral shaped mesh generation is crucial for the finite element community, but the automatic generation of such a high quality all-hexahedral mesh for a general three-dimension domain is still not available, especially for meshing a large scale complex geometry containing faults (discontinuities).

Due to difficulties in automatic generation of such a high quality all-hexahedral mesh, several methods on special classes of geometry have been proposed and become the easier ways to go, such as (1) the Octree method (e.g. Schneiders et al. 1996; Loic 2001), which generates small cubes inside the geometric model and generates a mesh by mapping the surfaces of the boundary cubes onto the surfaces of the geometric model; Normally, it is difficult to fit the boundary well, thus its application is quite limited; (2) the mapped method, which firstly decomposes the whole geometry to one or several meshable blocks before meshing (Shin and Sakurai 1996; Taghavi 2000; Calvo and Idelsohm 2000); In addition, a shape recognition and boundary fitting based method is also proposed (Takahashi and Shimizu 1991; Chiba et al. 1998), which employs a unique shape-recognition technique to change a geometric model into an approximate one consisting of straight lines. Boundary fitting maps small cubes that are generated by dividing the approximate model, onto the geometric and generate hexahedral meshes. However, this sometimes can not be always successfully applied in case of some complicated models, such as (a) If the geometric model contains surfaces which has three or fewer edges, a recognition model that is topologically equal to the geometric model cannot be generated; and (b) If the assigned edge directions are not correct, a recognition model cannot be generated even if the geometric model is topologically correct. This method has then been improved by automating the model-editing task using feature line extraction (Hariya et al. 2006), but it still just works case-by-case; (3) sweeping method (e.g. Scott et al. 2005). To ensure that the geometry be meshed with all-hexahedral finite elements, sweeping requires that the geometry should be 2.5D or decomposable into 2.5D sub-geometries. It includes the following two ways: One-to-One sweeping and Many-to-One or Many-to-Many Sweeping. As for One-to-One Sweeping (Scott et al. 2005), sweeping of "One-to-One" geometry begins by identifying "source", "target", and connected "guiding" curves/surfaces. The source surface is then usually meshed with quadrilaterals using an unstructured scheme such as paving (Blacker and Stephenson 1991). Each guiding curves/surfaces must also be meshed with a mapped or sub-mapped mesh. The surface mesh on the source is then swept or extruded one layer at a time along the mapped mesh on the guiding curves/surfaces toward the target mesh. This type of sweep is termed as "One-to-One" because of the One-to-One correspondence between the source and target surface. Due to the One-to-One sweeping's strict requirements, few geometry satisfy the topological constraints required to generate a swept mesh. The Many-to-One or Many-to-Many Sweeping methods have been proposed to decompose more complex geometry into 2.5D sweep "blocks" or "barrels" which then be sweepable (e.g. Lai et al. 2000; Shepherd et al. 2000; Knupp 1998; White et al. 2004). Sweepable geometries or geometries that may be decomposed into sweepable parts can be detected automatically with a fair amount of success (White et al. 2000); (4) converting from the tetrahedral shaped mesh. Normally, the tetrahedron mesh is much easier to be generated and can be converted to the hexahedron mesh, but such a kind of hexahedron mesh is normally in poor shape quality and with dramatically increased node and element numbers, thus not always acceptable for further finite element analysis. Related efforts have been attempted to re-generate the domain occupied by tetrahedral shaped mesh to the hexahedron mesh with high quality, but not always achievable (e.g. Owen et al. 1997, 2000).

The above methods are applied here to mesh the following practical geoscience problems. It is important to note that geometry decomposition is widely used as above for the hexahedral shaped mesh generation. For a continuous geometry, the actual decomposition of the geometry does not occur, only an internal characterization of sweep/mapped blocks; but for a discontinuous geometry containing faults, the discontinuous boundaries (faults) will be used as a constraint for the geometry decomposition for further mesh generation.

## 3.2 Fracture Dominated Reservoir System

Figure 1 shows a fault system in a certain fracture dominated gas reservoir. The faults at the upper surface are depicted as 19 curves, and the fault surfaces are straight along the vertical direction. Therefore, the sweeping method can be applied to generate the hexahedral shaped mesh for such a 2.5 dimension case. Normally, one may use all the faults at the upper surface as the constrained curves to generate the quadrilateral shaped mesh without additional geometrical operation, and then generate the hexahedral shaped mesh using the sweeping method, i.e. take the quadrilateral shaped surface mesh as front and then advance inward. It may work well for a normal continuous domain. However, for such a complicated fracture dominated reservoir system, this will make all the meshes continuous without taking all the faults as the real discontinuous boundaries (curves/surfaces) for the quadrilateral/hexahedral shaped mesh in 2D/3D, and the meshing result even fails when the constrained curves are complicated and interacted with each other, as shown in Fig. 2 with the current fault system. Therefore, the whole geometry is firstly decomposed into a number of simple meshable shaped geometries (i.e. triangular or/and quadrilateral shape) with the constraint of the fault curves as shown in Fig. 3 and put into different groups; secondly, the quadrilateral shaped mesh is generated as above but sharing the same mesh seed at the common edges (curves) between the neighbour groups; thirdly, the hexahedral shaped mesh is generated using the sweeping method for the different groups; finally, the above hexahedral meshes are assembled together with "welding" operation (i.e. node equivalent), while keeping the meshes along the faults being discontinuous, which are taken as the frictional contact interfaces in the further finite element analysis. More regular and fine meshes are used around the faults, while coarse meshes are used in the other regions. The current model consists of about 46,000 hexahedron elements with 65,000 nodes and all the 19 faults (see Fig. 4).

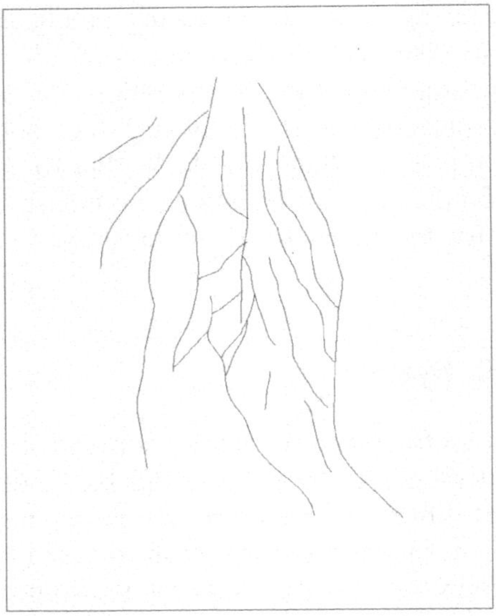

**Fig. 1** The geometry and distribution of 19 faults in a certain fracture dominated gas reservoir

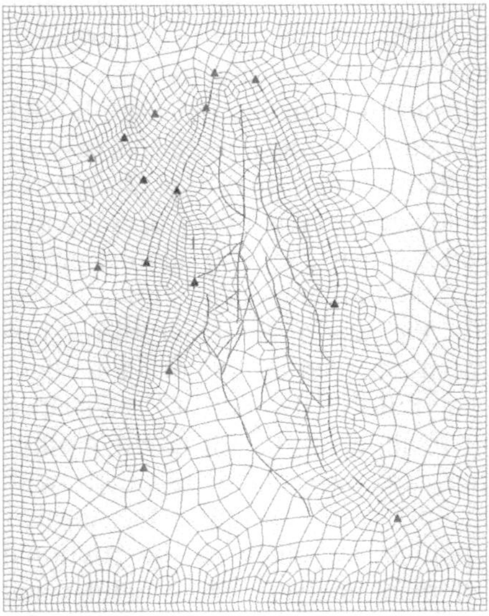

**Fig. 2** A failed example of the quadrilateral shaped mesh generation on the *top* surface using all the 19 complicated faults as the *constrained curves*, because only the faults marking with *black triangular* are really involved for such an automatic operation

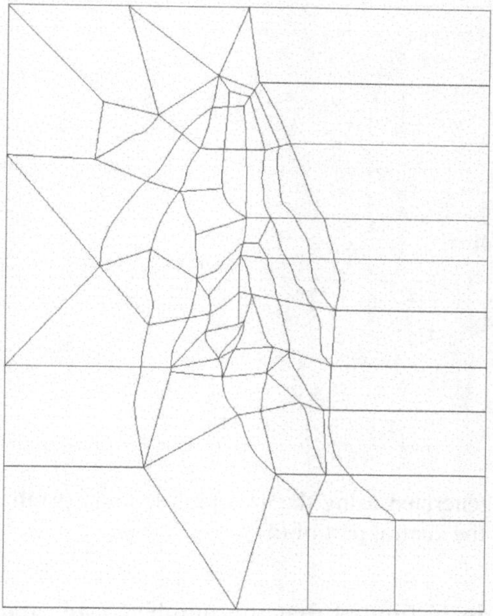

**Fig. 3** The *top* surface of the above fault system is decomposed into the 2D meshable geometries for the quadrilateral shaped mesh generation with the constraint of all the 19 *fault curves*

(a)

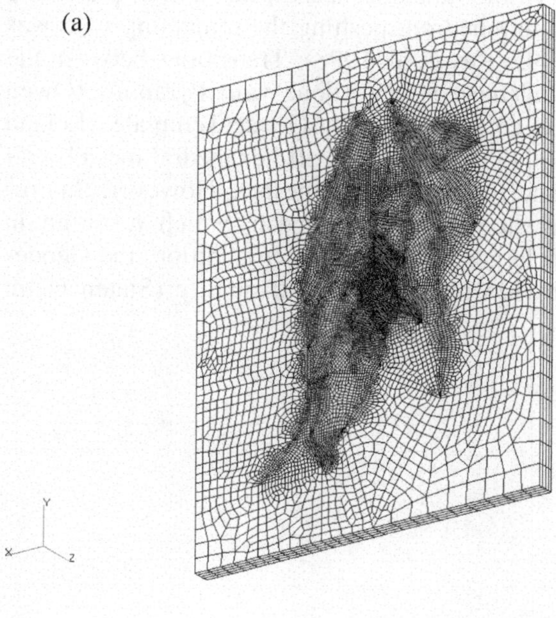

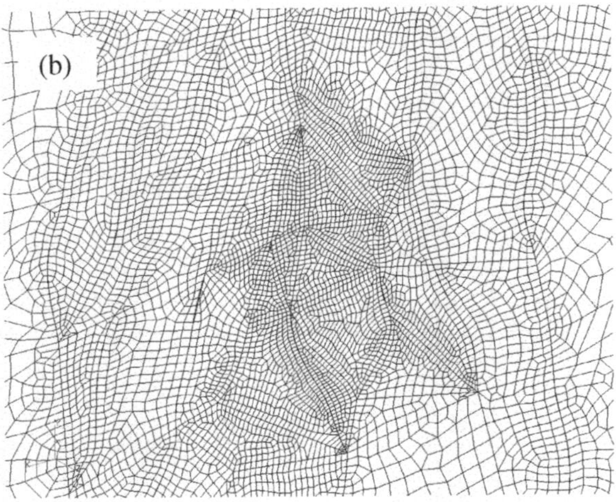


**Fig. 4** The hexahedral shaped mesh generated using the sweeping method (**a**) the whole mesh and (**b**) magnification of the central part of (**a**)

The sweeping method used here is limited that the number, size, and orientation of the quadrilateral faces on opposing fronts direction should match, thus it is rarely able to resolve the unmeshed voids and real general 3D problems. Once the opposing fronts collide, the algorithm frequently has deficiencies. Many creative attempts have been made to resolve this unmeshed void left behind by plastering. For examples, since arbitrary 3D voids can be robustly filled with tetrahedral meshes, the idea of plastering in a few layers, followed by tetrahedron-meshing the remaining void was attempted (Dewhirst et al. 1995; Ray et al. 1998); Transitions between the tetrahedral and the hexahedral meshes were done with Pyramids (Owen et al. 1997) and multi-point constraints; The Geode-Template (Leland et al. 1998) provided a method of generating an all-hexahedral mesh by refining both the tetrahedral and the hexahedral meshes. However, this requires an additional refinement of the entire mesh, which resulting in node/element numbers much larger than required. In addition, the Geode-Template was unable to provide reasonable mesh quality (Staten et al. 2005).

## 3.3 Meshing Interacting Fault System of South Australia with Mapped Block Method

The South Australian interacting fault system was chosen as a representative intraplate fault system. This choice was based on South Australia being reasonably representative of an active fault system in Australia and the availability of sufficient data and geological expertise for this region to enable a mesh be constructed. With detailed fault data based on advanced digital images (e.g. Fig. 5a) and geological knowledge of the region, provided by Professor Mike Sandiford of Melbourne University, a few of faults along the vertical direction are not straight along the vertical direction, it is a real 3D case, thus the mapped block method rather than the sweeping method is applied here. A 3D fault geometry model within a block with dimensions of about $530 \times 350 \times 60\ km^3$ (Fig. 5b) was firstly constructed. This involved editing and smoothing the related curves/surfaces defining the faults. In order to easily generate the hexahedral mesh and specify the conditions necessary for the finite element simulation (i.e. boundary conditions and information about faults), the entire geometric model of faults was firstly divided into several different geometrical components/blocks representing components of the solid model (e.g. Fig. 6a, b, c, d, e, and f). All of them are meshable using the mapped block meshing method, and a few of them can also be meshed by sweeping method (such as that in Fig. 6c and 7c); and these blocks were then used to generate finite element meshes (e.g. see Fig. 7a, b, c, d, e and f). Finally, the hexahedral meshes generated for the different components were assembled together with "welding" operation (i.e. node equivalent), or our stick contact algorithm after meshing (Xing et al. 2007a) (see Fig. 8). As shown in Fig. 8, more regular and fine meshes are used around the faults, while coarse meshes are used in the other regions. This approach enables more accurate computational results be obtained with reasonable finite computational resources. The discretised model after optimization currently includes 504,471 nodes and 464,620 8-node hexahedron elements together with 9 contact interfaces (i.e. faults). It was further analyzed by using our finite element code (Xing and Mora 2006).

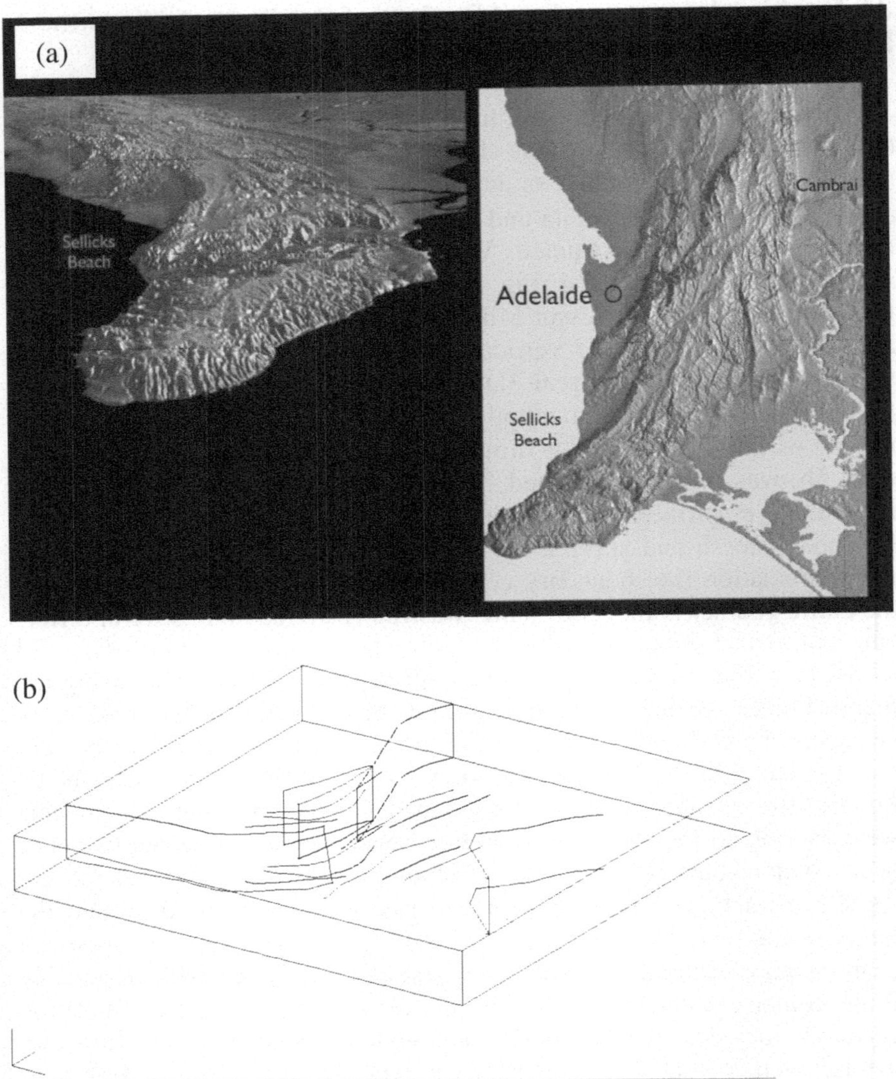


**Fig. 5** The South Australian interacting fault system (**a**) Image of the region of South Australia being considered; (**b**) 3-D fault model to be analyzed

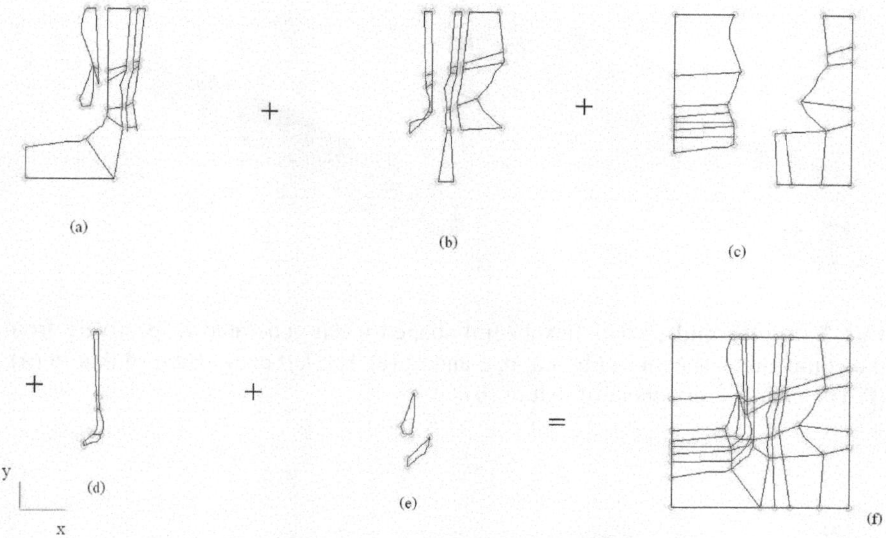


**Fig. 6** The entire geometric model of SA fault system is decomposed into several different geometrical components/blocks and shown in the XY plane

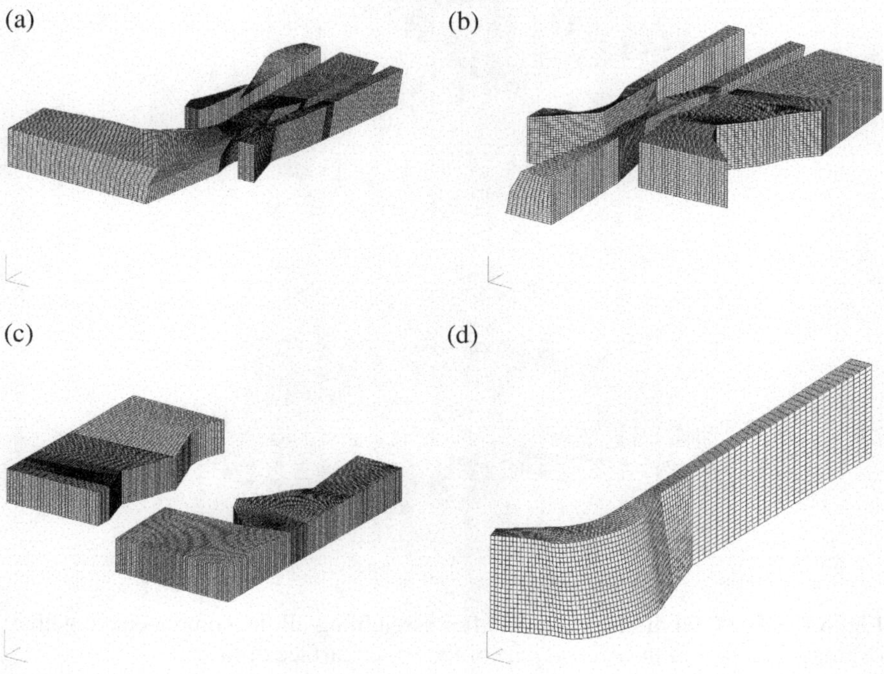

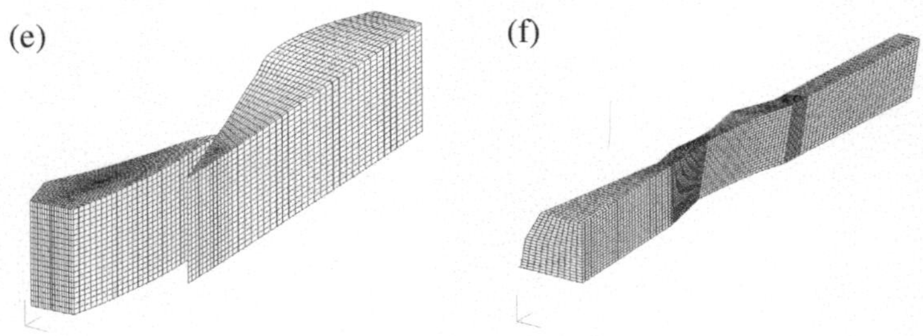


**Fig. 7** (**a**, **b**, **c and d**) The hexahedral shaped mesh generated respectively from the components shown in Fig. 6a, b, c and d; (**e**) The *left* component of that in (**a**); (**f**) The *middle* component of that in (**b**)

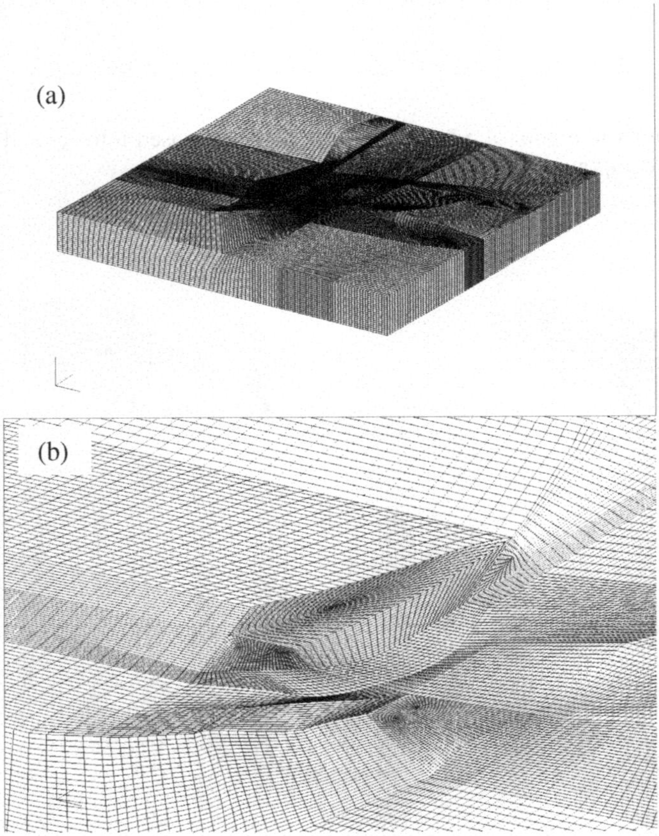


**Fig. 8** (**a**) The total mesh generated after assembling all the components together; (**b**) magnification of the *central part* of the *upper* surface of (**a**)

## 3.4 All Hexahedron Mesh Generation for a Whole-Earth Model

Mesh generation for a certain region in the reservoir and the fault system scale is described above, but sometimes the whole-earth model is highly required, such as for the tide deformation/stress analysis, deformable whole-earth rotation, the whole-earth system analysis and the interaction of a planetary system with the earth. Geophysical earth-models have advanced from the simplest Guttenburg-Bullen Model to the more complex Preliminary Reference Earth Model (PREM) (Dziewonski and Anderson 1981) and current 3D models including plate/fault boundaries. This provides more opportunities for the further analysis with a more accurate earth model, while it is a great challenge for the mesh generation before the finite element analysis could be carried out. Yin-Yang grid (Kagayama et al. 2004) used in the mantle convection simulation of the whole-earth, which is easily generated, but the meshes with Yin-Yang method are partly overlapped and not suitable for the finite element modeling. Our efforts on all-hexahedral shaped mesh generation of the whole-earth of model without/with considering the real plate boundaries are introduced here.

### *3.4.1 The PREM whole-Earth model*

The PREM model (Dziewonski and Anderson 1981) is a widely applied whole-earth model. It is composed of the inner core, the outer core, the mantle, the transition zone and the crust. For simplicity, we simplify it to a four-layered geophysical earth-model: the inner core, the outer core, the mantle and the crust (including the transition zone), as shown in Fig. 9. For convenience, the mapped block technique is applied here to decompose the whole spherical earth to different meshable blocks (groups). Each layer of the above whole-Earth model consists of 6 groups, thus 24 groups are generated in the whole-Earth model (Fig. 9), the model construction and mesh generation are thus much more simplified. The whole-Earth is discretised into 44,602 nodes and 43,008 hexahedron elements for the continuous case. For simplicity, both the same and the different material parameters can be assigned in each layer (but varying amongst the different layers) for the further finite element analysis.

For a certain case for simply analyzing the fault effects, such as the tidal deformation of the entire earth with a discontinuous outer layer (Xing et al. 2007b), a simplified fault (i.e. a discontinuous seismogenic interface) is enough, which can be assumed to exist in the outer layer between the groups g_1 and g_1_3 as shown in Fig. 9. But the practical plate boundaries with the detailed geometry and fault properties (i.e. frictional contact parameters along the plate boundaries) are necessary for such as earthquake dynamics analysis.

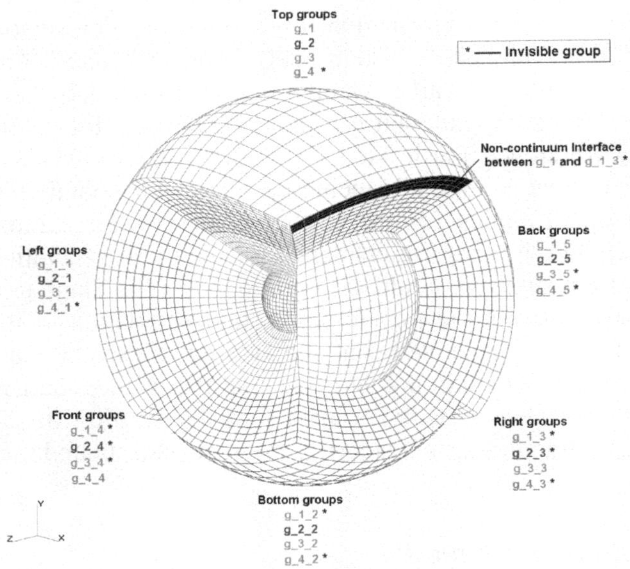


**Fig. 9** The entire geophysical Earth-model to be analysed. It is composed of four layers (from the *inside* to the *outside*): the *inner code*, the *outer core*, the *mantle* and *outer layer* (crust), and each layer consists of 6 groups

### *3.4.2 The Whole-Earth Crust with Plate Boundaries*

As described above, various earth models are available now. Here, the plate boundary data from the Plates Project at Texas University is applied to construct the whole-earth discontinuous crustal model (Fig. 10. See the detailed data at http://www.ig.utexas.edu/research/projects/plates/index.htm). It seems to be impossible to mesh the crustal layer with such complicated plate boundary geometries by directly using the above meshing methods.

For simplification, we firstly construct and edit the fault data in the longitude-latitude plane coordinate system (i.e. with the map Mercator projection), and create the curves with the input point data of the plate boundaries, then divide it into three following domains respectively: Domain A, ranging for the longitude [–180°, 180°] and the latitude [–90°, –85°]; Domain B: longitude [–180°, 180°] and latitude [85°, 90°]; and Domain C: longitude [–180°, 180°] and latitude [–85°, 85°]. The former two domains A and B are similar and the corresponding zone can be easily meshed, but the latter domain C are much more complicated due to the discontinuity from all the plate boundaries. Due to shortage of the dip angle data of the plate boundaries, we take all the faults are straight along the radial direction of the earth. The following procedure is then applied to generate the hexahedral shaped mesh of the whole model including the above plate boundaries: (1) The geometrical domain C is decomposed to several meshable sub-domains with the constraints of the plate boundaries as shown in Figs. 10 and 11a; (2) It is further grouped into different groups, and each group is meshed to the quadrilateral shaped mesh using paving/mapped mesh method and then transformed to an independent location for easy depiction, see Fig. 11b, c, d, e and f; (3) The above quadrilateral shaped mesh in Fig. 11b, c, d, e and f are respectively used as seed (fronts) to advance inward for the hexahedral shaped mesh generation with the sweeping method and then projected back to the 3D spherical space from the above projected plane coordinate system, as correspondingly shown in Fig. 12a, b, c, d and e. Here the crustal thickness is taken as 60 km; (4) Once the domain C is meshed, the blocks relevant with domains A and B can be easily meshed with the mapped method as described above but with the compatibility to the existing mesh/mesh seed along the common surfaces/edges shared with domain A and B, as denoted by A in Fig. 13a and b; (5) The hexahedral shaped meshes generated for the different groups/domains are assembled together with "welding" operation (i.e. node equivalent) except the nodes along the current plate boundaries and potential faults (that is treated as sticking frictional state when simulated using our finite element code (Xing et al. 2007a)), Figs. 13a, b and c. The generated meshes are optimized, and then the fault information could be extracted as shown in Fig. 14.

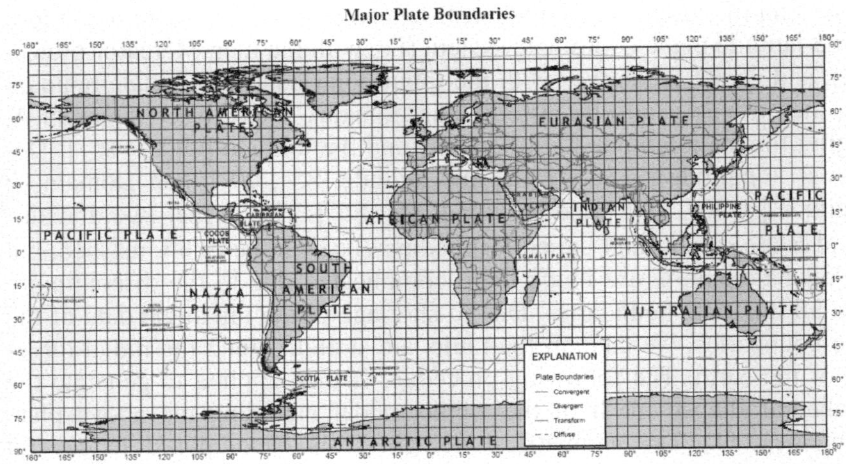


**Fig. 10** The major plate boundaries of the entire Earth to be analyzed (courtesy of the U.S. Geological Survey)

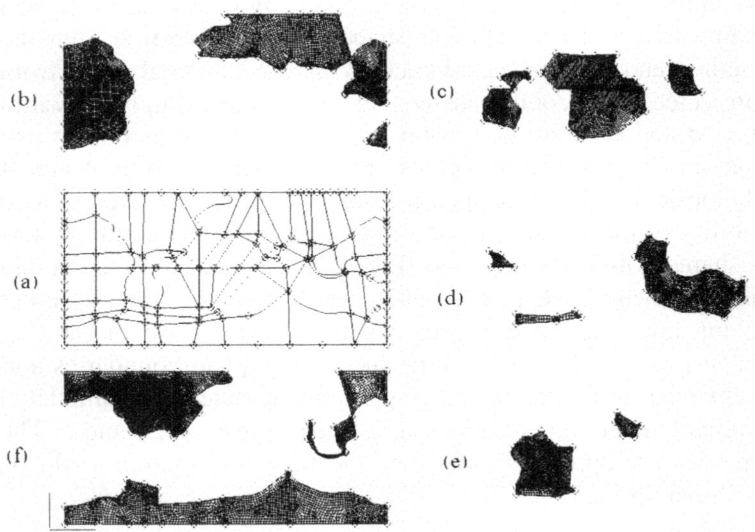


**Fig. 11** The whole earth is divided into 3 domains A, B and C (a) The geometrical domain C here is decomposed to several meshable sub-domains (groups) with the constraints of the plate boundaries; (**b**, **c**, **d**, **e** and **f**) The quadrilateral shaped mesh generated with each group within the geometrical domain C (transformation is applied to each group in the figures for easy depiction)

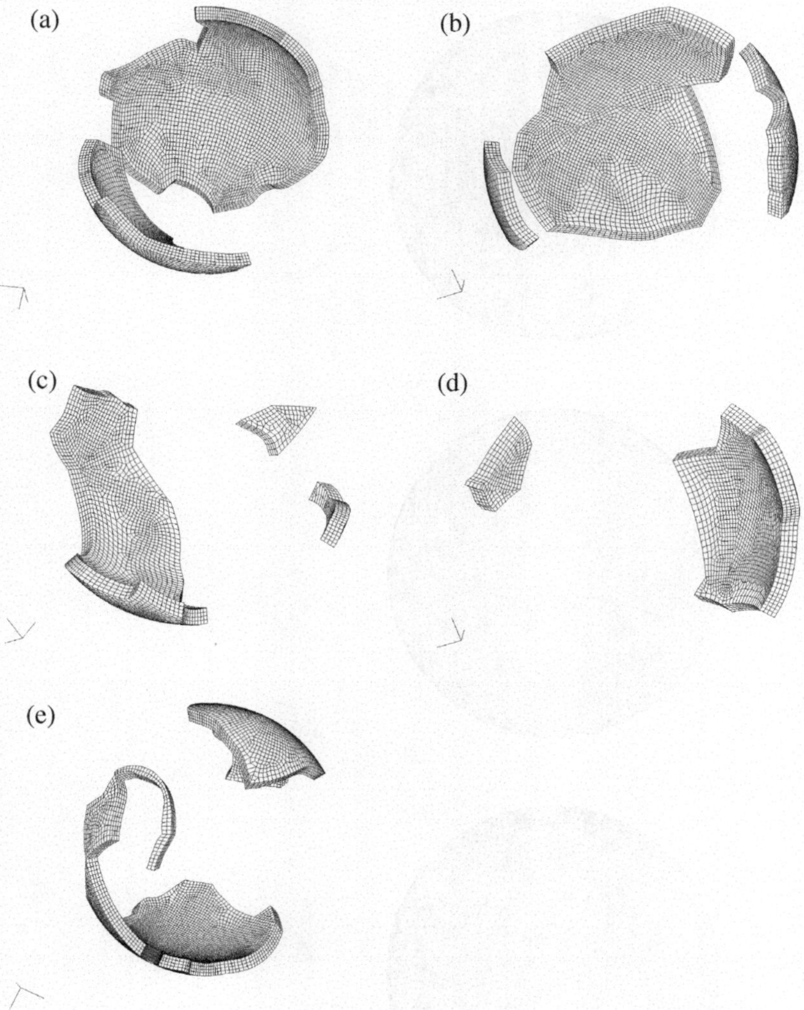


**Fig. 12** (**a**, **b**, **c**, **d** and **e**) The hexahedral shaped mesh is generated using the sweeping method with the above quadrilateral shaped mesh respectively in (**b**, **c**, **d**, **e** and **f**) as seed (*fronts*) and then transferred to the 3D XYZ space from the above projected longitude-latitude plane

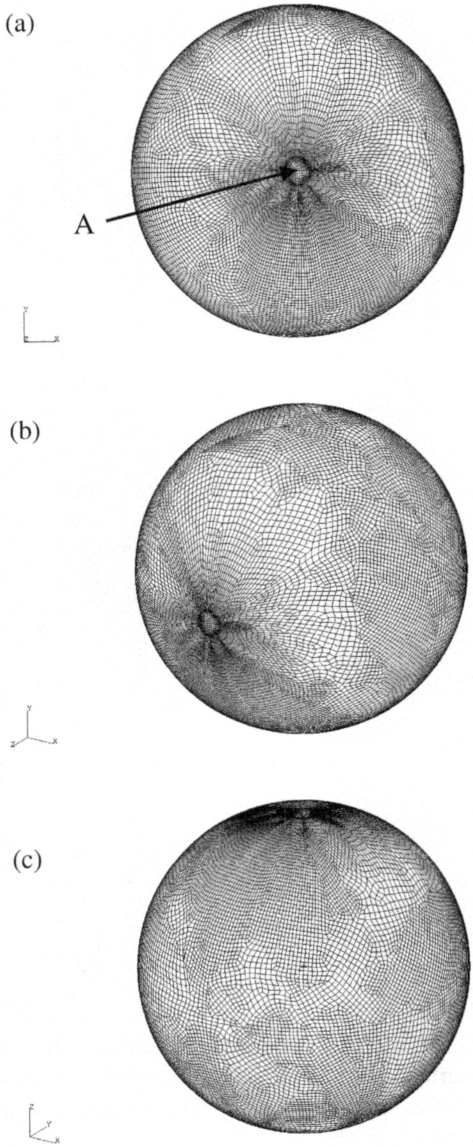


**Fig. 13** The hexahedral shaped mesh generated after assembling all the components together in the whole domain (A, B and C) and viewed in the following coordinate systems (**a**) XY plane, (**b**) ZXY space and (**c**) XYZ space. A in (**a**) and (**b**) shows the meshes generated in the subdomain A or B after assembling with the meshes generated in the domain C

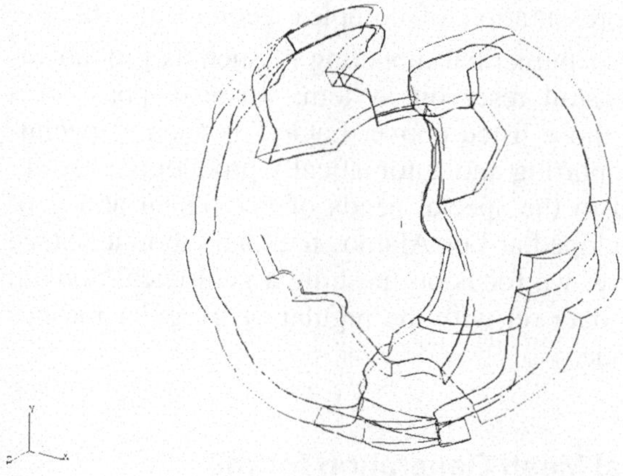

**Fig. 14** The geometry and shape of the discontinuous surfaces (including the plate boundaries) in the hexahedral shaped mesh generated

## 4 Tetrahedral Mesh Generation

### 4.1 Introduction

Besides the hexahedral mesh as above, the tetrahedron is also very popular for mesh generation in 3D. All the above methods for all-hexahedral mesh generation could be directly used for the tetrahedral mesh generation, because a hexahedral mesh can be easily divided into tetrahedral meshes. The great advantage of tetrahedral shaped mesh generation over the above hexahedron is that it could be much easier to be automatically generated for a general physical domain using the Delaunay algorithm and advancing front technique et al. (see Table 1). The Delaunay algorithm (Delaunay 1934) is most widely used and has two following important properties: (1) the empty circle (or empty sphere) criterion, which means no node should be contained in the circumcircle (or circumsphere) of any triangle (or tetrahedral) element; (2) the maximum–minimum angle criterion, which means the smallest angle in the Delaunay triangular mesh is the largest one in all possible triangular mesh of a given node set. Such criteria were used and extended in mesh generation by Bowyer-Watson (Watson 1981), Baker (1989), Weatherill (1994) and George (1991) et al. For more details, please refer to the related surveys as above.

Tetrahedral shaped mesh generation has a wide variety of geological applications for accurate representation of complex geological structure and stratigraphy for the further numerical modeling of such as groundwater resource development, gas/oil reservoir system, waste disposal in a geologic repository. Besides those listed above, LaGriT is such a special-purposed software tool for importing and automatically producing unstructured tetrahedral mesh tuned to the special needs of geological and geo-engineering applications developed at Los Alamos national laboratory (see http://meshing.lanl.gov/). Here, we focus on meshing a geological domain based on the available point data set with the regular or irregular meshes using our mesh generation code.

## 4.2 Automatic Tetrahedral Mesh Generation for the Stratigraphy Point Set

With the advanced measurement technique, the stratigraphy information in a certain geological domain is widely available, which may be described by a huge amount of point data. Therefore, the boundary/material surfaces can be extracted and defined by the relevant point data, which can be further described as triangular irregular networks (TINS). Figure 15a shows an example for a certain given domain to be analysed, which is composed of 3 different materials as denoted by different colours. The top surface and the setting of the different surfaces are shown in Fig. 15b and c, respectively. For a single stratigraphy, besides the above surfaces (i.e. the top and bottom surfaces), the other surfaces (the user defined normal regular plane to define the range of the above domain, i.e. the four vertical surfaces of this example) can be more easily described by triangular meshes which matching with the existing neighbour edges of the top and bottom surfaces), thus all these triangular surfaces could form a closed volume/representation of this single stratigraphy. And then a node distribution inside the closed volume domain is applied according to the prescribed characters, such as the material properties. Finally it is meshed into tetrahedral shaped meshes by using the Delaunay algorithm and further optimization. Here, assuming the material interfaces and the middle layer need to be specially addressed and thus the finer meshes generated around those areas correspondingly, Fig. 16a and b.

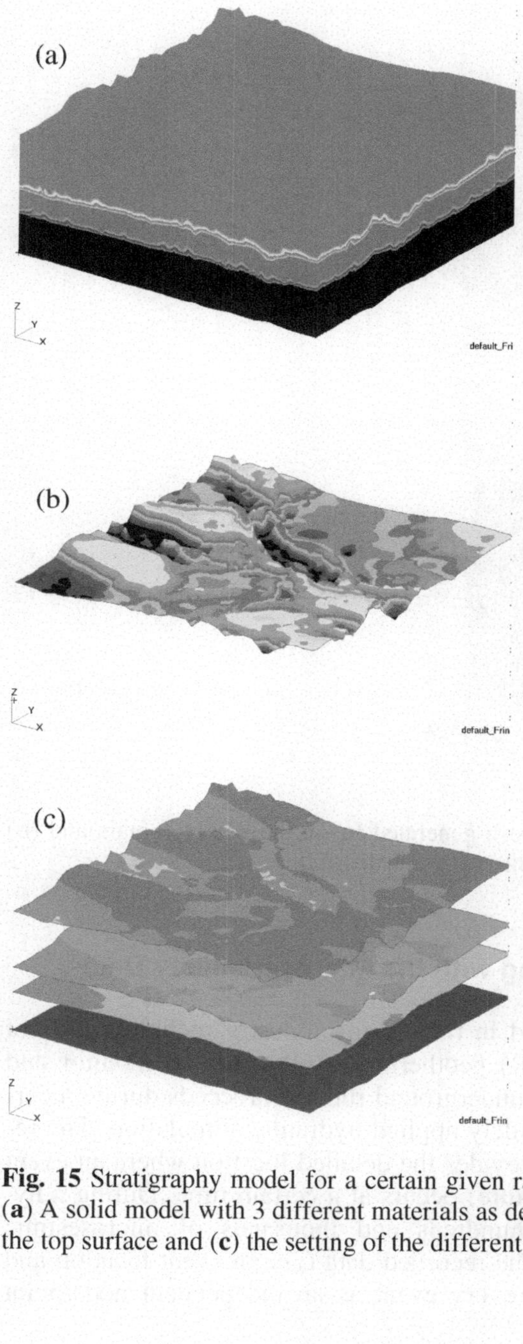


**Fig. 15** Stratigraphy model for a certain given range of the region to be analysed (**a**) A solid model with 3 different materials as denoted using different colours; (**b**) the top surface and (**c**) the setting of the different surfaces of stratigraphy model

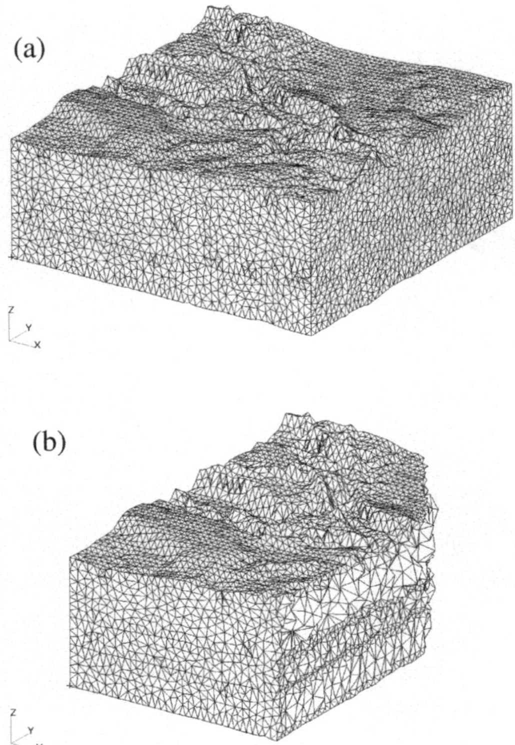


**Fig. 16** The tetrahedral shaped mesh generated (**a**) in the whole domain and (**b**) the specified domain to show the internal mesh distribution

## 4.3 Visualizing and Meshing with the Microseismicity Data

Microseismicity is widely used in the mining industry including the hot dry/fractured rocks (HDR/HFR) geothermal exploitation to monitor and determine where and how the underground rupture proceeds during a certain processing, such as the widely applied hydraulic stimulation. The recorded microseismicity data provides the detailed location where an event (i.e. underground dynamic rupture) occurs at a certain time. During a hydraulic stimulation process, hundreds and thousands of microsesimic events are recorded. With all the recorded data (i.e. an event location and its occurrence time), we take every event as an independent node/point with its location in 3D space as recorded and colour it with its occurrence time, thus we can directly know where the underground dynamic rupture locates and how it proceeds with the time from the above information. Fig. 17a, b, c, d, e, f and g show an example within a certain range, which includes 11,724 events over 52 days recorded during the hydraulic stimulation process by the Geodynamics Limited (see Movie 1 for more details).

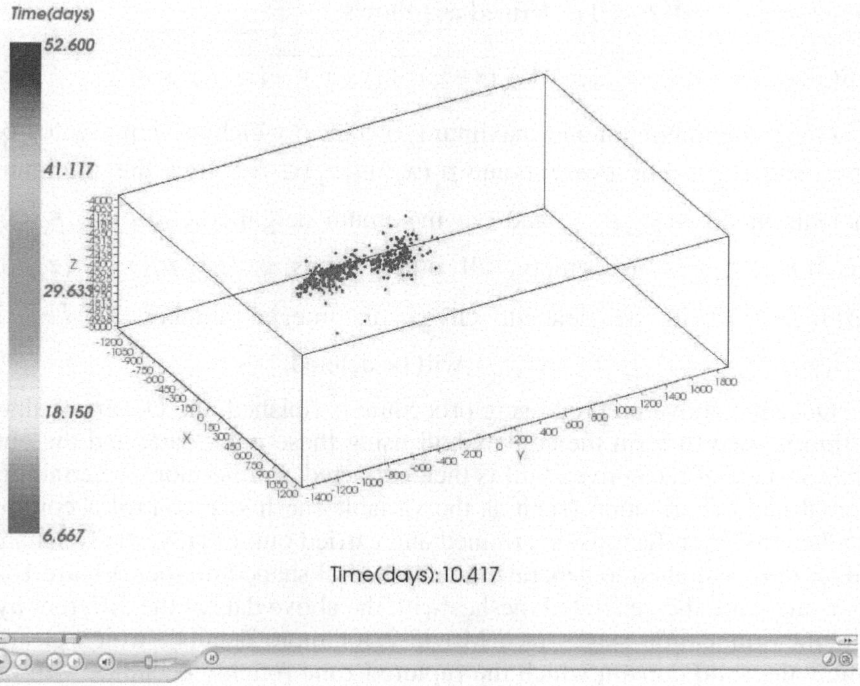


**Movie 1** Visualization of the micro-seismicity proceeding with time in a hydraulic stimulation process (available on accompanying DVD)

Moreover, we want to generate the mesh using the recorded data for determining the solid domain of the ruptured zone and the further numerical analysis. The Delaunay algorithm may be the most suitable mesh generation method for such a point set and thus is applied here. However, the recorded data are not suitable to be directly used for mesh generation, because there may be lots of coincided points (including those located too close) as well as some points which are far away from the main body data. A preprocess before mesh generation is taken as follows.

Define the recorded microseismicity data as the following scattered point set in 3D,

$$N = \{(x, y, z) \mid x, y, z \in \Re\} \tag{1}$$

Given a point $p_i(x_i, y_i, z_i) \in N$ and a tolerance $\varepsilon \in \Re$, the neighborhood point set $\delta(p_i, \varepsilon)$ is defined as follows:

$$\delta(p_i, \varepsilon) = \{(x, y, z) \mid (x, y, z) \in N, \sqrt{(x - x_i)^2 + (y - y_i)^2 + (z - z_i)^2} < \varepsilon\} \tag{2}$$

Given the minimum and maximum tolerance, which are represented by $\varepsilon_{\min}$ and $\varepsilon_{\max}$. For every point $p_i(x_i, y_i, z_i) \in N$, find the minimum neighborhood $\delta(p_i, \varepsilon_{\min})$ and the maximum neighborhood $\delta(p_i, \varepsilon_{\max})$. If $\delta(p_i, \varepsilon_{\min})$ is not empty, all other points except $p_i(x_i, y_i, z_i)$ in $\delta(p_i, \varepsilon_{\min})$ will be deleted. Given an integer number $\lambda \in I^+$, if $|\delta(p_i, \varepsilon_{\max})| < \lambda$, $p_i(x_i, y_i, z_i)$ will be deleted.

Once the above preprocessing procedure is finished, the Delaunay algorithm is used to form the convex hull using these point data, and the outside surface of the convex hull is then extracted. Furthermore, a certain internal node distribution (such as the variable mesh size control according to the outside surface) is determined and carried out. Finally, the Delaunay algorithm is applied to generate the tetrahedral shaped meshes. Figure 18a, b, c and d are the generated meshes with the above data at the different hydraulic stimulation stages (see Movie 2 for more details), which clearly show the solid domain which the ruptured zone roughly occupies with the microseismicity rupture proceeding and can also be used for the further finite element analysis. Here, the related parameters are set as: $\varepsilon_{\min} = 10$, $\varepsilon_{\max} = 100$, $\lambda = \log(|N|)$.

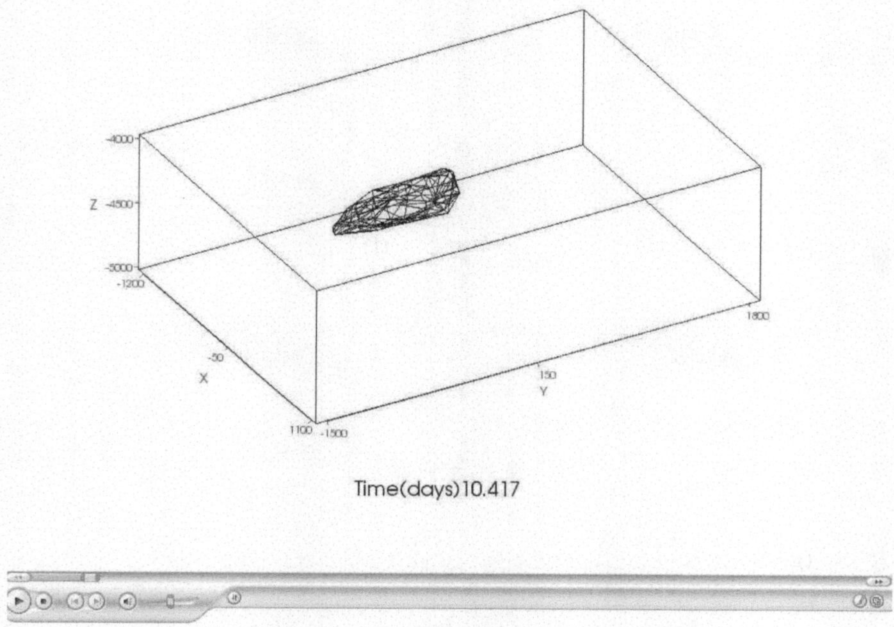


**Movie 2** Visualization of the solid domain of the micro-seismicity rupture zone proceeding with time in a hydraulic stimulation process (available on accompanying DVD)

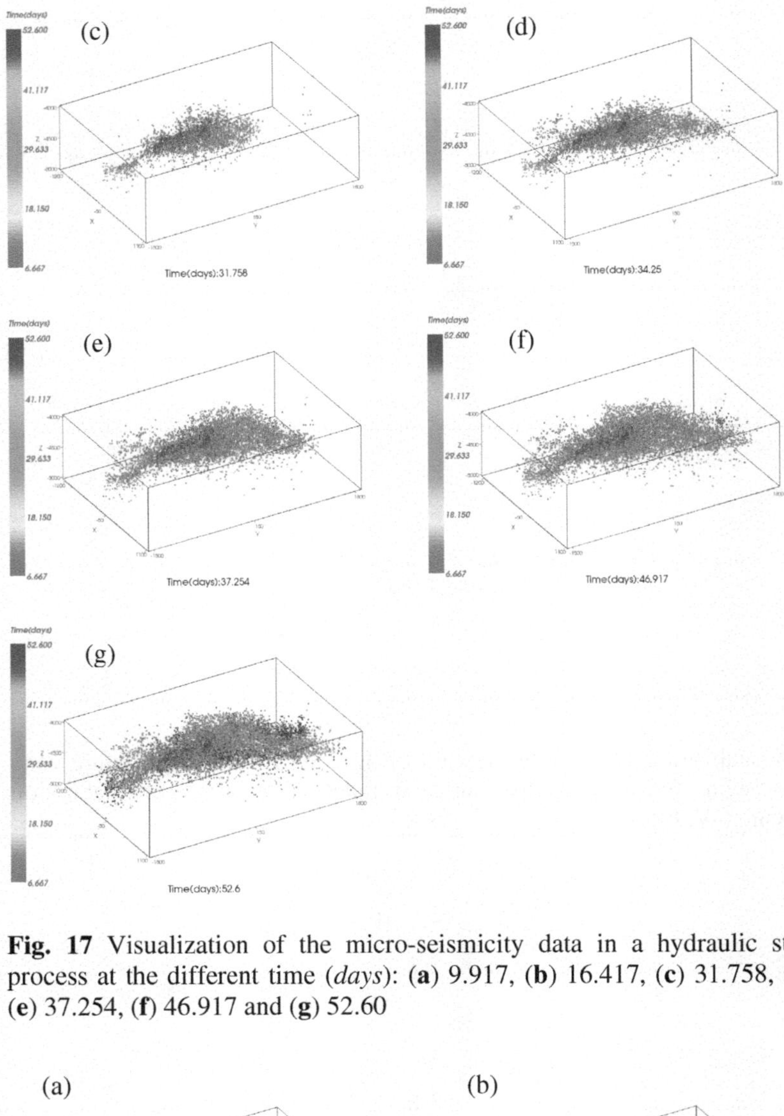


**Fig. 17** Visualization of the micro-seismicity data in a hydraulic stimulation process at the different time (*days*): (**a**) 9.917, (**b**) 16.417, (**c**) 31.758, (**d**) 34.25, (**e**) 37.254, (**f**) 46.917 and (**g**) 52.60

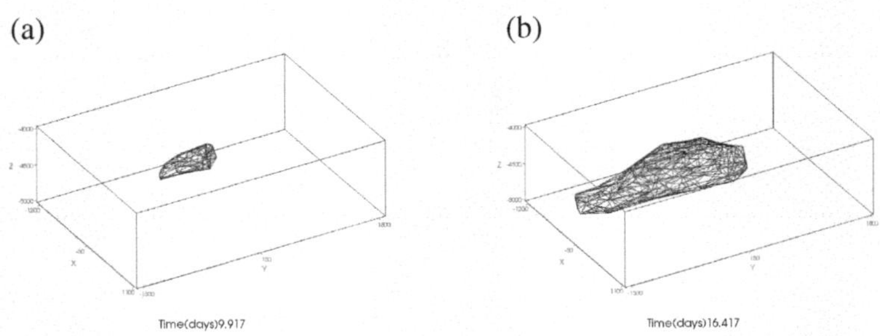

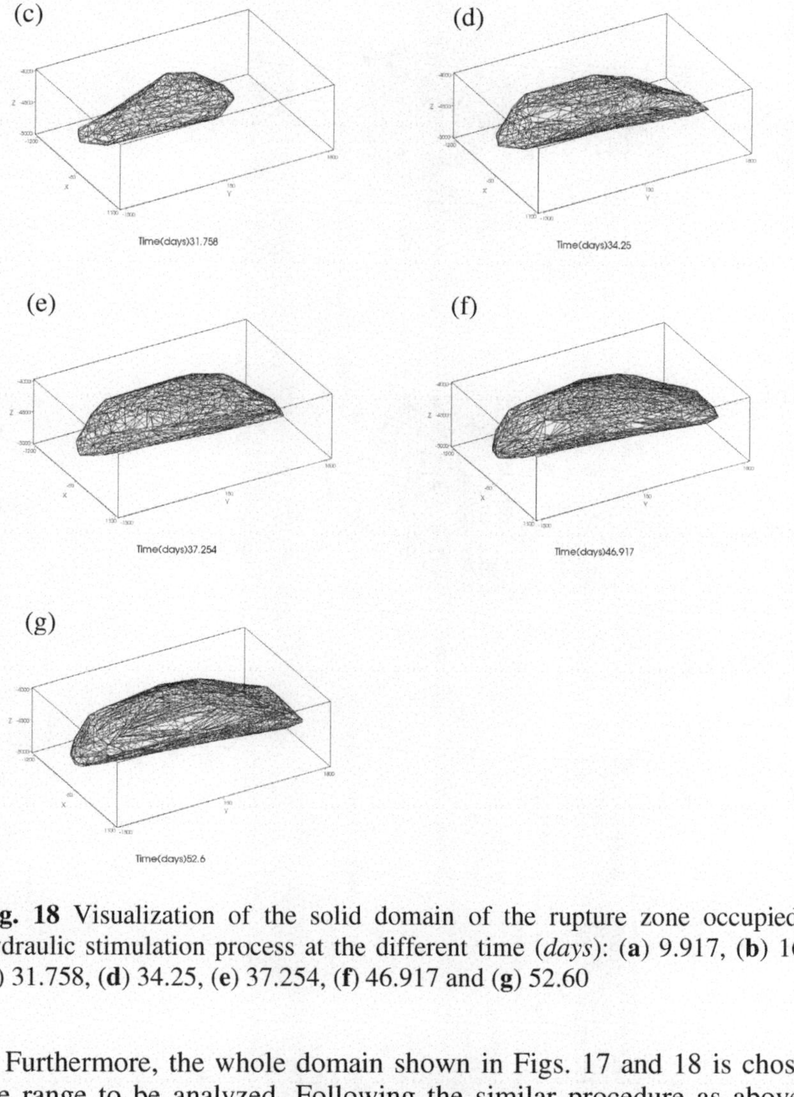


**Fig. 18** Visualization of the solid domain of the rupture zone occupied in a hydraulic stimulation process at the different time (*days*): (**a**) 9.917, (**b**) 16.417, (**c**) 31.758, (**d**) 34.25, (**e**) 37.254, (**f**) 46.917 and (**g**) 52.60

Furthermore, the whole domain shown in Figs. 17 and 18 is chosen as the range to be analyzed. Following the similar procedure as above, the tetrahedral meshes generated using Delaunay algorithm are shown in Fig. 19a, b and c. Here, the mesh size on the 6 outer surfaces is taken as the same, but the nodal positions inside (i.e. mesh size) are controlled by the above microseismic data. Therefore, the inside mesh is quite heterogeneously distributed and its mesh size appears much smaller around the microseismicity concentrated zone, see Figs. 19 b and c.

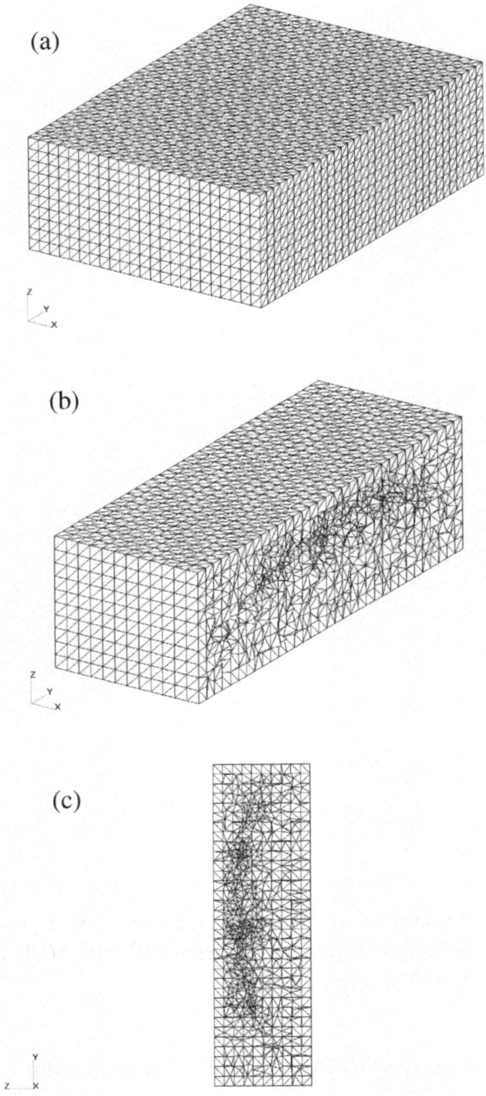


**Fig. 19** The tetrahedral shaped mesh generated (**a**) in the full domain; (**b**) and (**c**) the internal mesh distribution seeing through a cross-section depicted in the XYZ space and the YZ plane respectively

## 5 Conclusions

Mesh generation is widely and successfully applied in the engineering computing, but it is relatively "new" for the geoscience community. This paper briefly introduces the relevant progresses and then discusses the possibility for applying and extending the above successful stories in engineering computing to geo-computing. For geoscience, the available input data are normally a large quantity of point data in the 3D space rather than the defined shapes and dimensions with reasonable tolerances provided by the industrial designers, thus quite different from the engineering cases. To deal with such geoscience data, this paper focuses on applying and/or extending the related methods to generate all hexahedral or tetrahedral shaped mesh in 3D for the different practical geoscience application examples based on the relevant progresses on geometrical modeling and hexahedral/tetrahedral shaped mesh generation. That includes all-hexahedral shaped mesh generation for a fracture dominated reservoir system, the South Australia interacting fault system and the entire earth models with the simplified/practical plate boundaries, and all-tetrahedral shaped mesh generation for a multi-layer underground geological model, and meshing with the microseismicity data recorded during a hydraulic stimulation process in a geothermal reservoir. It lays the foundation for the future research on such as the adaptive meshing and remeshing, parallel mesh generation as required from various different application cases.

**Acknowledgements** Support is gratefully acknowledged by the Australian Research Council Linkage project LP05609326 and Discovery project DP0666203, AuScope, the Queensland State Government, The University of Queensland, and SGI. The authors are grateful to Dr Doone Wyborn from the Geodynamics Limited, Professor Mike Sandiford from Melbourne University and Professor John Rundle from University of California at Davis for providing the related data used in this paper.

## References

Baker TJ (1989) Automatic Mesh Generation for Complex Three-Dimensional Regions Using a Constrained Delaunay Triangulation. Engineering with Computers, Vol 5 pp 161–175

Blacker T and Stephenson MB (1991) Paving: A New Approach to Automated Quadrilateral Mesh Generation. International Journal for Numerical Methods in Engineering, Vol 32 pp 811–847

Borouchaki H and Frey PJ (1998) Adaptive Trangular-Quadrilateral Mesh Generation. International Journal for Numerical Methods in Engineering, Vol 41 pp 915–934

Borouchaki H and George PL (1997) Aspects of 2-D Delaunay Mesh Generation. International Journal for Numerical Methods in Engineering, Vol 40 pp 1957–1975

Borouchaki H, George PL and Lo SH (1996) Optimal Delaunay Point Insertion. International Journal for Numerical Methods in Engineering, Vol 39 pp 3407–3437

Borouchaki H, Laug P and George PL (2000) Parametric Surface Meshing Using a Combined Advancing-Front Generalized Delaunay Approach. International Journal for Numerical Methods in Engineering, Vol 49 pp 233–259

Borouchaki H and Lo SH (1995) Fast Delaunay Triangulation in Three Dimensions. Computer Methods in Applied Mechanics and Engineering, Vol 128 pp 153–167

Calvo NA and Idelsohm SR (2000) All-Hexahedral Element Meshing: Generation of the Dual Mesh by Recurrent Subdivision. Computer Methods in Applied Mechanics and Engineering, Vol 182 pp 371–378

Cheng GD and Li H (1996) New Method for Graded Mesh Generation of Quadrilateral Finite Elements. Computers and Structures, Vol 59 Num 5 pp 823–829

Chiba N, Nishigaki I, Yamashita Y, Takizawa C and Fujishiro K (1998) A Flexible Automatic Hexahedral Mesh Generation by Boundary-Fit Method. Computer Methods in Applied Mechanics and Engineering, Vol 161 pp 145–154

Delaunay BN (1934) Sur la Sphere. Vide. Izvestia Akademia Nauk SSSR, VII Seria, Otdelenie Matematicheskii i Estestvennyka Nauk, Vol 7 pp 793–800

Dewhirst D, Vangavolu S, Wattrick H (1995) The Combination of Hexahedral and Tetrahedral Meshing Algorithms. Proceeding 4th International Meshing Roundtable pp 291–304

Du Q and Wang DS (2003) Tetrahedral Mesh Generation and Optimization Based on Centroidal Voronoi Tessellations. International Journal for Numerical Methods in Engineering, John Wiley & Sons, Ltd., Vol 56 Num 9 pp 1355–1373

Dziewonski AD and Anderson DL (1981) Preliminary Reference Earth Model. Physics of the Earth and Planetary Interiors, Vol 25 Num 2 pp 97–356

Frey P, Benoit S and Gautherie M (1994) Fully Automatic Mesh Generation for 3-D Domains Based Upon Voxel Sets. International Journal for Numerical Methods in Engineering, John Wiley, Num 37 pp 2735–2753

George PL (1997) Improvements on Delaunay-Based Three-Dimensional Automatic Mesh Generator. Finite Elements in Analysis and Design, Elsevier, Vol 25 pp 297–317

George PL, Hecht F and Saltel E (1991) Automatic Mesh Generator with Specified Boundary. Computer Methods in Applied Mechanics and Engineering, North-Holland, Vol 92 pp 269–288

George PL and Seveno E (1994) The Advancing-Front Mesh Generation Method Revisited. International Journal for Numerical Methods in Engineering, Wiley, Vol 37 pp 3605–3619

Hariya M, Nishigaki I, Kataoka I, Hiro Y (2006) Automatic Hexahedral Mesh Generation with Feature Line Extraction. Proceeding 15th Round Table Meshing pp 453–468

Joe B (1991a) GEOMPACK – A Software Package for the Generation of Meshes Using Geometric Algorithms. Advances in Engineering Software, Elsevier, Vol 13 Num 5 pp 325–331

Joe B (1991b) Construction of Three-Dimensional Delaunay Triangulations Using Local Transformations. Computer Aided Geometric Design, Elsevier Science Publishers (North-Holland), Num 8 pp 123–142

Joe B (1991c) Delaunay Versus Max–Min Solid Angle Triangulations For Three-Dimensional Mesh Generation. International Journal for Numerical Methods in Engineering, John Wiley & Sons, Vol 31 pp 987–997

Joe B (1995) Construction of Three-Dimensional Improved-Quality Triangulations Using Local Transformations. SIAM Journal of Science Computing, Vol 16 pp 1292–1307

Kageyama A and Sato T (2004) The 'Yin-Yang Grid': An Overset Grid in Spherical Geometry. Geochem. Geophys. Geosyst., Q09005, doi:10.1029/2004GC000734

Kanai T, Makinouchi A, Oishi Y (2000) Development of Tectonic CAD/Database Systems. In Abstracts of International Workshop on Solid Earth Simulation and ACES WG Meeting (Ed. Mitsu'ura M et al.). Tokyo, Jan 17–21

Knupp PM (1998) Next-Generation Sweep Tool: A Method For Generating All-Hex Meshes on Two-And-One-Half Dimensional Geomtries. Proceedings the 7th International Meshing Roundtable pp 505–513

Knupp PM (1999) Applications of Mesh Smoothing: Copy, Morph, and Sweep on Unstructured Quadrilateral Meshes. International Journal for Numerical Methods in Engineering, Wiley, Vol 45 pp 37–45

Lai MW, Benzley S and White D (2000) Automated Hexahedral Mesh Generation by Generalized Multiple Source to Multiple Target Sweeping. International Journal for Numerical Methods in Engineering, John Wiley, Vol 49 Num 1 pp 261–275

Lau TS and Lo SH (1996) Finite Element Mesh Generation Over Analytical Surfaces. Computers and Structures, Elsevier Science Ltd., Vol 59 Num 2 pp 301–309

Lee CK (2003) Automatic Metric 3D Surface Mesh Generation Using Subdivision Surface Geometrical Model. Part 2: Mesh Generation Algorithm and Examples. International Journal for Numerical Methods in Engineering, John Wiley & Sons, Ltd., Vol 56 Num 11 pp 1615–1646

Lee CK and Hobbs RE (1999) Automatic Adaptive Finite Element Mesh Generation Over Arbitrary Two-Dimensional Domain Using Advancing Front Technique. Computers and Structures Pergammon, Vol 71 pp 9–34

Lee YK and Lee CK (2002) Automatic Generation of Anisotropic Quadrilateral Meshes on Three-Dimensional Surfaces Using Metric Specifications. International Journal for Numerical Methods in Engineering, John Wiley & Sons, Ltd., Vol 53 Num 12 pp 2673–2700

Lee YK, Lee CK (2003) A New Indirect Anisotropic Quadrilateral Mesh Generation Scheme with Enhanced Local Mesh Smoothing Procedures. International Journal for Numerical Methods in Engineering, John Wiley & Sons, Ltd., Vol 58 Num 2 pp 277–300

Lee CK and Lo SH (1994) A New Scheme for the Generation of a Graded Quadrilateral Mesh. Computers and Structures Pergammon Vol 52 Num 5 pp 847–857

Lee DT and Schacter BJ (1980) Two Algorithms for Constructing a Delaunay Triangulation. International Journal of Computer and Information Sciences, Vol 3 Num 9 pp 219–242

Leland RW, Melander D, Meyers R, Mitchell S, Tautges T (1998) The Geode Algorithm: Combining Hex/Tet Plastering, Dicing and Transition Elements for Automatic, All-Hex Mesh Generation. Proceeding 7th International Meshing Roundtable pp 515–521

Lo SH (1991a) Volume Discretization into Tetrahedra-I. Verification and Orientation of Boundary Surfaces. Computers and Structures, Pergamon Press, Vol 39 Num 5 pp 493–500

Lo SH (1991b) Volume Discretization into Tetrahedra – II. 3D Triangulation by Advancing Front Approach. Computers and Structures Pergamon, Vol 39 Num 5 pp 501–511

Lo SH (1992) Generation of High Quality Gradation Finite Element Mesh. Engineering Fracture Mechanics, Pergamon, Vol 2 Num 41 pp 191–202

Lohner R (1996a) Progress in Grid Generation via the Advancing Front Technique. Engineering with Computers, Springer-Verlag, Vol 12 pp 186–210

Lohner R (1996b) Extensions and Improvements of the Advancing Front Grid Generation Technique. Communications in Numerical Methods in Engineering, John Wiley & Sons, Ltd., Vol 12 pp 683–702

Lohner R (1996c) Regridding Surface Triangulations. Journal of Computational Physics, Academic Press, Vol 126 pp 1–10

Lohner R and Cebral JR (2000) Generation of Non-Isotropic Unstructured Grids via Directional Enrichment. International Journal for Numerical Methods in Engineering, John Wiley, Vol 49 Num 1 pp 219–232

Lohner R and Eugenio O (1998) An Advancing Front Point Generation Technique. Communications in Numerical Methods in Engineering, Wiley, Vol 14 pp 1097–1108

Loic M (2001) A New Approach to Octree-Based Hexahedral Meshing. Proceedings 10th International Meshing Roundtable, pp 209–221

Owen SJ, Canann S, Siagal S (1997) Pyramid Elements for Maintaining Tetrahedra to Hexahedra Conformability. Trends in Unstructured Mesh Generation AMD, Vol 220 pp 123–129

Owen SJ, Saigal S (2000) H-Morph: An Indirect Approach to Advancing Front Hex Meshing. International Journal for Numerical Methods in Engineering, John Wiley & Sons, Ltd.. Vol 49 Num 1–2 pp 289–312

Owen SJ, Staten ML, Canann SA and Saigal S (1999) Q-Morph: An Indirect Approach to Advancing Front Quad Meshing. International Journal for Numerical Methods in Engineering, Wiley, Vol 9 Num 44 pp 1317–1340

Ray M, Tautges T, Tuchinsky P (1998) The "Hex-Tet" Hex-Dominant Meshing Algorithm as Implemented in CUBIT. Proceeding 7th International Meshing Roundtable pp 151–158

Rosenbaum G, Lister GS and Duboz C (2002) Reconstruction of the tectonic evolution of the western Mediterranean since the Oligocene. In: Rosenbaum, G. and Lister, G. S. 2002. Reconstruction of the evolution of the Alpine-Himalayan Orogen. Journal of the Virtual Explorer, Vol 8 pp 107–126

Schneiders R (1996) A Grid-Based Algorithm for the Generation of Hexahedral Element Meshes. Engineering with Computers, Vol 12 pp 168–177

Schneiders R (1997) An Algorithm for the Generation of Hexahedral Element Meshes Based on an Octree Technique. Proceedings 6th International Meshing Roundtable pp 183–194

Schneiders R and Bunten R (1995) Automatic Generation of Hexahedral Finite Element Meshes. Computer Aided Geometric Design, Elsevier, Vol 12 pp 693–707

Schneiders R, Schindler R and Weiler F (1996) Octree-Based Generation of Hexahedral Element Meshes. Proceedings 5th International Meshing Roundtable pp 205–216

Scott MA, Earp MN, Benzley SE and Stephenson MB (2005) Adaptive Sweeping Techniques. Proceedings the 14th International Meshing Roundtable pp 417–432

Shephard MS and Marcel KG (1991) Automatic Three-Dimensional Mesh Generation by the Finite Octree Technique. International Journal for Numerical Methods in Engineering, Wiley, Vol 32 pp 709–749

Shephard MS and Marcel KG (1992) Reliability of Automatic 3D Mesh Generation. Computer Methods in Applied Mechanics and Engineering, North-Holland, Vol 101 pp 443–462

Shepherd J, Mitchell SA, Knupp P, and White D (2000) Methods for Multisweep Automation. Proceedings of the 9th International Meshing Roundtable, Sandia National Laboratories, pp. 77–87

Shewchuk JR (2002) Delaunay Refinement Algorithms for Triangular Mesh Generation, Computational Geometry: Theory and Applications, Vol 22 Num 1–3 pp 21–74, May 2002

Shin BY and Sakurai H (1996) Automated Hexahedral Mesh Generation by Swept Volume Decomposition and Recomposition. Proceeding 5th International Meshing Roundtable pp 273–280

Staten ML, Canann SA and Owen SJ (1998) BMSWEEP: Locating Interior Nodes During Sweeping. Proceeding the 7th International Meshing Roundtable pp 7–18

Staten ML, Owen SJ, Blacker TD (2005) Unconstrained Paving & Plastering: A New Idea for all Hexahedral Mesh Generation. Proceedings the 14th International Meshing Roundtable pp 399–416

Taghavi R (2000) Automatic Block Decomposition Using Fuzzy Logic Analysis. Proceeding 9th International Meshing Roundtable pp 187–192

Takahashi H and Shimizu H (1991) A General Purpose Automatic Mesh Generation Using Shape Recognition Technique. Compt. Engrg. ASME, Vol 1 pp 519–526

Tu T and O'Hallaron DR (2004) Extracting Hexahedral Mesh Structures from Balanced Linear Octrees. Proceedings 13th International Meshing Roundtable Williamsburg VA pp 191–200

Watson DF (1981) Computing the Delaunay Tesselation with Application to Voronoi Polytopes. The Computer Journal, Vol 24 Num 2 pp 167–172

Weatherill NP and Hassan O (1994) Efficient Three-dimensional Delaunay Triangulation with Automatic Point Creation and Imposed Boundary Constraints. International Journal for Numerical Methods in Engineering, Wiley, Num 37 pp 2005–2039

White D, Saigal S and Owen S (2004) CCSweep: Automatic Decomposition of Multi-Sweep Volumes. Engineering with Computers, Vol 20 pp 222–236

White D, Tautges T and Timothy J (2000) Automatic Scheme Selection for Toolkit Hex Meshing. International Journal for Numerical Methods in Engineering, Vol 49 pp 127–144

Wright JP and Alan GJ (1994) Aspects of Three-Dimensional Constrained Delaunay Meshing. International Journal for Numerical Methods in Engineering, Wiley, Num 37 pp 1841–1861

Xing HL, Miyamura T, Makinouchi A, Homma T, Kanai T, Oishi Y (2001) Development of High Performance Finite Element Software System for Simulation of Earthquake Nucleation and Development, Exploration Geodynamics. (Eds. Moresi L, Muller D and Hobbs B). Western Australia, pp 178–188

Xing HL, Makinouchi A and Mora P (2007a) Finite Element Modeling of Interacting Fault System, Physics of the Earth and Planetary Interiors, Vol 163 pp 106–121. DOI 10.1016/j.pepi.2007.05.006

Xing HL and Mora P (2003) Mesh Generation Software Survey, Technical Report of ACcESS (ESSCC, The University of Queensland). pp 1–20

Xing HL and Mora P (2006) Construction of an Intraplate Fault System Model of South Australia, and Simulation Tool for the iSERVO Institute Seed Project. Pure and Applied Geophysics, Vol 163 pp 2297–2316. DOI 10.1007/s00024-006-0127-x

Xing HL, Zhang J and Yin C (2007b) A Finite Element Analysis of Tidal Deformation of the Entire Earth with a Discontinuous Outer Layer. Geophysical Journal International, Vol 170 Num 3 pp 961–970. DOI 10.1111/j.1365-246X.2007.03442.x

Yamakawa S and Shimada K (2003) Anisotropic Tetrahedral Meshing via Bubble Packing and Advancing Front. International Journal for Numerical Methods in Engineering, John Wiley & Sons, Ltd., Vol 57 Num 13 pp 1923–1942

Yerry MA and Shephard MS (1984) Automatic Three-Dimensional Mesh Generation by the Modified Octree Technique. International Journal for Numerical Methods in Engineering, John Wiley, Num 20 pp 1965–1990

# III. Strategies for Preconditioning Methods of Parallel Iterative Solvers for Finite-Element Applications in Geophysics

Kengo Nakajima

Information Technology Center, The University of Tokyo,
2-11-16 Yayoi, Bunkyo-ku, Tokyo 113-8658, Japan.

## 1 Background

### 1.1 Why Preconditioned Iterative Solvers?

Solving large-scale systems of linear equations $[A]\{x\}=\{b\}$ is one of the most expensive and critical processes in scientific computing. In particular, for simulation codes based on the finite-element method (FEM), most of the computational time is devoted to solving linear equation systems with sparse coefficient matrices. For this reason, a significant proportion of scalable algorithm research and development is aimed at solving these large, sparse linear systems of equations on parallel computers. Sparse linear solvers can be broadly classified as being either *direct* or *iterative*.

Direct solvers, such as Gaussian elimination and LU factorization, are based on a factorization of the associated sparse matrix. They are extremely robust and yield the exact solution of $[A]\{x\}=\{b\}$ after a finite number of steps without round-off errors. However, their memory requirements grow as a nonlinear function of the matrix size because initially zero components of the original matrix fill in during factorization. In contrast, iterative methods are memory scalable. Iterative methods are therefore the only choice for large-scale simulations by massively parallel computers. While iterative methods are memory scalable, a disadvantage is that their convergence can be slow or they can fail to converge. The rate of convergence of iterative methods depends strongly on the spectrum of the coefficient matrix. Hence, iterative methods usually involve a second matrix that

H. Xing, *Advances in Geocomputing*, Lecture Notes in Earth Sciences 119,
DOI 10.1007/978-3-540-85879-9_3

transforms the coefficient matrix to a matrix with a more favorable spectrum. This transformation matrix is called a *preconditioner*. The improvement in the convergence of an iterative method yielded by the application of an effective preconditioner outweighs the extra cost of constructing and applying it. Indeed, without a preconditioner the iterative method may even fail to converge.

In preconditioned iterative methods, the original linear equation:

$$[A]\{x\}=\{b\} \tag{1}$$

is transformed into the following Eq. (2) using the preconditioner (or preconditioning matrix) $[M]$:

$$[\tilde{A}]\{x\}=\{\tilde{b}\},\ [\tilde{A}]=[M]^{-1}[A],\ \{\tilde{b}\}=[M]^{-1}\{b\} \tag{2}$$

Equation (2) has the same solution as Eq. (1), but the spectral properties of the coefficient matrix $[\tilde{A}]=[M]^{-1}[A]$ may be more favorable, facilitating a faster convergence.

Various types of preconditioners have been proposed, developed and applied. The simplest method is called *diagonal scaling* or the *point Jacobi* method, where $[M]$ is given by the diagonal components of the original coefficient matrix $[A]$. Jacobi, Gauss-Seidel and SOR type stationary iterative methods are also well-known preconditioners, while preconditioners using various types of polynomials have also been widely used (Barrett et al. 1994).

The incomplete lower-upper (*ILU*) and incomplete Cholesky (*IC*) factorization methods are the most popular preconditioning techniques for accelerating the convergence of Krylov iterative methods (Barrett et al. 1994). These ILU/IC methods are based on *LU/Cholesky factorization* used in direct solution techniques. LU factorization is applicable to general un-symmetric matrices, while Cholesky is applicable to symmetric matrices. In *LU/Cholesky factorization*, many *fill-ins* are introduced during the factorization process, and so the factorized matrix can be dense even if the original matrix is sparse. **ILU(p)/IC(p)** is an incomplete factorization in which *p*-th-order fill-ins are allowed. Larger values of *p* provide a more accurate factorization and usually lead to robust preconditioning, but are more expensive in both memory and CPU time. In many engineering applications, **ILU(0)/IC(0)** is widely used where there are no fill-ins and the non-zero pattern of the original coefficient matrix is maintained in the factorized matrix.

## 1.2 *Selective Blocking* Preconditioning for Contact Problems

### 1.2.1 GeoFEM Project

From 1999 to 2002, the author developed parallel iterative solvers and preconditioning methods for geophysics problems in the *GeoFEM* project,[1] which develops a parallel finite-element platform for solid earth simulation on the *Earth Simulator*.[2] One of the most important applications of GeoFEM is the simulation of the stress accumulation process at plate boundaries (faults), which is critical for estimating the earthquake generation cycle (Figs. 1 and 2). A fine resolution (less than 1 km) is required around zones with higher stress accumulations, and so more than hundreds of millions of meshes may be required for detailed simulations. In this type of simulation, material, geometric and boundary nonlinearity should be considered. Among these, boundary nonlinearity due to the contact of faults is the most critical. In GeoFEM, the augmented Lagrange method (ALM) and penalty method are implemented, with a large penalty number, $\lambda$, introduced for constraint conditions around faults (Iizuka et al. 2000). The nonlinear process is solved iteratively by the Newton-Raphson (NR) method. A large $\lambda$ ($\sim 10^3 \times$ Young's modulus) can provide an accurate solution and fast nonlinear convergence for NR processes, but the condition number of the coefficient matrices of the corresponding linear equations is large, and several iterations are required for the convergence of iterative solvers (Fig. 3). Therefore, a robust preconditioning method is essential for such ill-conditioned problems.

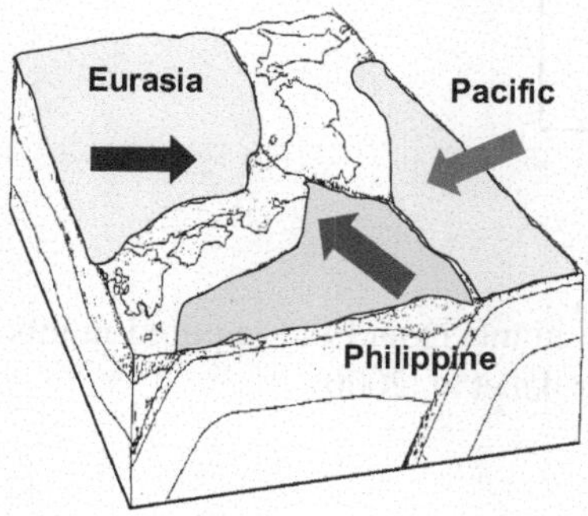


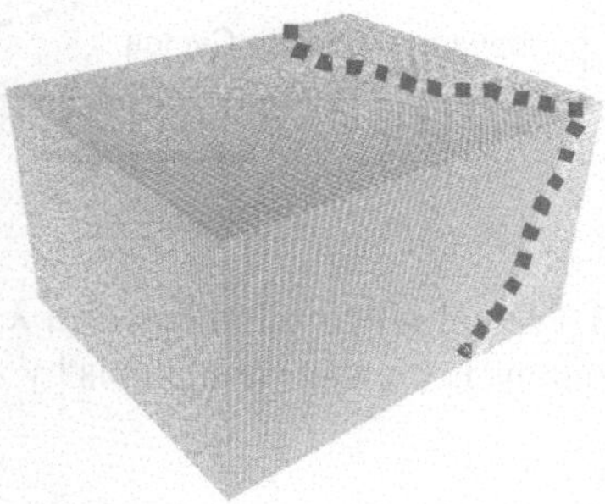

**Fig. 1** Subductive plate boundaries (faults) around Japanese Islands and an example of the finite-element model

[1] http://geofem.tokyo.rist.or.jp/
[2] http://www.es.jamstec.go.jp/

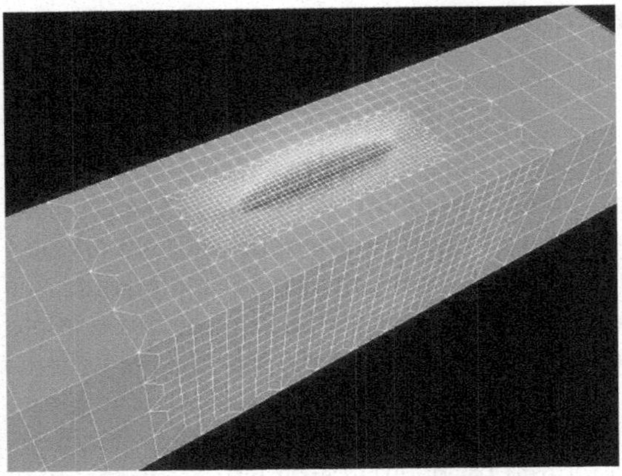

**Fig. 2** Example of the finite-element model with locally refined meshes for a transcurrent fault (movie available on accompanying DVD)

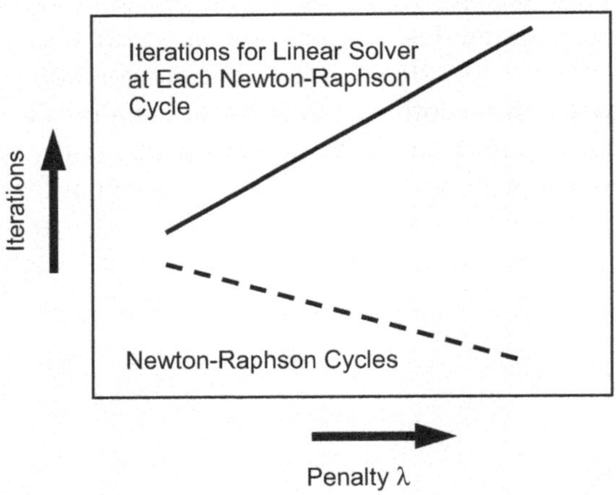


**Fig. 3** Typical relationship between λ (penalty number) and the required number of iterations in contact simulations by ALM (Iizuka et al. 2000)

### 1.2.2 Selective Blocking

*Selective blocking* is a special preconditioning method developed for this type of application by the author on GeoFEM's framework (Nakajima 2003, Nakajima and Okuda 2004). In this method, finite element nodes in the same contact group coupled through penalty constraints are placed into a large block (selective block or super node) (Fig. 4). For symmetric positive definite matrices, preconditioning with *block* incomplete Cholesky factorization *using selective blocking* (SB-BIC) yields an excellent performance and robustness (Nakajima 2003, Nakajima and Okuda 2004). Details of the parallel iterative solvers of GeoFEM and algorithm of selective blocking are described in Appendices 1 and 2 of this chapter.

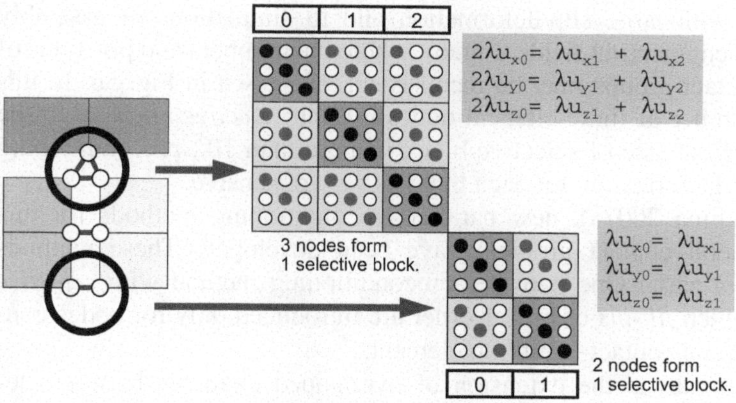


**Fig. 4** Matrix operation of nodes in contact groups for *selective blocking* preconditioning

## 1.3 Overview of this Work

Contact phenomena are one of the most important and critical issues in various types of scientific and engineering problems. In previous works (Nakajima 2003, Nakajima and Okuda 2004), the numbers of nodes in contact groups are consistent, and conditions for infinitesimal deformation have been also assumed, as shown in Fig. 5a. With this approach, the positions of nodes do not change and a consistent relationship among nodes in contact groups is maintained during the simulation. Moreover, a special partitioning method, where all nodes in the same contact group are located in the same domain, has been applied, as shown in Appendix 2. However, this approach is not therefore flexible, and cannot be applied to fault contact simulations with large slip/deformation and to simulations of assembly structures in engineering fields (Fig. 6), where the numbers and positions of nodes in contact groups may be inconsistent, as shown in Fig. 5b. In this situation, number of finite-element nodes in each *selective block* might be very large. If the size of selective block is more than $10^3$, preconditioning with full LU factorization for each block is very expensive.

In (Nakajima 2007a), new parallel preconditioning methods for this type of general contact problem have been developed. These methods comprise two parts: One part is a preconditioning method with *selective fill-ins*, in which *fill-ins* of higher order are introduced only for nodes connected to special contact-condition elements.

The other part is the extension of overlapped elements between domains. It is widely known that convergence of parallel finite-element applications with preconditioned iterative solvers strongly depends on method of domain decomposition. In (Nakajima 2007a), the *selective overlapping* method was proposed, which extends the layers of overlapped elements according to the information of the special elements used for contact conditions. Both methods are based on the idea of *selective blocking*, but are more general and flexible than that approach.

These methods are very unique, because dropping rules of the preconditioning matrices are defined according to properties of individual finite-element and features of finite-element applications before assembling entire coefficient matrices.

In addition, the following two methods are further introduced in this work:

- Local reordering in distributed data
- Hierarchical Interface Decomposition (HID) (Henon and Saad 2007)

HID provides a method of domain decomposition with robustness and scalability for parallel ILU/IC preconditioners. The robustness and efficiency of HID and *selective overlapping* are compared in this work.

In the following part of this chapter, these four methods (selective fill-ins, selective overlapping, local reordering and HID) are reviewed in detail. These methods are implemented as preconditioners of iterative solvers for parallel finite-element applications for ill-conditioned problems. Parallel codes are based on the framework for parallel FEM procedures of GeoFEM, and the GeoFEM's local data structure (ref. Appendix 1) is applied.

The results of example problems with contact conditions using 64-core PC clusters are shown. Finally, the developed methods are applied to general ill-conditioned problems for problems with heterogeneous material properties.

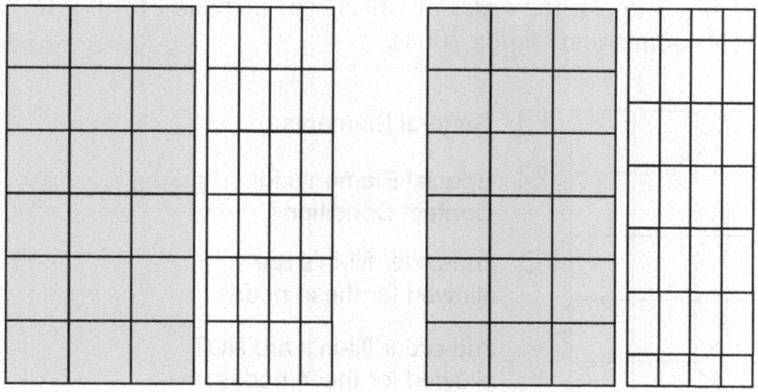


**Fig. 5** Consistent and inconsistent node numbers at contact surfaces in FEM models applied to contact simulations

**Fig. 6** Example of an assembly structure: Jet Engine

## 2 Various Approaches for Parallel Preconditioning Methods in Ill-Conditioned Problems

### 2.1 Selective Fill-Ins

The *selective blocking* preconditioning method (Nakajima and Okuda 2004) is a robust and efficient preconditioning method for contact problems. However, it can only be applied in a very limited number of situations, as shown in the previous section.

Incomplete LU factorization with p-th-order fill-in **(ILU(p))** preconditioning methods are widely used for various types of applications (Saad 2003). The higher the order of fill-ins (**p**) is, the more robust the preconditioner will be, but this normally comes at the cost of being computationally more expensive. The required memory for coefficient matrices increases by a factor of from 2 to 5 if the order of fill-ins (**p**) increases from 0 to 1, or from 1 to 2 (Nakajima and Okuda 2004).

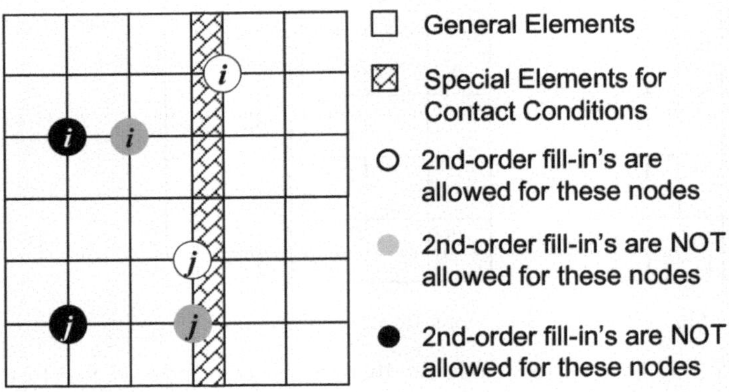


**Fig. 7** Example of the **ILU(1+)** preconditioning technique

In (Nakajima 2007a), new preconditioning methods for general contact problems have been developed. The first approach is a preconditioning method with *selective fill-ins*, called **ILU(p+)**. Figure 7 describes the principle of **ILU(p+)**. Denoting the i,j-th component of the preconditioner matrix by $m_{ij}$ in **ILU(p+)**, (p+1)-th order fill-ins are allowed for $m_{ij}$ such that both the *i-th* and *j-th* nodes are connected to special contact-condition elements, such as *master-slave* type elements (Iizuka et al. 2000). In Fig. 7, second-order fill-ins can be allowed for all three *i-j* pairs, according to graphical connectivity information. However, in the **ILU(p+)** preconditioning approach, only the white circles are allowed to generate second-order fill-ins.

This approach closely resembles that of *selective blocking*, in which full LU factorization is applied to nodes in contact groups, but is much more general and flexible. Since constraint conditions are applied to the nodes that are connected to special elements through penalty terms, *selective* ILU factorization with higher order fill-ins for these nodes is expected to provide robust convergence with efficiency. In (Washio et al. 2005), a preconditioning method with block ILU factorization is proposed for coupled equations of incompressible fluid flow and solid structure. Different orders of fill-ins are applied to velocity and pressure components to generate block ILU factorization of coefficient matrices. **ILU(p+)** is very similar to this idea.

Figure 8 describes the model used for the validation of the developed preconditioning methods. This problem simulates general contact conditions, in which the positions and number of nodes on contact surfaces are inconsistent. In this model there are four blocks of elastic material that are discretized into cubic tri-linear type finite-elements. Each block is connected through elastic truss elements generated at each node on the contact surfaces. The truss elements together take up the form of a cross, as shown in Fig. 8. In the present case, the elastic coefficient of the truss elements is set to $10^3$ times that of the solid elements, which corresponds to the coefficient $\lambda$ ($=10^3$) for constraint conditions of the augmented Lagrangian method (ALM). Poisson's ratio is set to 0.25 for the cubic elements.

Symmetric boundary conditions are applied at the x=0 and y=0 surfaces, while a Dirichlet fixed condition for deformation in the direction of the z-axis is applied to z=0 surfaces. Finally, a uniform distributed load in the direction of the z-axis is applied to $z=Z_{max}$ surfaces. This problem lies in the area of linear elasticity, but the coefficient matrices are particularly ill-conditioned, and so this problem provides a good simulation of nonlinear contact problems (Nakajima 2003, Nakajima and Okuda 2004).

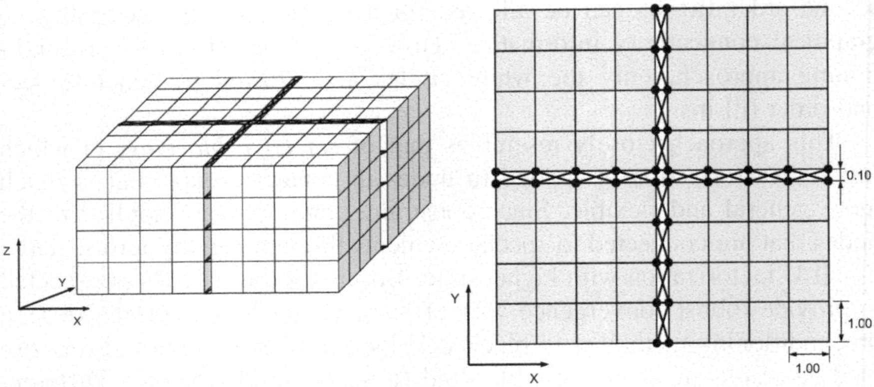


**Fig. 8** Elastic blocks connected through truss elements

The plots in Fig. 9 display results describing the performance of four preconditioning techniques for the problem in linear elasticity described above. All calculations were performed using a single core of AMD Opteron 275 (2.2 GHz)[3] with PGI FORTAN90 compiler.[4] Each block in Fig. 8 has 8,192 (=16×16×32) cubes, where the total problem size has 107,811 degrees of freedom (DOF). Generalized product-type methods based on Bi-CG (GPBi-CG) (Zhang 1997) for general coefficient matrices have been applied as an iterative method, although the coefficient matrices for this problem are positive indefinite. Each node has three DOF in each axis in 3D solid mechanics; therefore, ***block*** **ILU (BILU)** type preconditioning (Nakajima 2003, Nakajima and Okuda 2004) has been applied.

**BILU(1+)**, in which additional *selective fill-ins* to **BILU(1)** have been applied for nodes connected to special elements (elastic truss elements in Fig. 8), provides the most robust and efficient convergence. **BILU(p)** provides faster convergence the larger the value of p, as shown in Fig. 9b, but is also more computationally expensive with increasing p, as shown in Fig. 9c, where the number of off-diagonal components in preconditioning matrices [*M*] is described. **BILU(1)** and **BILU(1+)** are competitive, but **BILU(1+)** provides a better convergence rate.

[3] http://www.amd.com/

[4] http://www.pgroup.com/

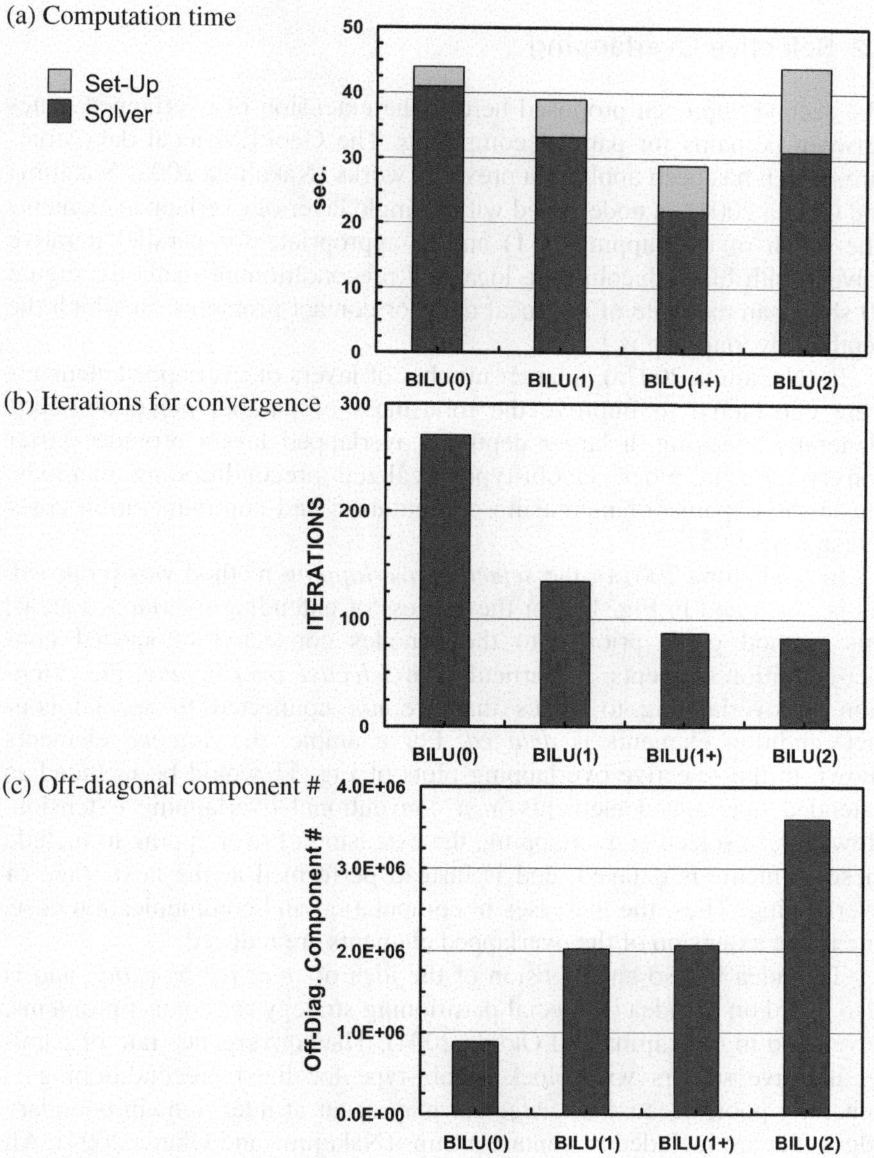


**Fig. 9** Results for the problem in linear elasticity, whose configuration is given in Fig. 8, considering simple cube geometries of 107,811 DOF with contact conditions on a single core of AMD Opteron 275 (2.2 GHz) with the PGI FORTRAN90 compiler

## 2.2 Selective Overlapping

The second approach proposed here is the extension of overlapped zones between domains for parallel computing. The GeoFEM local data structure, which has been applied in previous works (Nakajima 2003, Nakajima and Okuda 2004), is node-based with a single layer of overlapped elements (the depth of overlapping is 1) and is appropriate for parallel iterative solvers with block Jacobi-type localized preconditioning methods. Figure 10 shows an example of the local data for contact problems, in which the depth of overlapping is 1.

In (Nakajima 2007a), a larger number of layers of overlapped elements were considered to improve the robustness of parallel preconditioners. Generally speaking, a larger depth of overlapped layers provides faster convergence in block Jacobi-type localized preconditioning methods, but at the expense of increasing computation and communication costs (Nakajima 2005).

In (Nakajima 2007a), the *selective overlapping* method was proposed. As is illustrated in Fig. 11, for the process of extending overlapped areas, this method gives priority to those nodes connected to special contact-condition elements. In particular, in *selective overlapping*, the extension of overlapping to nodes that are *not* connected to special contact-condition elements is *delayed*. For example, the *hatched* elements shown in the selective overlapping plots of Fig. 11 would be included as extended overlapped elements in a conventional overlapping extension. However, in selective overlapping, the extension of overlapping to include these elements is delayed, and is instead performed at the next stage of overlapping. Thus, the increases in computation and communication costs due to the extension of the overlapped elements are reduced.

This idea is also an extension of the idea of *selective blocking*, and is also based on the idea of special partitioning strategy for contact problems, developed in (Nakajima and Okuda 2004). The convergence rate of parallel iterative solvers with block Jacobi-type localized preconditioning is generally poor, because the *edge-cut* may occur at inter-domain boundary edges that are included in contact groups (Nakajima and Okuda 2004). All nodes in the same contact group should be in the same domain in order to avoid such edge-cuts. Because the constraint conditions are applied to those nodes that are connected to special elements through penalty terms, the *selective* extension of overlapping for these nodes is expected to provide robust convergence with efficiency.

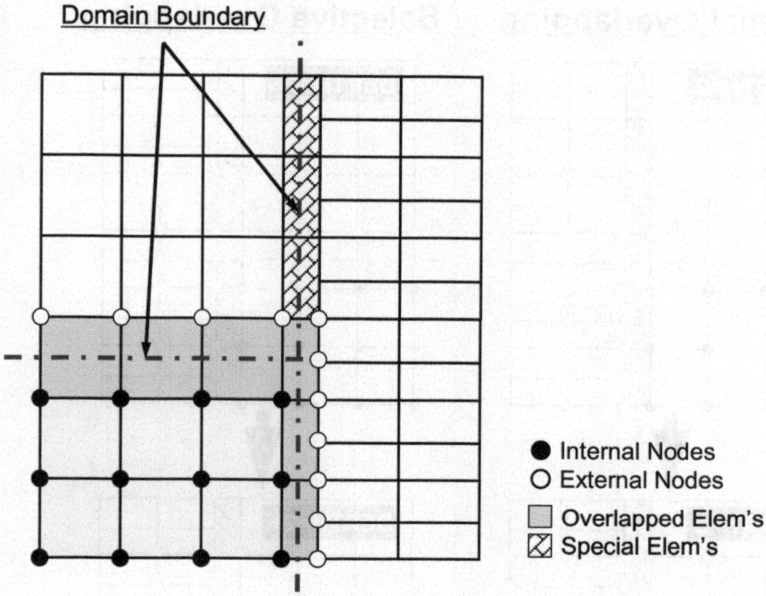


**Fig. 10** Example of GeoFEM's local data structure for contact problems

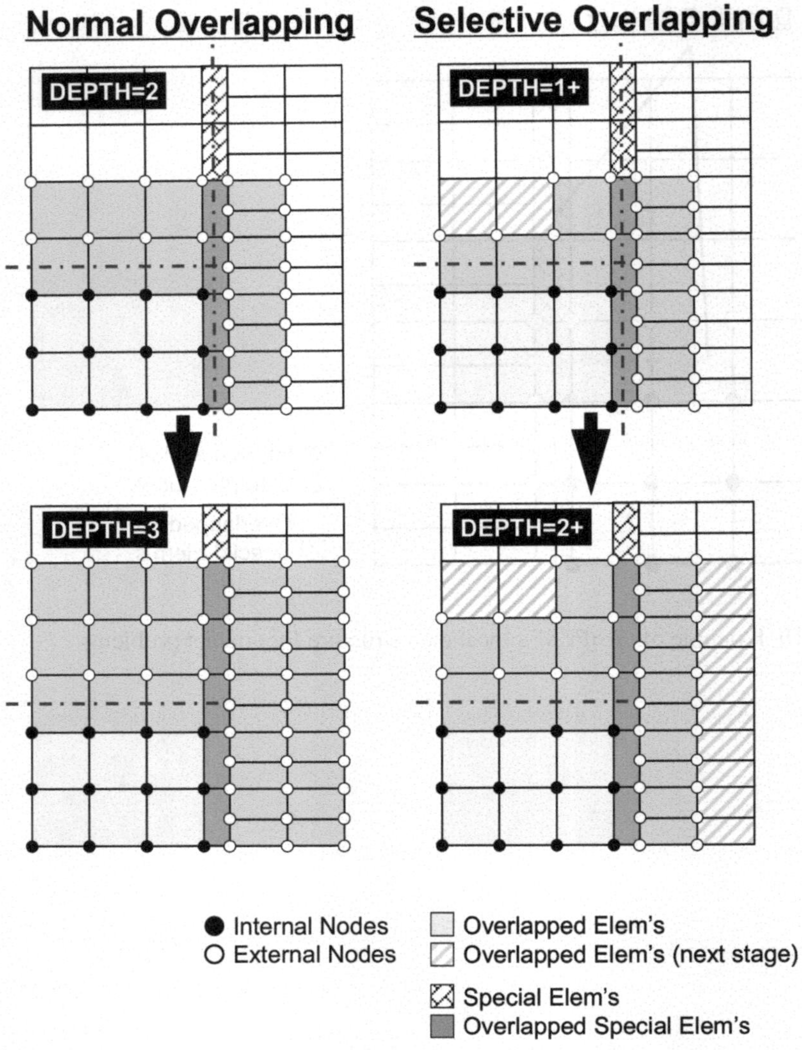


**Fig. 11** Example of *selective overlapping*, precedence for extensions of overlapped layers is given to nodes connected to special contact-condition elements

## 2.3 Local Reordering in Distributed Data

It is widely known that the reordering of vertices strongly affects the convergence of iterative solvers with ILU-type preconditioners (Nakajima 2007b).

As shown in Appendix 1, nodes in each local mesh data of the GeoFEM data structure are classified into the following *three* categories from the viewpoint of message passing:

- Internal nodes (originally assigned to the domain)
- External nodes (forming an element in the domain, but are from external domains)
- Boundary nodes (external nodes of other domains)

Figure 12 describes a very simple example, where an initial entire mesh comprising 18 nodes and 10 quadrilateral elements is partitioned into two domains. Local numbering starts from the internal nodes. The external nodes are numbered after all the internal nodes have been numbered, as shown in Fig. 12. According to this numbering method, the bandwidth of local sparse coefficient matrices including the external nodes is relatively large, as shown in Fig. 12. Coefficient matrices with larger bandwidth usually provide slower convergence for iterative solvers with ILU-type preconditioning methods (Saad 2003). As long as block Jacobi-type fully localized preconditioning methods in Nakajima (2003) are adopted, this effect of bandwidth is rather smaller. But if overlapping among domain is applied, this effect may be more significant.

In this work, a *global* numbering is introduced, where both the internal and external nodes in each domain are reordered according to their original global ID. Figure 13 shows local data meshes obtained through this *global* numbering, and corresponding local coefficient matrices. It may be noted that the bandwidth of local coefficient matrices is smaller than that of original matrices in Fig. 12.

The Reverse Cuthill-Mckee (RCM) method is a well-known reordering technique, which is also suitable for reducing the bandwidth of sparse matrices (Nakajima 2003, Saad 2003). In the previous works (Nakajima 2003, Nakajima 2007b), RCM reordering has been applied to internal nodes for parallel efficiency. In this work, the following two types of approaches have been applied:

- RCM is applied only to internal nodes (Fig. 14)
  - local numbering with *RCM-internal*
- RCM is applied to both internal and external nodes (Fig. 15)
  - local numbering with *RCM-entire*

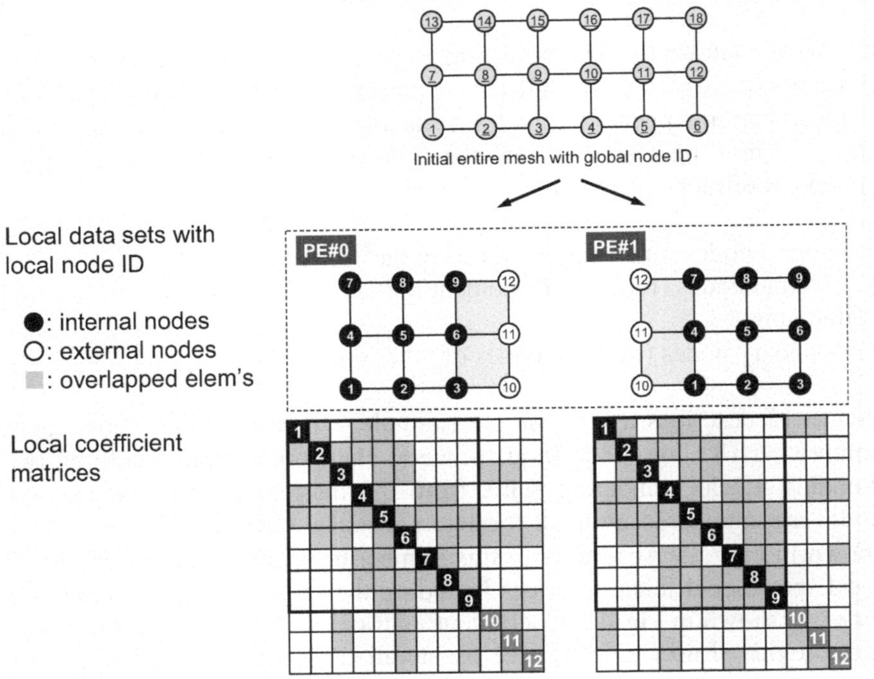


**Fig. 12** An initial entire mesh with global node ID, local domains with local node ID and local coefficient matrices for two domains

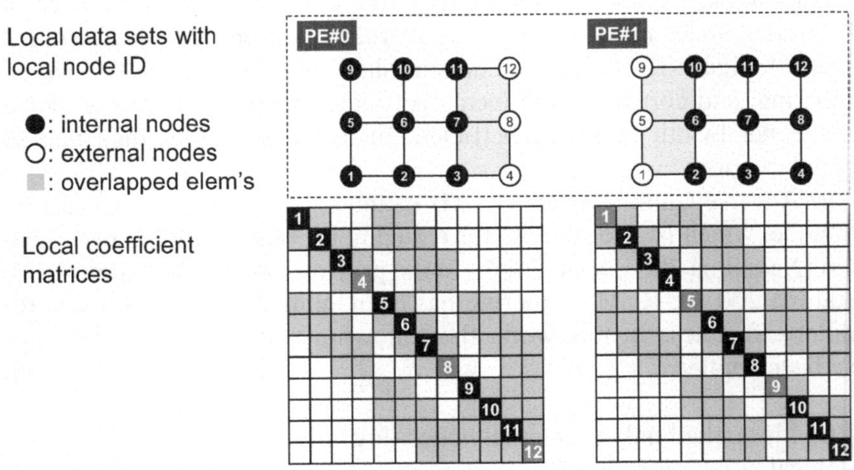


**Fig. 13** Local domains and coefficient matrices according to *global* numbering

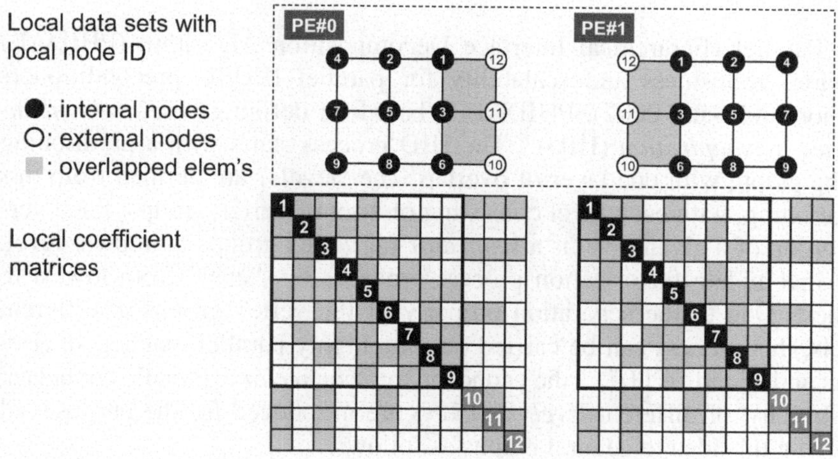


**Fig. 14** Local domains and coefficient matrices with *RCM-internal* reordering

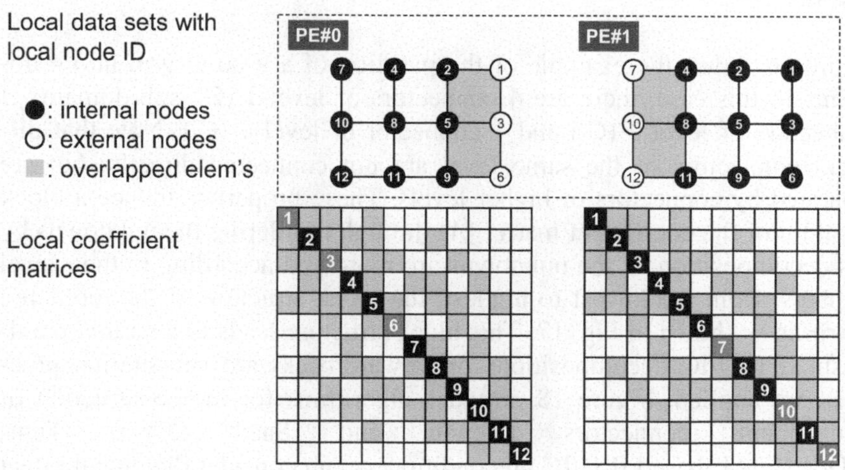


**Fig. 15** Local domains and coefficient matrices with *RCM-entire* reordering

## 2.4 HID (Hierarchical Interface Decomposition)

The Parallel Hierarchical Interface Decomposition Algorithm (PHIDAL) provides robustness and scalability for parallel ILU/IC preconditioners (Henon and Saad 2007). PHIDAL is based on defining "*hierarchical interface decomposition* (HID)". The HID process starts with a partitioning of the graph, with one layer of overlap. The "levels" are defined from this partitioning, with each level consisting of a set of vertex groups. Each vertex group of a given level is a *separator* for vertex groups of a *lower* level. The incomplete factorization process proceeds by "level" from lowest to highest. Due to the separation property of the vertex groups at different levels, this process can be carried out in a highly parallel manner. In (Henon and Saad (2007), the concept of *connectors* (small connected sub-graphs) of different *levels* and *keys* are introduced for the purposes of applying this idea to general graphs as follows:

- Connectors of *level-1* ($C^1$) are the sets of interior points. Each set of interior points is called a *sub-domain.*
- A connector of *level-k* ($C^k$) (k>1) is adjacent to k *sub-domains.*
- No $C^k$ is adjacent to any other connector of level-k.
- *Key*($u$) is the set of sub-domains (connectors of level-1, $C^1$) connected to vertex $u$.

Figure 16 shows the example of the partition of a 9-point grid into 4 domains. In this case, there are 4 connectors of level-1 ($C^1$, sub-domain), 4 connectors of level-2 ($C^2$) and 1 connector of level-4 ($C^4$). Note that different connectors of the same level are not connected directly, but are separated by connectors of higher levels. These properties induce a block structure of the coefficient matrix [$A$] through reordering the unknowns by this decomposition. If the unknowns are reordered according to their level numbers, from the lowest to highest, the block structure of the reordered matrix is as shown in Fig. 17. This block structure leads to a natural parallelism if ILU/IC decompositions or forward/backward substitution processes are applied. Figure 18 provides algorithms for the construction of independent connectors (Henon and Saad 2007). Thus, HID/PHIDAL-based ILU/IC preconditioners can consider the global effect of external domains in parallel computations, and are expected to be more robust than block Jacobi-type localized ones. In this work, HID and *selective overlapping* are compared from this point of view.

In (Nakajima 2007b), GeoFEM's original partitioner for domain decomposition was modified so that it could create a distributed hierarchical data structure for HID. Each *sub-domain* (interior vertices, connectors of level-1) is assigned to an individual *domain*, which corresponds to each MPI process. Higher-level connectors are distributed to each domain so that load-balancing can be attained and communications can be minimized. Figure 19 shows an example of the final partition of a 9-point grid into 4 sub-domains.

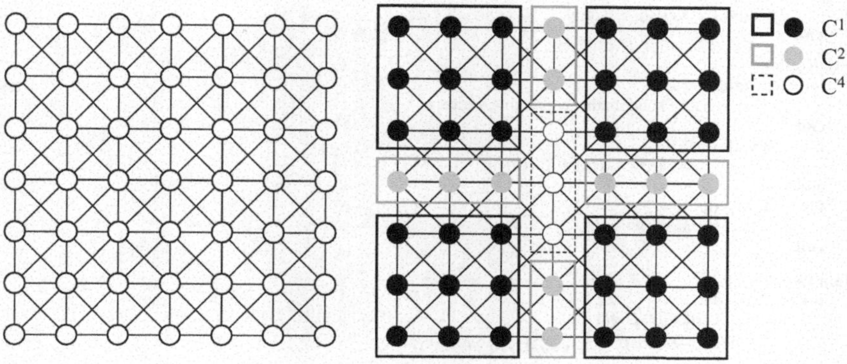


(a) Initial entire grid (b) *Connectors* and *levels*

**Fig. 16** HID partitioning of a 9-point grid into 4 sub-domains

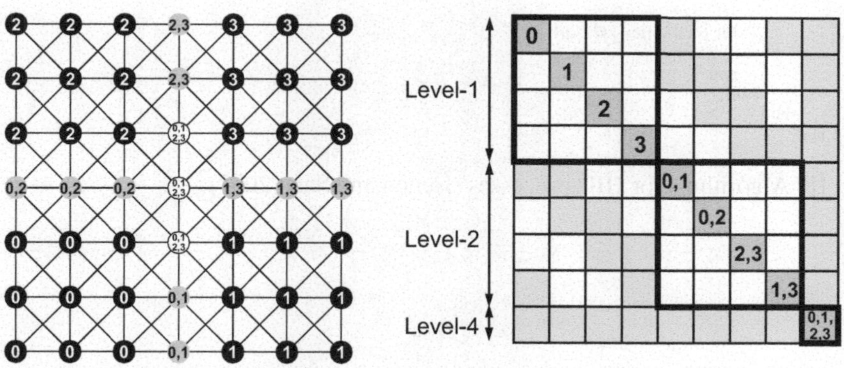


(a) Domain decomposition (*connectors* and *keys*) (b) matrix and non-empty blocks

**Fig. 17** Domain/block decomposition of the coefficient matrix according to HID reordering

**Initialization:**

**for** each vertex $u \in V$
  $Key(u)$ := list of subdomains containing vertex $u$
**end**
**for** each vertex $u \in V$
  $Kdeg_u := |\{v \in V^l(u) \, / \, Key(v) \neq Key(u)\}|$
**end**
**for** $l = 1$, to $p$ **do:**
  $L^l = \{u \in V \, / \, |Key(u)| == l\}$
**end**

**Initialization:**

**for** each vertex $u \in V$
  $Key(u)$ := list of subdomains containing vertex $u$
**end**
**for** each vertex $u \in V$
  $Kdeg_u := |\{v \in V^l(u) \, / \, Key(v) \neq Key(u)\}|$
**end**
**for** $l = 1$, to $p$ **do:**
  $L^l = \{u \in V \, / \, |Key(u)| == l\}$
**end**

**Main Loop:**

**for** $l = 2$ to $p$ **do:**
  **for** each vertex $u \in L^l$ **do:**
    $Key(u) := Key(u) \cup \bigcup_{k=1,\dots,l-1} \left( \bigcup_{v \in V^k(u)} Key(v) \right)$
  **end**
  **while** all vertices in $L^l$ have not been processed;
    get $u$ the vertex in $L^l$ such that $Kdeg_u$ is maximum.
    $Key(u) := Key(u) \cup \bigcup_{v \in V^k(u)} Key(v)$
    $m = |Key(u)|$
    **if** $m > l$ **then**
      $L^l := L^l \setminus \{u\}$
      $L^m := L^m \cup \{u\}$
      **for** each vertex $v \in V^l(u)$
        $Kdeg_v := Kdeg_v - 1$
      **end**
    **endif**
  **end**
**end**

**Fig. 18** Algorithms for HID processes (Henon and Saad 2007)

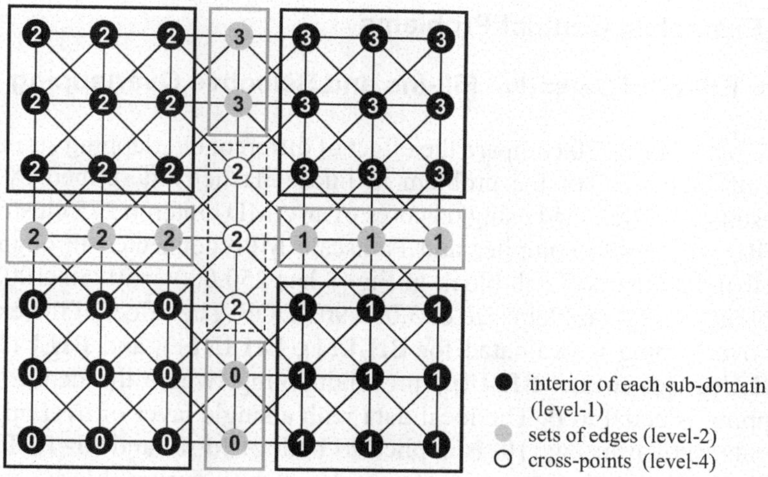


(a) Final partition (number's correspond to ID of partition (0-3))

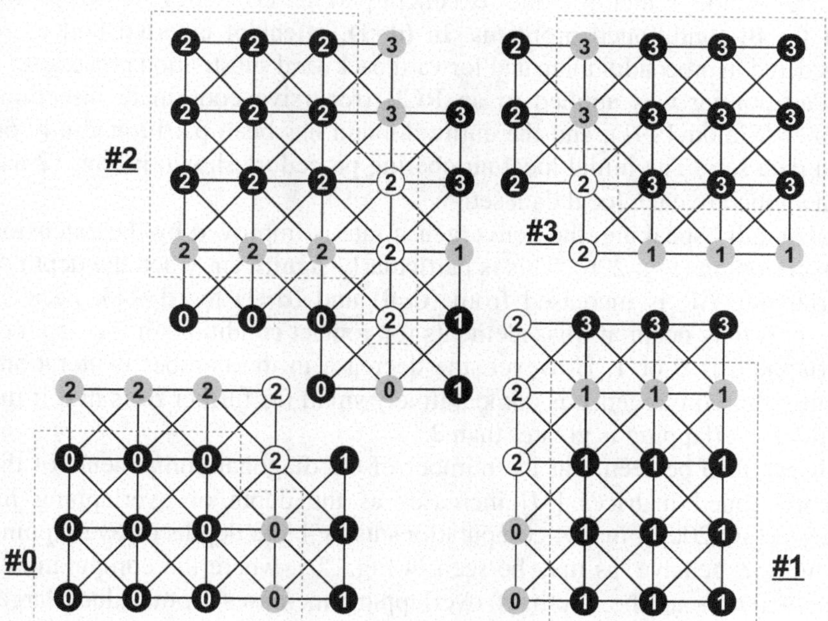


(b) Distributed local data sets with external vertices

**Fig. 19** The final partition of a 9-point grid into 4 domains

## 3 Examples: Contact Problems

### 3.1 Effect of Selective Fill-Ins and Selective Overlapping

The plots in Fig. 20 compare the effect of different overlapping strategies on results obtained for the problem in linear elasticity described by Fig. 8. Results were obtained using 64 cores of an AMD Opteron 275 cluster with a PGI FORTAN90 compiler and Pathscale MPI[5] connected through an Infiniband[6] network. Each block in Fig. 8 has 250,000 (=50×50×100) cubes, yielding a total problem size of 3,090,903 DOF. The effect of the extension of overlapping is evaluated for **BILU(1)**, **BILU(1+)**, and **BILU(2)**. Here **BILU(p)-(d)** means **BILU(p)** preconditioning, where the depth of overlapping is equal to **d**. The local data with a single layer of overlapped elements, shown in Fig. 10, is applied to both of **(d=0)** and **(d=1)**. In **(d=0)**, effect of *external nodes* are *not* considered at all during ILU/IC decompositions and forward/backward substitution processes. Therefore, **BILU(p)-(0)** corresponds to *pure* block Jacobi-type localized preconditioning method, which provides excellent parallel efficiency, but is not robust for ill-conditioned problems. In **(d=1)**, effect of external nodes are considered in preconditioning and forward/backward substitution processes.

Partitioning was applied in an RCB (recursive coordinate bisection) manner (Simon 1991), and the entire domain has been partitioned into 64 local data sets. The initial local numbering procedure shown in Fig. 12 has been applied to each local data set.

Generally speaking, the convergence rate is improved by the extension of overlapping (Fig. 20b). This is particularly significant when the depth of overlapping (**d**) is increased from **(d=0)** and **(d=1)** to **(d=1+),** because *edge-cuts* may occur at truss elements for contact conditions if the depth of overlapping is 0 or 1. However, the decrease in the number of iterations required for convergence is comparatively small for further rises in **d** if the depth of overlapping is greater than 2.

It can also be seen that the number of off-diagonal components of the preconditioned matrices [*M*] increases as the depth of overlapping increases (Fig. 20c). Finally, computations using large depths of overlapping are more expensive, as may be seen in Fig. 20a, where the computational time increases as the depth of overlapping increases from values larger than 2 (Fig. 20a). Methods **BILU(1)-(1+)** and **BILU(1+)-(1+)** offer the best performance, and these two methods are closely matched.

[5] http://www.pathscale.com/

[6] http://www.infinibandta.org/

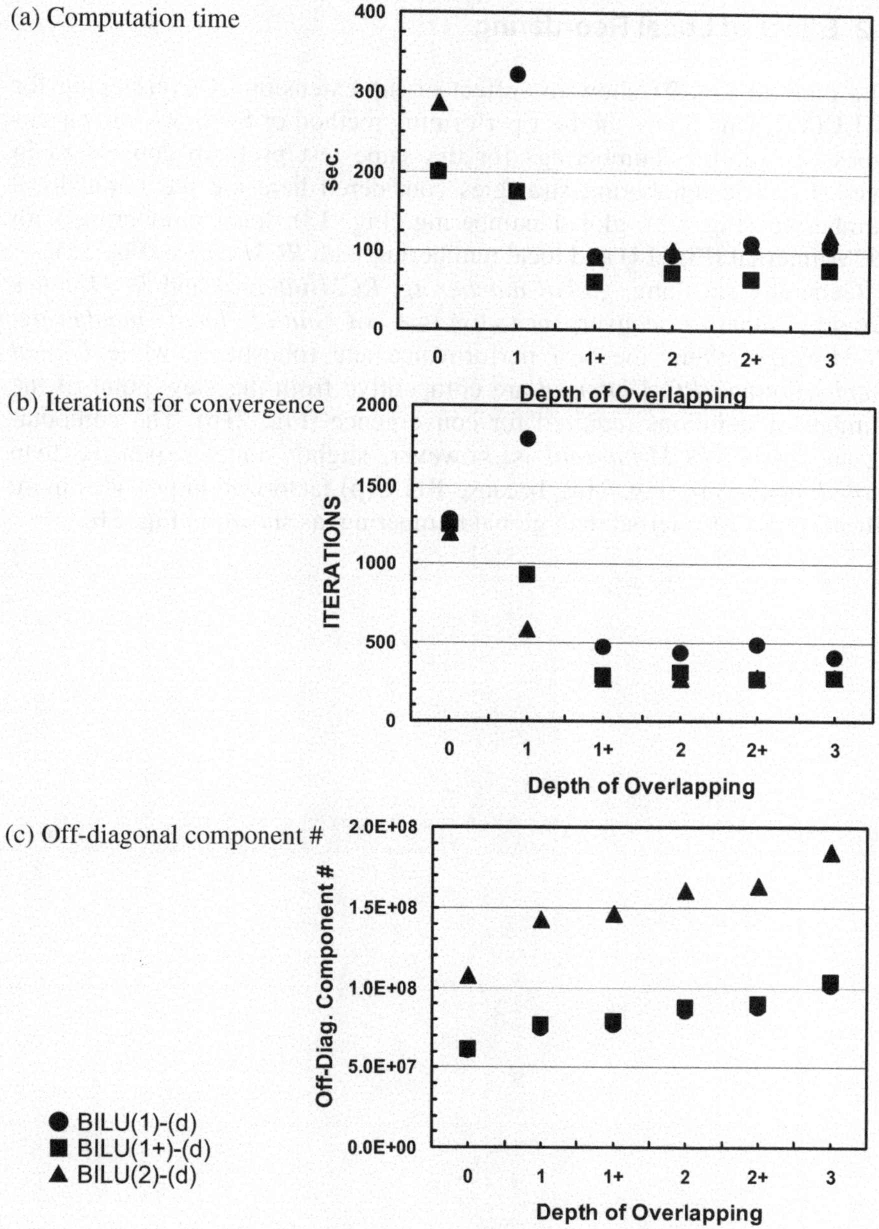


**Fig. 20** Results detailing the effect of overlapping for the model problem considering simple cube geometries of 3,090,903 DOF with contact conditions in linear elasticity described by Fig. 8 on 64 cores of AMD Opteron 275 cluster using the PGI compiler

### 3.2 Effect of Local Reordering

The plots in Fig. 21 show the effect of the extension of overlapping for **BILU(1+)**, which was the best performing method of Sect. 3.1, on various types of orderings/numberings for the same test problem considered in Sect. 3.1, The numbering strategies considered here are the initial local numbering (Fig. 12), global numbering (Fig. 13), local numbering with RCM-internal (Fig. 14) and local numbering with *RCM-entire* (Fig. 15).

Generally speaking, *global numbering*, *RCM-internal* and *RCM-entire* provide superior convergence to that of *initial local numbering*. *RCM-entire* attains the best performance and robustness, while *Global numbering* and *RCM-internal* are competitive from the view point of the number of iterations required for convergence (Fig. 21b). The computational cost of *RCM-internal* is, however, slightly more expensive than *global numbering (*Fig. 21a), because **BILU(p)** factorization provides more fill-ins in RCM-internal than global numbering, as shown in Fig. 21c.

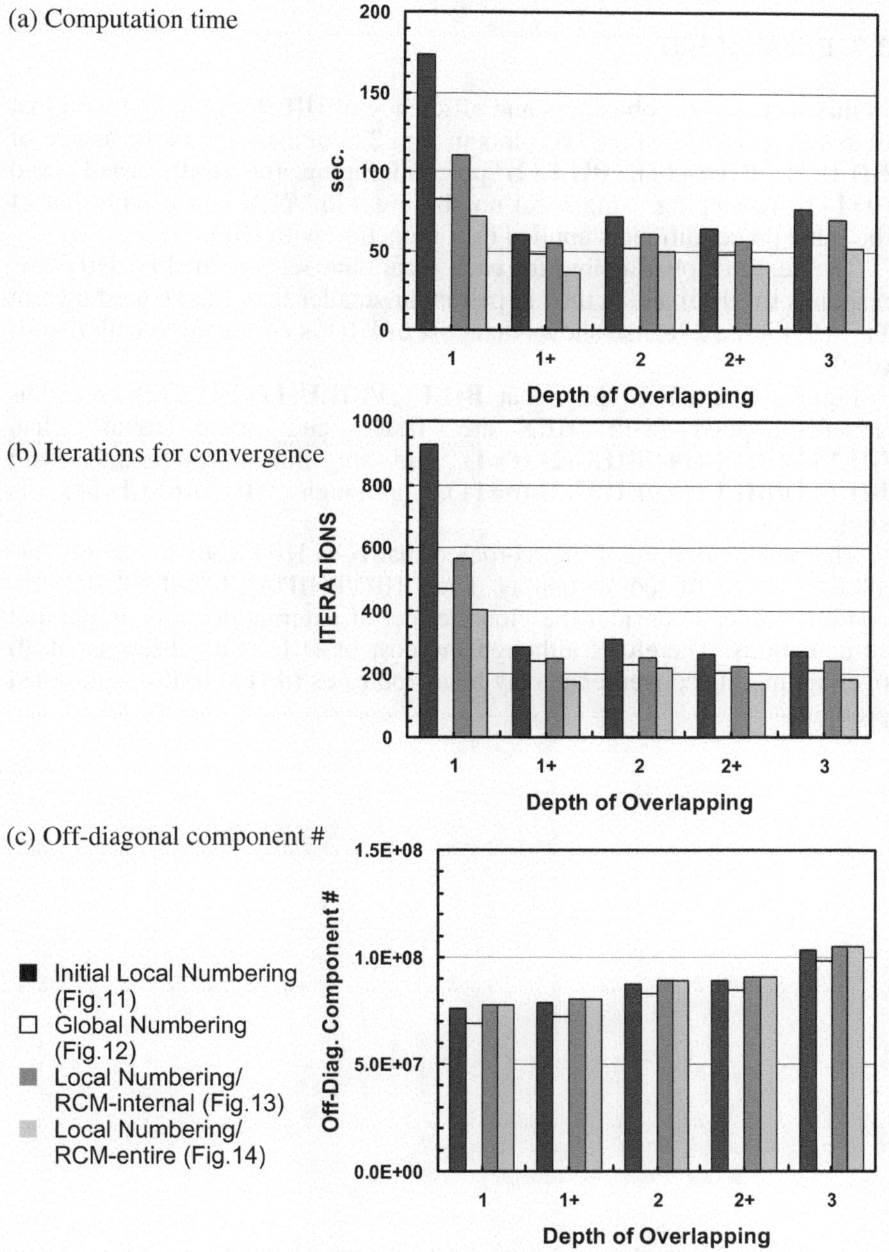


**Fig. 21** Effect of overlapping and local reordering for **BILU(1+)** in linear-elastic problem for simple cube geometries of 3,090,903 DOF with contact conditions in Fig. 8 on 64 cores of AMD Opteron 275 cluster with PGI compiler

### 3.3 Effect of HID

In this section, the robustness and efficiency of HID is compared with that of *selective overlapping*. The plots in Fig. 22 compare the performance of **BILU(1)**, **BILU(1+)**, **BILU(2)** preconditioning for **(d=0)**, **(d=1)** and **(d=1+)** overlapping using local numbering with *RCM-entire* with that of the same preconditioners applied in conjunction with HID.

The depth of overlapping for each local data set provided by HID corresponds to **(d=0)** and is thus in particular smaller than **(d=1)**, as shown in Fig. 19. Figure 22c also shows that cost of HID is competitive with that of **(d=0)**

Figures 22a and b show that **BILU(1)/BILU(1+)/BILU(2)** preconditioners applied with HID are faster and more robust than **BILU(1)/BILU(1+)/BILU(2)-(d=1)**, and are almost competitive with **BILU(1)/BILU(1+)/BILU(2)-(d=1+)**, although **BILU(p)-(d=1+)** is slightly better.

The block structure of the reordered matrix in HID leads to natural parallelism in ILU/IC computations. Thus, HID/PHIDAL-based ILU/IC preconditioners can consider the global effect of external domains in parallel computations. Therefore, although the cost of HID is as cheap as **(d=0)** overlapping, its convergence may be as robust as **(d=1+)** in ill-conditioned problems.

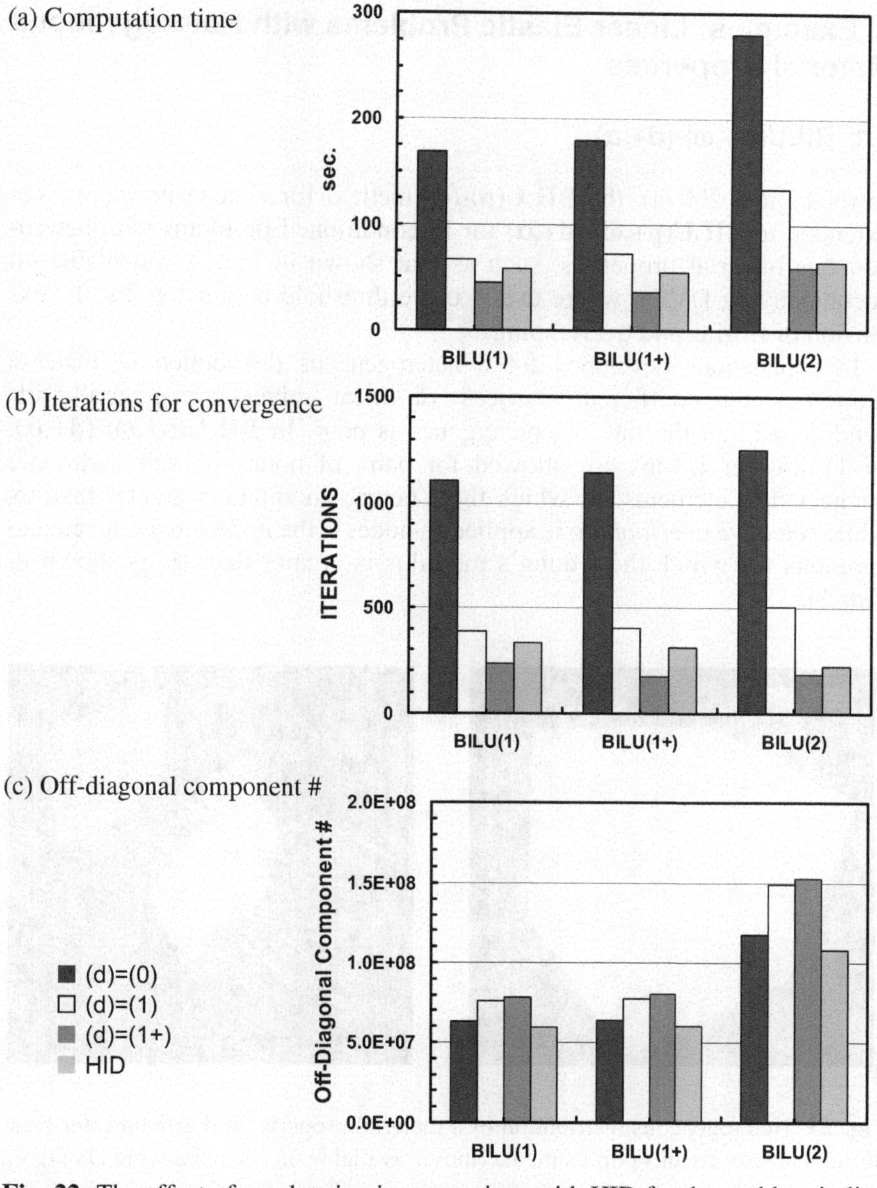


**Fig. 22** The effect of overlapping in comparison with HID for the problem in linear elasticity considering simple cube geometries of 3,090,903 DOF with contact conditions as shown in Fig. 8, on 64 cores of AMD Opteron 275 cluster using the PGI compiler, *RCM-entire* reordering applied

# 4 Examples: Linear-Elastic Problems with Heterogeneous Material Properties

## 4.1 BILU(p+,ω)-(d+,α)

In (Nakajima 2007a), the **BILU(p)-(d)** method for contact problems was extended to **BILU(p+,ω)-(d+,α)** for ill-conditioned problems with heterogeneous material properties, such as that shown in Fig. 23 (available on accompanying DVD), where **ω** and **α** are threshold parameters for the extension of fill-ins and overlapping.

In applications developed for a heterogeneous distribution of material properties, the coefficient matrices of linear solvers are generally ill-conditioned and the rate of convergence is poor. In **BILU(p+,ω)-(d+,α)**, (p+1)-th-order fill-ins are allowed for pairs of nodes if both nodes are connected to elements for which the Young's modulus is greater than **ω**, while *selective overlapping* is applied to nodes if the nodes are connected to elements for which the Young's modulus is greater than **α**, as shown in Fig. 24.

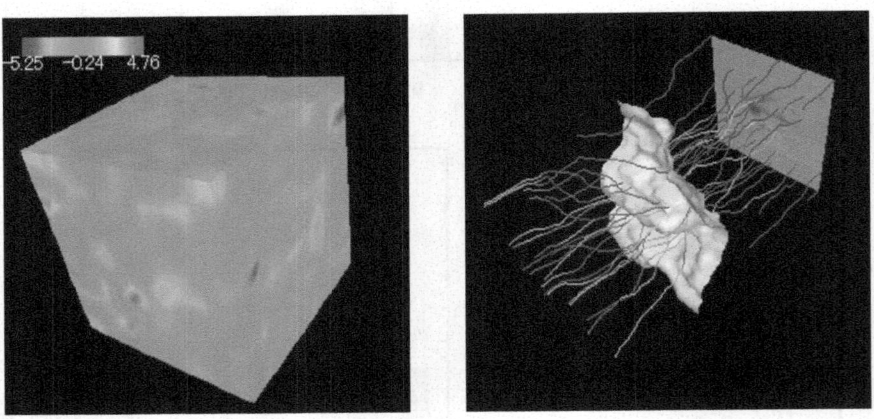


**Fig. 23** Heterogeneous distribution of a material property, and groundwater flow through heterogeneous porous media (movie available on accompanying DVD)

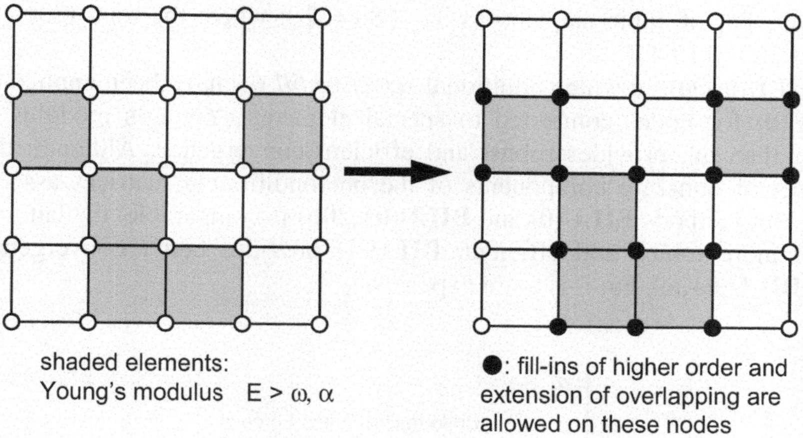


**Fig. 24** Selective *fill-ins* and *overlapping* for a heterogeneous field

## 4.2 Problem Description

Figure 25 describes the boundary conditions for a model problem in linear elasticity considering heterogeneous material of simple cubic geometry. Each element is a cubic tri-linear type finite-element. Poisson's ratio is set to 0.25 for all elements, while the heterogeneous distribution of Young's modulus in each tri-linear element is calculated by a sequential Gauss algorithm, which is widely used in the area of geo-statistics (Deutsch and Journel 1998). The minimum and maximum values of Young's modulus are $10^{-3}$ and $10^{3}$, respectively, where the average value is 1.0.

Symmetric boundary conditions are applied to the x = 0 and y = 0 surfaces, and the Dirichlet fixed condition for deformation in the direction of the z-axis is applied at z = 0. Finally, a uniform distributed load in the direction of the z-axis is applied at the $z = Z_{max}$ surface. This problem is linearly elastic, but the coefficient matrices are particularly ill-conditioned.

The GPBi-CG method for general coefficient matrices is used here as the iterative solution technique, although the coefficient matrices of this problems are positive indefinite. Each node has three DOF in each axis in 3D solid mechanics; therefore, **block ILU (BILU)** type preconditioning has been applied.

The plots of Fig. 26 show results obtained from computations using **BILU** preconditioning applied to the linearly elastic model problem shown in Fig. 24, using a single core of AMD Opteron 275 with PGI compiler.

The number of cubic elements is 32,768 (=$32^3$), where the total problem size is 107,811 DOF.

**BILU(0+,ω)**, in which additional *selective fill-ins* have been applied to **BILU(0)** for nodes connected to special elements (Young's modulus is larger than **ω**), provides robust and efficient convergence. Although the number of non-zero components of the preconditioning matrices associated with methods **BILU(0)** and **BILU(0+,200)** is comparable, the latter is much more robust and efficient. **BILU(1)** provides better convergence than **BILU(0+,ω)**, but it is more expensive.

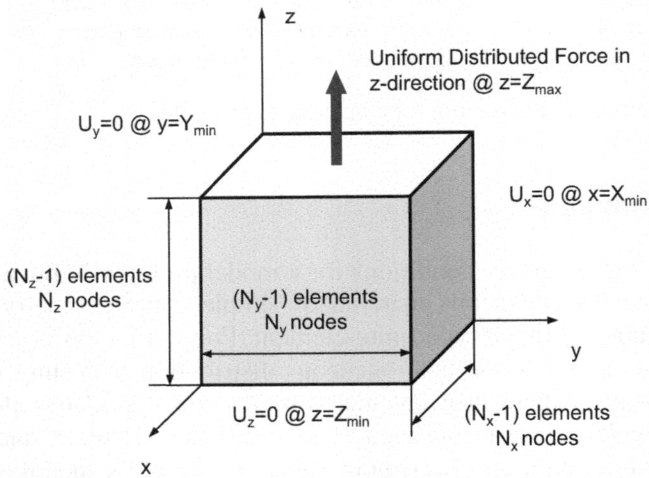


**Fig. 25** Boundary conditions of a model problem in linear elasticity considering simple cubic geometries with heterogeneity as shown in Fig. 24

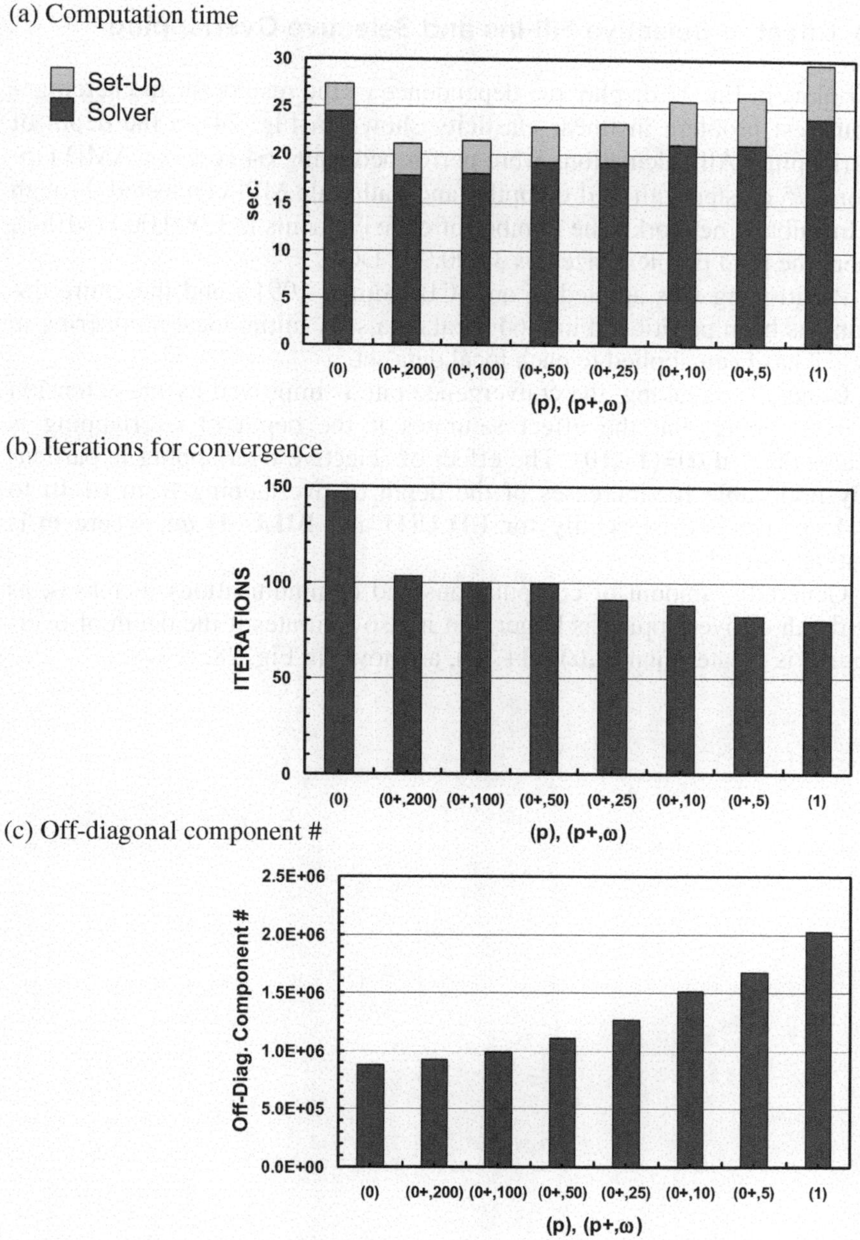


**Fig. 26** Results for the problem in linear elasticity considering simple cube geometries of 107,811 DOF with heterogeneity as shown in Fig. 24 on a single core of AMD Opteron 275 using the PGI compiler

## 4.3 Effect of Selective Fill-Ins and Selective Overlapping

The plots in Fig.27 display the dependence of the results for the heterogeneous test problem in linear elasticity shown in Fig. 24 on the depth of overlapping. All calculations were performed using 64 cores of AMD Opteron 275 cluster with PGI compiler and Pathscale MPI connected through an Infiniband network. The number of cube elements is 1,000,000 (=$100^3$), where the total problem size has 3,090,903 DOF.

Partitioning was applied in an RCB (Simon 1991), and the entire domain has been partitioned into 64 local data sets. Initial local numbering in Fig. 12 has been applied to each local data set.

Generally speaking, the convergence rate is improved by the extension of overlapping, but the effect saturates if the depth of overlapping is greater than **(d,α)=(1+,10)**. The effect of selective overlapping is particularly noticeable for increases of the depth of overlapping from **(d=0)** to **(d=1)** or **(d=1+)**, especially for **BILU(1)** and **BILU(0+,ω)**, where **ω** is relatively small.

Generally, amount of computations and communications increases, as the depth of overlapping is larger, but it also saturates if the depth of overlapping is greater than **(d,α)=(1+,10)**, as shown in Fig. 28.

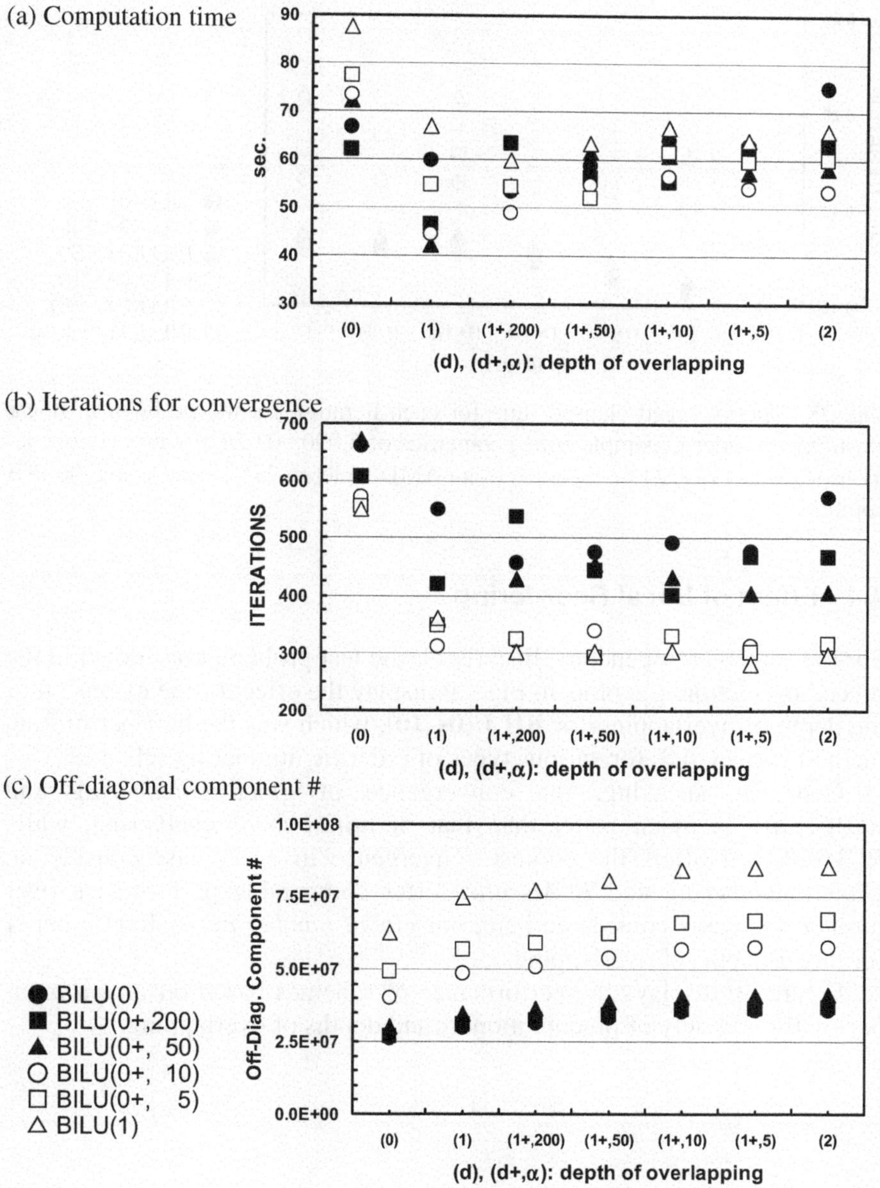


**Fig. 27** The effect of overlapping on the results for the problem in linear elasticity considering simple cube geometries of 3,090,903 DOF with heterogeneity as shown in Fig. 24 on 64 cores of an AMD Opteron 275 cluster using the PGI compiler

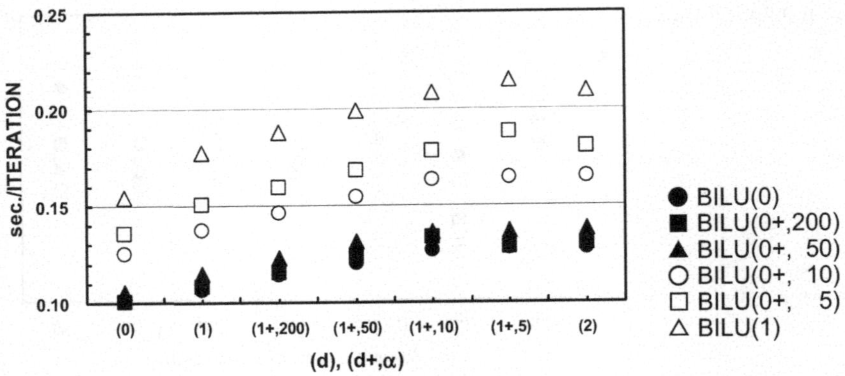


**Fig. 28** The averaged elapsed time for each iteration for the problem in linear elasticity considering simple cube geometries of 3,090,903 DOF with heterogeneity as shown in Fig. 24 on 64 cores of an AMD Opteron 275 cluster using the PGI compiler

## 4.4 Effect of Local Reordering

For the same heterogeneous, linearly elastic test problem considered in the preceding section, the plots in Fig. 29 display the effect of the extension of the depth of overlapping for **BILU(0+,10)**, which was the best performing method in Sect. 4.3, for various types of ordering/numbering schemes.

Generally speaking, the convergence of *global numbering* and *RCM-entire* is much better than that of *initial local numbering*, while *RCM-internal* offers the poorest convergence in every case considered. *Global numbering* and *RCM-entire* offer comparable performance over most of the cases considered here, but *global numbering* is slightly better for larger depths of overlapping.

Figure 30 displays the performance of schemes based on *global numbering* for a variety of preconditioners and depths of overlapping.

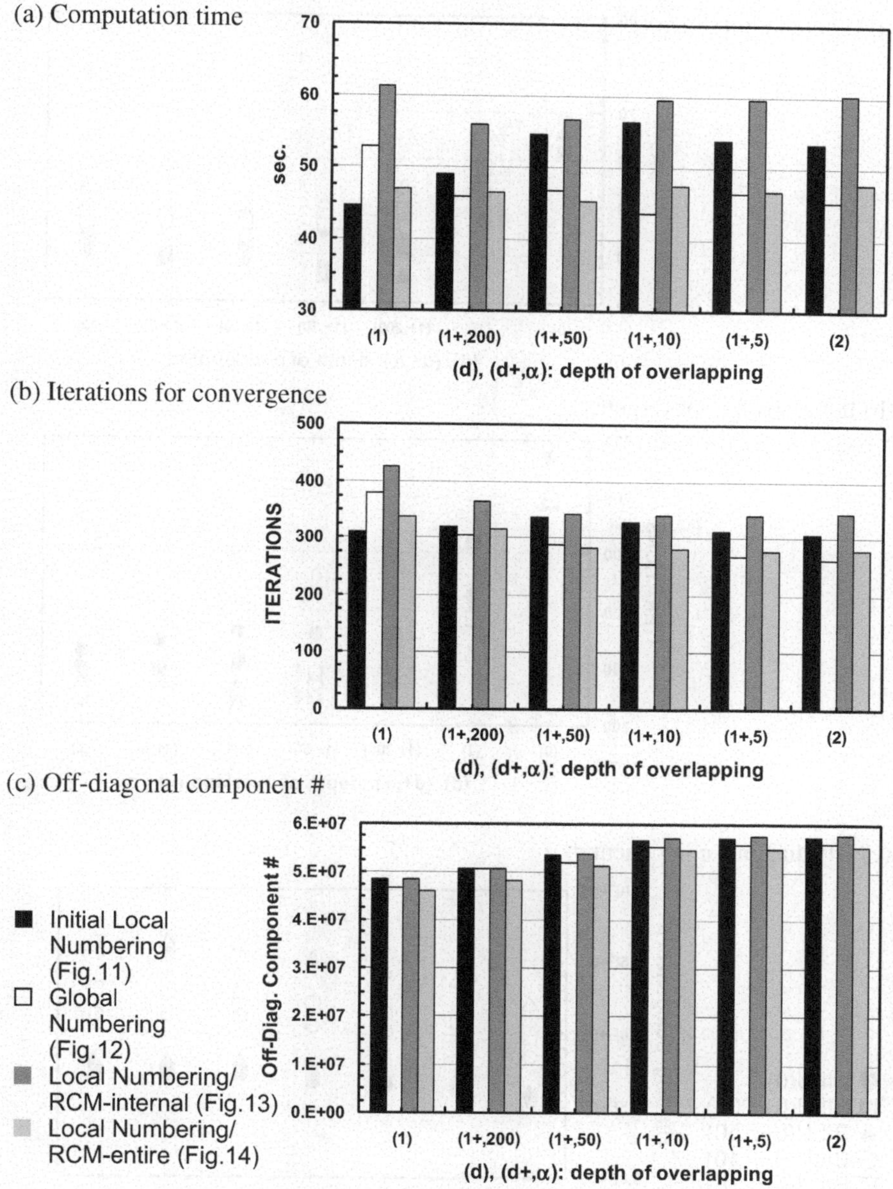


**Fig. 29** The effect of overlapping and local reordering on the results for **BILU(0+,10)** applied to the problem in linear elasticity simple cube geometries of 3,090,903 DOF with heterogeneity as shown in Fig. 24 on 64 cores of AMD Opteron 275 cluster using the PGI compiler

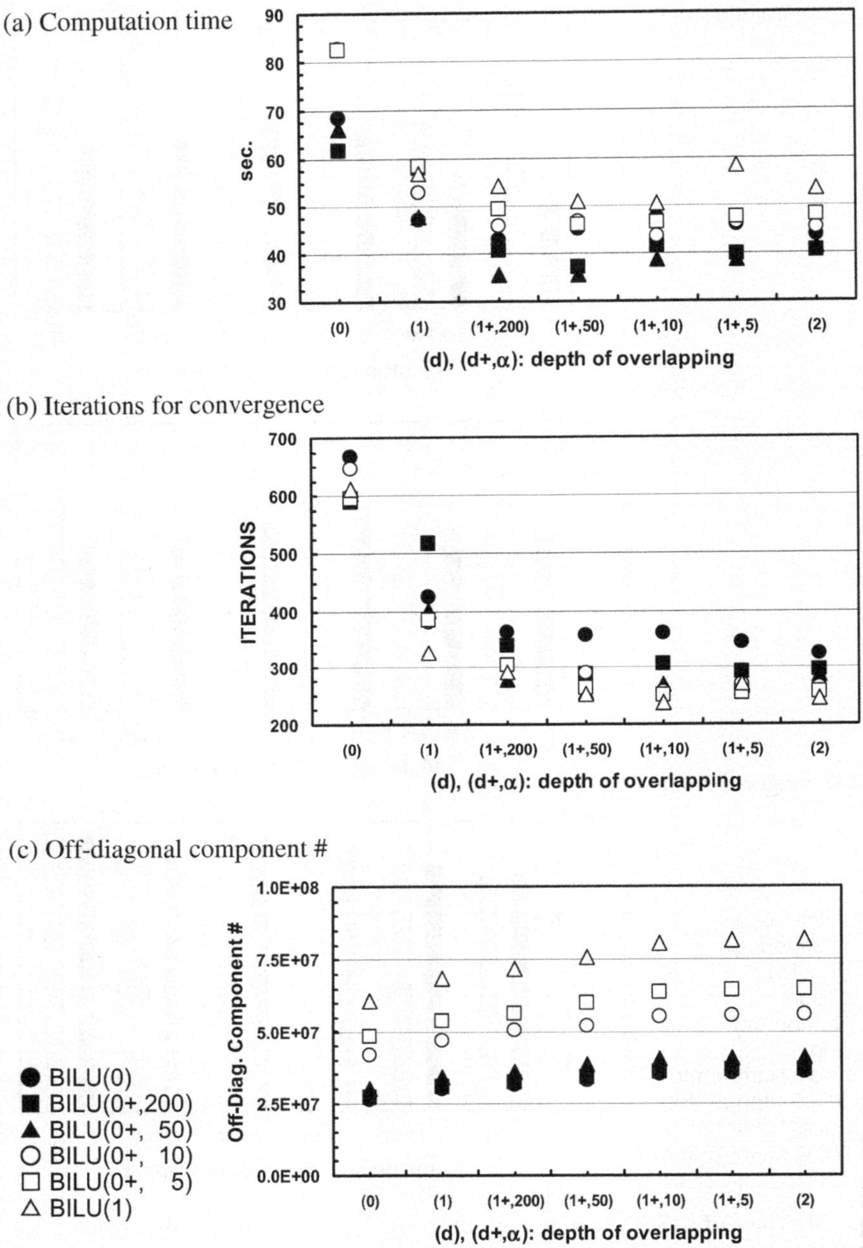


**Fig. 30** The effect of overlapping on results for the problem in linear elasticity considering simple cube geometries of 3,090,903 DOF with heterogeneity as shown in Fig. 24 on 64 cores of an AMD Opteron 275 cluster using the PGI compiler, *global numbering* applied

## 4.5 Effect of HID

In this section the robustness and efficiency of HID is compared with that of *selective overlapping*. The plots in Fig. 31 evaluate the performance of **BILU(0)**, **BILU(0+)**, **BILU(1)** for **(d=0)**, **(d=1)** and **(d=1+,α)** overlapping using *global numbering* (Fig. 13) together with **BILU(1)**, **BILU(1+)**, **BILU(2)** applied with HID.

The depth of overlapping for each local data set provided by HID corresponds to **(d=0),** and is thus in particular smaller than **(d=1)**, as shown in Fig. 19. Figure 31c shows that cost of HID is competitive with that of **(d=0)**.

Figure 31a and b show that **BILU(0)/BILU(0+)/BILU(1)** with HID are faster and more robust than **BILU(0)/BILU(0+)/BILU(1)-(d=1)**, and **BILU(0)/BILU(0+)/BILU(1)-(d=1+,α)** in most of the cases considered.

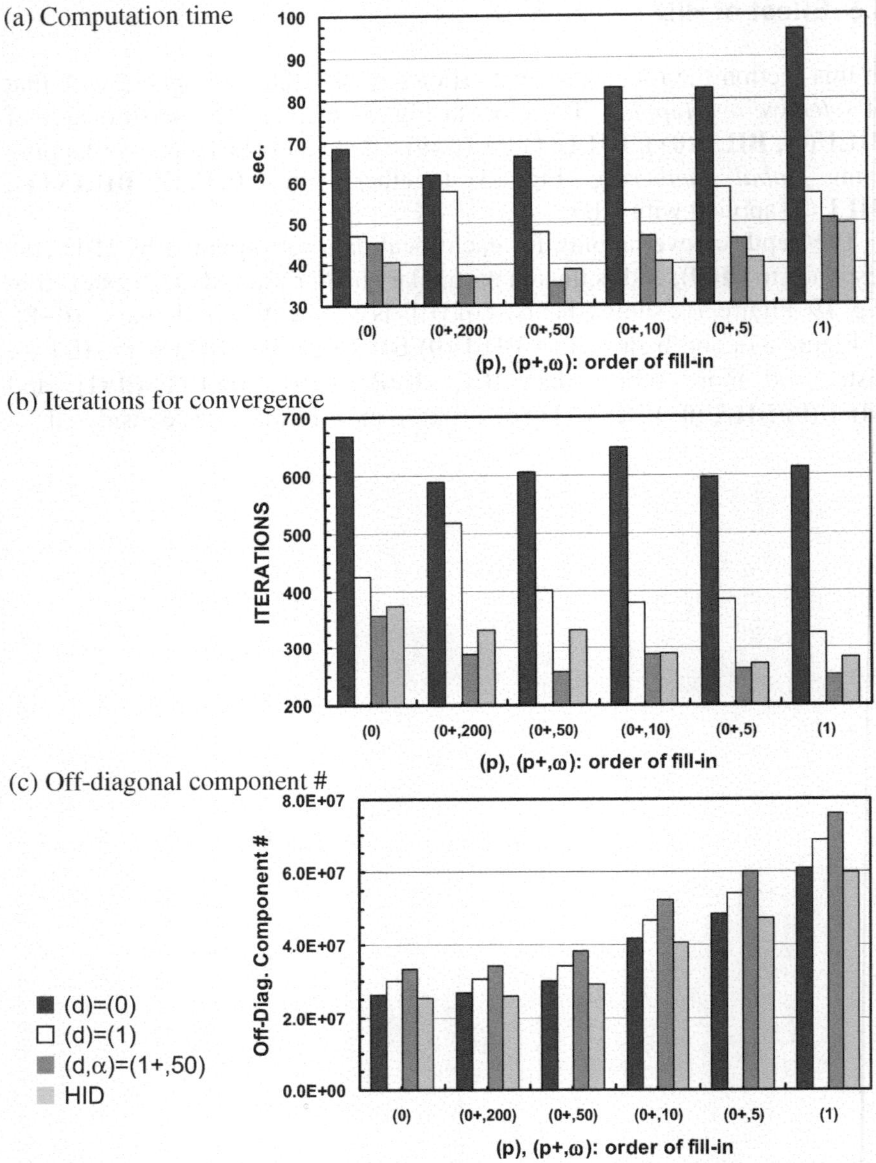


**Fig. 31** The effect of overlapping compared with HID on results for the problem in linear elasticity considering simple cube geometries of 3,090,903 DOF with heterogeneity as shown in Fig. 24 on 64 cores of an AMD Opteron 275 cluster using the PGI compiler, *global numbering* applied

## 5 Concluding Remarks

In this work, the following four approaches have been proposed and introduced as parallel preconditioning methods for ill-conditioned problems:

- Selective fill-ins
- Selective overlapping
- Local reordering
- HID

These methods have been implemented to parallel iterative solvers for finite-element applications, and applied to two types of 3D linear elasticity problems with ill-conditioned coefficient matrices. The first problem includes contact conditions, while the other problem concerns a medium with heterogeneous material properties.

*Selective fill-ins* and *selective overlapping* are very unique methods, because the dropping rules of the preconditioning matrices are defined according to the properties of individual finite-elements and features of the finite-element applications before assembling entire coefficient matrices.

Generally speaking, **BILU(1+)-(1+)** with selective fill-ins **(p=1+)** and selective overlapping **(d=1+)**, provides the best performance with robustness for contact problems, while **BILU(0+,ω)-(1+,α)** with selective fill-ins **(p=0+)** and selective overlapping **(d=1+)** offers the best performance for heterogeneous cases. The effect of *selective overlapping* is particularly marked for increases in the depth of overlapping from **(d=0)** or **(d=1)** to **(d=1+)**.

In this work, the effect of reordering of local nodes on convergence has also been evaluated. Although the optimum method was different for the two test problems considered here, both *global numbering* and *local numbering with RCM-entire* provide better convergence than the other methods considered. These two methods apply renumbering on both the internal and external nodes in each local data set. If a deeper overlapping of domains is employed in the preconditioning processes, both the internal and external nodes should be reordered for better convergence.

Furthermore, HID was compared with *selective overlapping*. Through reordering the unknowns according to their level numbers, the properties of HID ensure that the coefficient matrix $[A]$ has a block structure. This block structure of the reordered matrix in turn leads to a natural parallelism in ILU/IC computations Thus, HID/PHIDAL-based ILU/IC preconditioners can consider the global effect of external domains in parallel computations. Although HID is as cheap in terms of computational costs as **(d=0)** overlapping, it is as robust as **(d=1+)** and **(d=1+,α)** even for ill-conditioned problems, as shown in this work. Of the two schemes, HID and *selective overlapping,* it is difficult at this stage to definitively favor one over the other. Further investigation and comparison of these two methods should be undertaken over various types of real applications.

The selection of optimum preconditioning methods with appropriate parameters for parallel computing is a difficult task, especially for ill-conditioned problems, the focus of this work. Usually, there are many parameters to be selected. For example we have order of fill-ins, depth of overlapping, threshold parameters for fill-ins and overlapping, and the method of local reordering in this work. First of all, further investigation of the effect of each parameter on convergence is required for various types of real applications.

There have been some projects considering the automatic selection of preconditioners and parameters, such as the *I-LIB* (*Intelligent Library*) project.[7] They are mainly focusing on the evaluation of the features of coefficient matrices derived from applications. In real applications, convergence of parallel iterative solvers is often affected by local heterogeneity and/or discontinuity of the field, as shown in this paper. Our strategy is to utilize both the global information obtained from derived coefficient matrices and also very local information, such as information obtained from each mesh in finite-element applications.

The strategy of domain decomposition strongly affects the convergence of the method. In this work, it has been shown that optimum methods for reordering of local data differ according to the particular application, although *RCM-entire* generally provides robust convergence. Furthermore, finite-element models for practical simulations contain various sizes and shapes of elements, although only uniform cubic elements were considered in this work.

The first step towards an automatic selection of parameters in parallel preconditioning methods for ill-conditioned problems is the development of an intelligent domain decomposer (partitioner). According to our experiences in this field, convergence declines if domain boundaries are on elements that provide *strong* connections, such as elements with higher values of Young's modulus in heterogeneous cases, and truss elements in contact cases. The intelligent partitioner should also include some rules for the distortion of elements.

If the HID approach is adopted, we do not have to consider the depth of overlapping, but the strategy for domain decomposition is of critical importance, especially for complicated geometries.

---

[7] http://www.super-computing.org/~kuroda/nadia.html

**Acknowledgements** This work is supported by the 21st Century Earth Science COE Program at the University of Tokyo, and CREST/Japan Science and Technology Agency.

## References

Barrett R, Berry M, Chan TF, Demmel JW, Donato J, Dongarra JJ, Eijkhout V, Pozo R, Romine C, van der Horst H (1994) Templates for the Solution of Linear Systems: Building Blocks for Iterative Methods, SIAM

Deutsch CV, Journel AG (1998) GSLIB Geostatistical Software Library and User's Guide, Second Edition, Oxford University Press

Henon P, Saad Y (2007) A Parallel Multistage ILU Factorization Based on a Hierarchical Graph Decomposition. SIAM Journal for Scientific Computing 28, 2266–2293

Iizuka M, Okuda H, Yagawa G (2000) Nonlinear Structural Subsystem of GeoFEM forFault Zone Analysis. Pure and Applied Geophysics 157, 2105–2124

Liou J, Tezduyar TE (1992) Clustered Element-by-Element Computations for Fluid Flow. In Simon HD (ed) Parallel Computational Fluid Dynamics (Implementations and Results), The MIT Press, pp. 167–187

Nakajima K (2003) Parallel Iterative Solvers of GeoFEM with Selective Blocking Preconditioning for Nonlinear Contact Problems on the Earth Simulator, ACM/IEEE Proceedings of SC2003

Nakajima K, Okuda H (2004) Parallel Iterative Solvers for Simulations of Fault Zone Contact using Selective Blocking Reordering. Numerical Linear Algebra with Applications 11, 831–852

Nakajima K (2005) Parallel Preconditioned Iterative Solvers for Contact Problems (in Japanese), Proceedings of Annual Meeting of Japan Society for Applied Mathematics (JSIAM), 18–19

Nakajima K (2007a) Parallel Preconditioning Methods with Selective Fill-Ins and Selective Overlapping for Ill-Conditioned Problems in Finite-Element Methods. Lecture Notes in Computer Science 4489, 1085–1092

Nakajima K (2007b) Parallel Multistage Preconditioners Based on a Hierarchical Graph Decomposition for SMP Cluster Architectures with a Hybrid Parallel Programming Model. Lecture Notes in Computer Science 4782, 384–395

Saad Y (2003) Iterative Methods for Sparse Linear Systems, Second Edition, SIAM

Simon HD (1991) Partitioning of Unstructured Problems for Parallel Processing. Computing Systems in Engineering 2, 135–148

Washio T, Hisada T, Watanabe H, Tezduyar TE (2005) A Robust and Efficient Iterative Linear Solver for Strongly Coupled Fluid-Structure Interaction Problems. Computer Methods in Applied Mechanics and Engineering 194, 4027–4047

Zhang SL (1997) GPBi-CG: Generalized Product-Type Methods Based on Bi-CG for Solving Nonsymmetric Linear Systems. SIAM Journal of Scientific Computing 18, 537–551

# 6 Appendix 1: Parallel Iterative Solvers in GeoFEM

## 6.1 Distributed Data Structure

GeoFEM adopts domain decomposition for parallel computing where the entire model is divided into domains, and each domain is assigned to a processing element (PE). A proper definition of the layout of the distributed data structures is an important factor determining the efficiency of parallel computations with unstructured meshes. The local data structures in GeoFEM are node-based with overlapping elements, and as such are appropriate for the preconditioned iterative solvers used in GeoFEM.

Although MPI provides subroutines for communication among processors during computation for structured grids, it is necessary for users to design both the local data structure and communications for unstructured grids. In GeoFEM, the entire region is partitioned in a *node-based* manner and each domain contains the following local data:

- Nodes originally assigned to the domain
- Elements that include the assigned nodes
- All nodes that form elements but are from external domains
- A communication table for sending and receiving data
- Boundary conditions and material properties

Nodes are classified into the following three categories from the viewpoint of message passing:

- Internal nodes (originally assigned to the domain)
- External nodes (forming the element in the domain but are from external domains)
- Boundary nodes (*external nodes* of other domains)

Communication tables between neighboring domains are also included in the local data. Values on *boundary* nodes in the domains are *sent* to the neighboring domains and are *received* as *external* nodes at the *destination* domain. This data structure, described in Fig. 32, and the communication procedure described in Fig. 33 provide excellent parallel efficiency. This type of communication occurs in the procedure for computing the matrix-vector product of Krylov iterative solvers described in the next subsection. The partitioning program in GeoFEM works on a single PE, and divides the initial entire mesh into distributed local data.

In GeoFEM, coefficient matrices for linear solvers are assembled in each domain according to FEM procedures. This process can be performed without communication among processors using the information of overlapping elements.

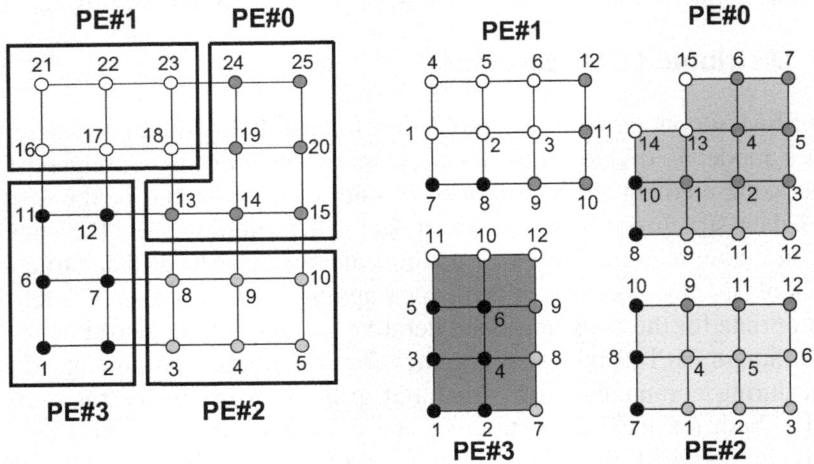


Fig. 32 Node-based partitioning into four PEs

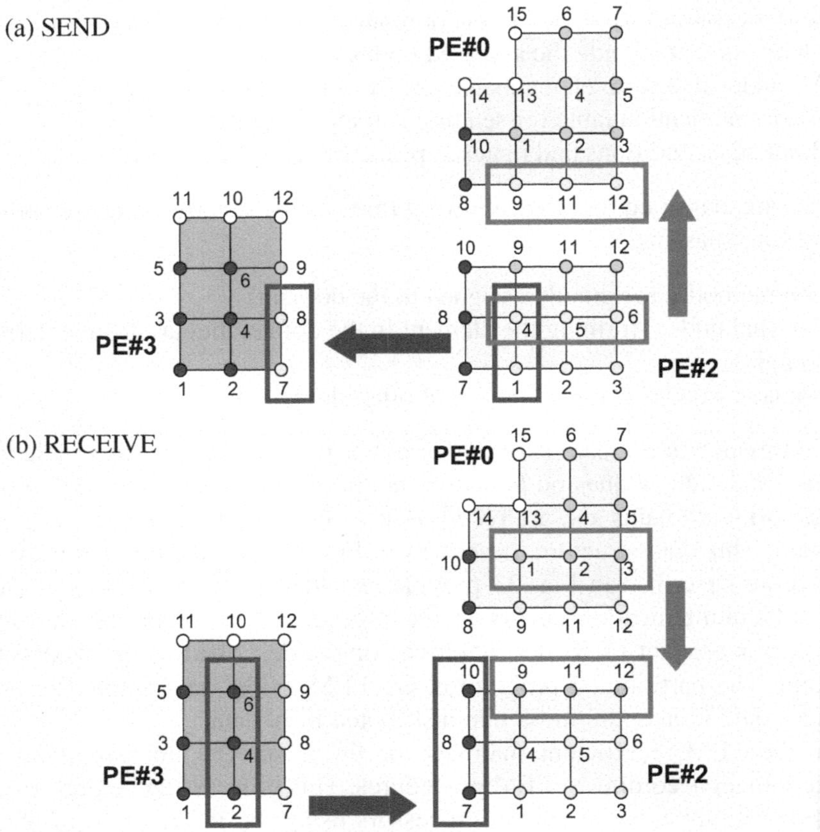


Fig. 33 Communication among processors

## 6.2 Localized Preconditioning

The incomplete lower-upper (ILU) and incomplete Cholesky (IC) factorization methods are the most popular preconditioning techniques for accelerating the convergence of Krylov iterative methods.

Of the range of ILU preconditioning methods, ILU(0), which does not allow fill-in beyond the original non-zero pattern, is the most commonly used. Backward/forward substitution (BFS) is repeated at each iteration. BFS requires global data dependency, and this type of operation is not suitable for parallel processing in which locality is of utmost importance. Most preconditioned iterative processes are a combination of the following four processes:

- matrix-vector products
- inner dot products
- DAXPY (linear combination of vectors) operations and vector scaling
- preconditioning operations

The first three operations can be parallelized relatively easily. In general, preconditioning operations such as BFS represent almost 50 % of the total computation if ILU(0) is implemented as the preconditioning method. Therefore, a high degree of parallelization is essential for the BFS operation.

The *localized* ILU(0) used in GeoFEM is a *pseudo* ILU(0) preconditioning method that is suitable for parallel processors. This method is not a *global* method, rather, it is a *local* method on each processor or domain. The ILU(0) operation is performed locally for a coefficient matrix assembled on each processor by zeroing out components located outside the processor domain. This is equivalent to solving the problem within each processor with zero Dirichlet boundary conditions during the preconditioning. This *localized* ILU(0) provides data locality on each processor and good parallelization because no inter-processor communications occur during ILU(0) operation. This idea is originally from the incomplete block Jacobi preconditioning method.

However, localized ILU(0) is not as powerful as the global preconditioning method. Generally, the convergence rate degrades as the number of processors and domains increases. At the critical end, if the number of processors is equal to the number of degrees of freedom (DOF), this method performs identically to diagonal scaling.

Table 1 shows the results of a homogeneous solid mechanics example with $3 \times 44^3$ DOF solved by the conjugate gradient (CG) method with localized IC(0) preconditioning. Computations were performed on the

Hitachi SR2201, which was operated by the Information Technology Center of the University of Tokyo.[8] Although the number of iterations for convergence increases according to the domain number, this increase is just 30% from 1 to 32 PEs.

Figure 34 shows the work ratio (real computation time/elapsed execution time including communication) for various problem sizes of simple 3D elastic problems with homogeneous boundary conditions. In these computations, the problem size for 1 PE was fixed. The largest case was 196,608,000 DOF on 1024 PEs. Figure 34 shows that the work ratio is higher than 95% if the problem size for 1 PE is sufficiently large. In this case, code was vectorized and a performance of 68.7 GFLOPS was achieved using 1024 PEs. Peak performance of the system was 300 GFLOPS with 1024 PEs; 68.7 GFLOPS corresponds to 22.9% of the peak performance. This good parallel performance is attributed largely to the reduced overhead provided by the use of communication tables as part of the GeoFEM's local data structure.

**Table 1** Homogeneous solid mechanics example with $3 \times 44^3$ DOF on Hitachi SR2201 solved by CG method with localized IC(0) preconditioning (convergence criteria $\varepsilon = 10^{-8}$)

| PE # | Iter. # | Sec. | Speed up |
|---|---|---|---|
| 1 | 204 | 233.7 | – |
| 2 | 253 | 143.6 | 1.63 |
| 4 | 259 | 74.3 | 3.15 |
| 8 | 264 | 36.8 | 6.36 |
| 16 | 262 | 17.4 | 13.52 |
| 32 | 268 | 9.6 | 24.24 |
| 64 | 274 | 6.6 | 35.68 |

[8] http://www.cc.u-tokyo.ac.jp

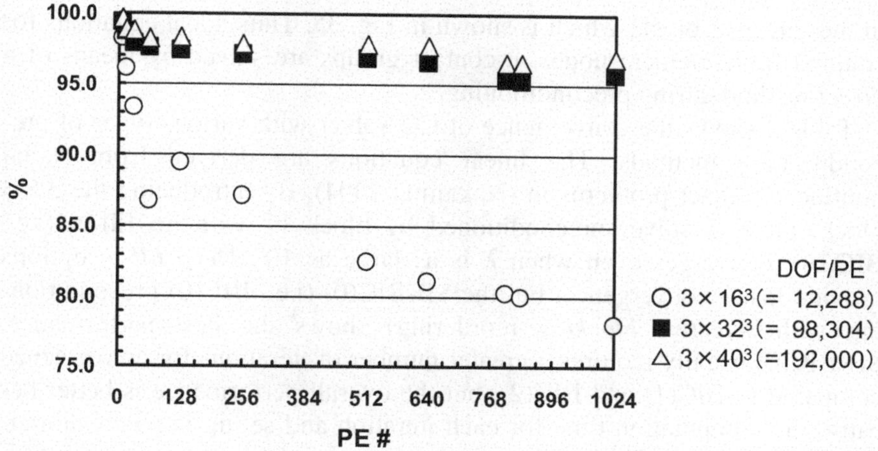


**Fig. 34** Parallel performance for various problem sizes for simple 3D elastic solid mechanics on Hitachi SR2201, problem size/PE is fixed, largest case is 196,608,000 DOF on 1024 PEs

# 7 Appendix 2: Selective Blocking

## 7.1 Robust Preconditioning Methods for Ill-Conditioned Problems

The IC/ILU factorization methods are the most popular preconditioning techniques for accelerating the convergence of Krylov iterative methods. The typical remedies using an IC/ILU type of preconditioning method for ill-conditioned matrices, which appear in nonlinear simulations using penalty constraints, are as follows:

- Blocking
- Deep Fill-in
- Reordering.

In addition to these methods, a special method called *selective blocking* was also developed for contact problems in Nakajima (2004). In the *selective blocking* method, strongly coupled finite-element nodes in the same contact group coupled through penalty constraints are placed into the same large block (*selective block* or *super* node) and all of the nodes involved are reordered according to this blocking information. Full LU factorization is applied to each selective block. The size of each block is $(3 \times NB) \times (3 \times NB)$ in 3D problems, where NB is the number of finite-element nodes in the selective block, which is shown in Fig. 35. Thus, local equations for coupled finite-element nodes in contact groups are solved by means of a *direct* method during preconditioning.

Table 2 shows the convergence of CG solver with various types of preconditioning methods. The linear equations are derived from actual nonlinear contact problems in (Nakajima 2004). By introducing the $3 \times 3$ block, the CG solver preconditioned by block IC with no fill-in (i.e., BIC(0)), converges even when $\lambda$ is as large as $10^6$. *Deep fill-in* options provide faster convergence, but the SB-BIC(0) (i.e., BIC(0) preconditioning with *selective blocking* reordering) shows the best performance. SB-BIC(0) usually requires a greater number of iterations for convergence compared to BIC(1) and BIC(2), but the overall performance is better because the computation time for each iteration and set-up is much shorter. As is also shown in Table 2, because no *inter-block* fill-in is considered for SB-BIC(0), the memory requirement for this method is usually as small as that in BIC(0) with no fill-in. Only the *inter-node* fill-in in each *selective block* is considered in SB-BIC(0).

The CG solver with SB-BIC(0) preconditioning can be considered to be a hybrid of iterative and direct methods. Local equations for coupled finite-element nodes in contact groups are solved by means of a direct method during preconditioning. This method combines the efficiency and scalability of iterative methods with the robustness of direct methods.

This idea of selective blocking is also related to the *clustered element-by-element method* (CEBE) (Liou and Tezduyar 1992). In CEBE, elements are partitioned into clusters of elements, with the desired number of elements in each cluster, and the iterations are performed in a cluster-by-cluster fashion. This method is highly suitable for both vectorization and parallelization, if it is used with proper clustering and element grouping schemes. Any number of elements can be brought together to form a cluster, and the number should be viewed as an optimization parameter to minimize computational cost. The CEBE method becomes equivalent to the direct method when the cluster size is equal to the total number of elements. Generally, larger clusters provide better convergence rates because a larger number of fill-in elements are taken into account during factorization, but the cost per iteration cycle increases according to the size of the cluster, as shown in Fig. 36. The trade-off between convergence and computational cost is not clear, but the results of examples by Liou and Tezduyar (1992) show that larger clusters provide better performance.

In *selective blocking*, clusters are formed according to information about the contact groups. Usually, the size of each cluster is much smaller than that in a general CEBE method. If a finite element node does not belong to any contact groups, it forms a cluster whose size is equal to one in the selective blocking.

**Table 2** Iterations/computation time for convergence ($\varepsilon=10^{-8}$) on a single PE of Intel Xeon 2.8 GHz by preconditioned CG for the 3D elastic fault-zone contact problem in (Nakajima 2004) (83,664 DOF), **BIC(p)**: Block IC with p-th-order fill-ins, **SB-BIC(0)**: BIC(0) with the selective blocking reordering

| Preconditioning method | $\lambda$ | Iterations | Set-up (sec.) | Solve (sec.) | Set-up + solve (sec.) | Single iter. (sec.) | Required memory (MB) |
|---|---|---|---|---|---|---|---|
| Diagonal Scaling | $10^2$ | 1,531 | <0.01 | 75.1 | 75.1 | 0.049 | 119 |
| | $10^6$ | N/A | – | – | – | – | |
| IC(0) (Scalar Type) | $10^2$ | 401 | 0.02 | 39.2 | 39.2 | 0.098 | 119 |
| | $10^6$ | N/A | – | – | – | – | |
| BIC(0) | $10^2$ | 388 | 0.02 | 37.4 | 37.4 | 0.097 | 59 |
| | $10^6$ | 2,590 | 0.01 | 252.3 | 252.3 | 0.097 | |
| BIC(1) | $10^2$ | 77 | 8.5 | 11.7 | 20.2 | 0.152 | 176 |
| | $10^6$ | 78 | 8.5 | 11.8 | 20.3 | 0.152 | |
| BIC(2) | $10^2$ | 59 | 16.9 | 13.9 | 30.8 | 0.236 | 319 |
| | $10^6$ | 59 | 16.9 | 13.9 | 30.8 | 0.236 | |
| SB-BIC(0) | $10^0$ | 114 | 0.10 | 12.9 | 13.0 | 0.113 | 67 |
| | $10^6$ | 114 | 0.10 | 12.9 | 13.0 | 0.113 | |

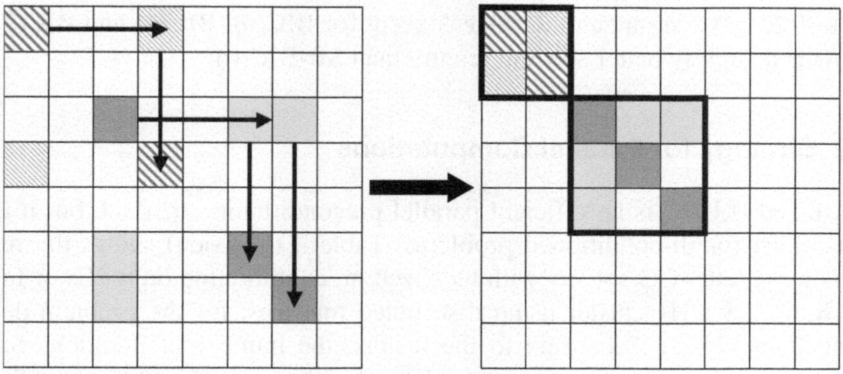

(a) (b)

**Fig. 35** Procedure of the *selective blocking*, strongly coupled elements are put into the same *selective block*, (**a**) searching for strongly coupled components and (**b**) reordering and selective blocking

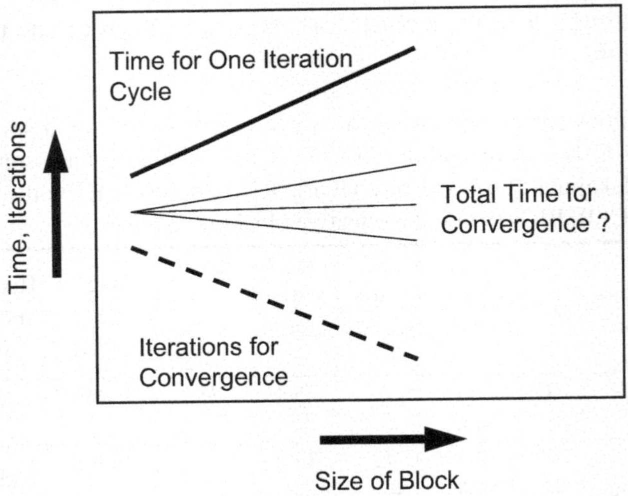


**Fig. 36** Trade-off between convergence and computational cost per on iteration cycle according to block size in CEBE type method, based on (Liou and Tezduyar 1992)

In (Nakajima 2003), the robustness of the preconditioning method was estimated according to the eigenvalue distribution of the $[M]^{-1}[A]$ matrix by the method in (Barrett et al. 1994), where $[A]$ is the original coefficient matrix and $[M]^{-1}$ is the inverse of the preconditioning matrix. According to the results, all of the eigenvalues are approximately constant and close to 1.00 for a wide range of $\lambda$ values except for BIC(0). BIC(1) and BIC(2) provide a slightly better spectral feature than SB-BIC(0).

## 7.2 Strategy for Parallel Computations

Localized ILU/IC is an efficient parallel preconditioning method, but it is not robust for ill-conditioned problems. Table 3 (left side) shows the results by parallel CG solvers with localized preconditioning on 8 PEs of Intel Xeon 2.8 GHz cluster using distributed matrices, for the problem described in Fig. 1. According to the results, the number of iterations for convergence increases by a factor of 10 in $\lambda=10^6$ cases. This is because the *edge-cuts* occur at inter-domain boundary edges that are included in contact groups.

In order to eliminate these edge-cuts, a partitioning technique has been developed so that all nodes which belong to the same contact group are in the same domain. Moreover, nodes are re-distributed so that load-balancing among domains should be attained for efficient parallel computing (Fig. 37).

In GeoFEM, there are several types of special elements for contact problems (types 411, 412, 421, 422, 511, 512, 521 and 522). Nodes included in the same elements of these types are connected through penalty constraints and form a contact group. In the new partitioning method, the partitioning process is executed so that these nodes in the same contact elements are on the same domain, or PE. These functions are added to the original domain partitioner in GeoFEM.

Table 3 (right side) shows the results obtained by this partitioning method. The number of iterations for convergence has been dramatically reduced for each preconditioning method although it is larger than that of the single PE cases shown in Table 2 due to localization.

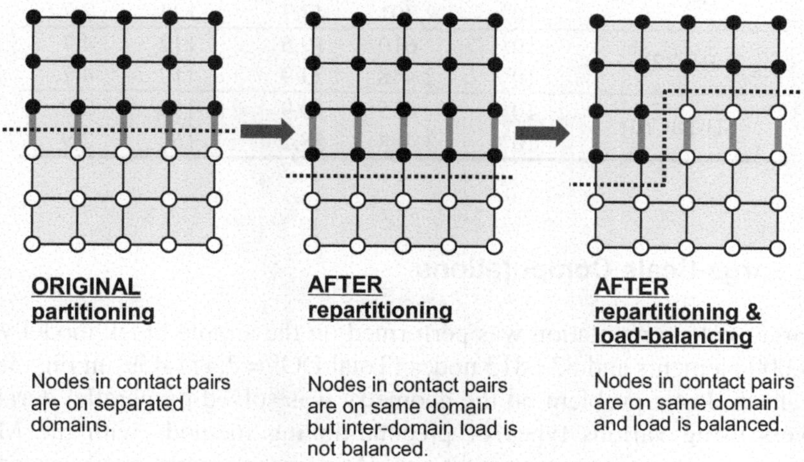


**Fig. 37** Partitioning strategy for the nodes in contact groups

**Table 3** Iterations/computation time for convergence ($\varepsilon=10^{-8}$) on 8 PEs of Intel Xeon 2.8 GHz cluster by preconditioned CG for the 3D elastic fault-zone contact problem in (Nakajima 2004) (83,664 DOF), **BIC(n)**: Block IC with n-level fill-in, **SB-BIC(0)**: BIC(0) with the selective blocking reordering, effect of repartitioning method in Fig. 37 is evaluated

| Preconditioning method | $\lambda$ | ORIGINAL partitioning Iterations | ORIGINAL partitioning Set-up + solve (sec.) | IMPROVED partitioning Iterations | IMPROVED partitioning Set-up + solve (sec.) |
|---|---|---|---|---|---|
| BIC(0) | $10^2$ | 703 | 7.5 | 489 | 5.3 |
| | $10^6$ | 4,825 | 50.6 | 3,477 | 37.5 |
| BIC(1) | $10^2$ | 613 | 11.3 | 123 | 2.7 |
| | $10^6$ | 2,701 | 47.7 | 123 | 2.7 |
| BIC(2) | $10^2$ | 610 | 19.5 | 112 | 4.7 |
| | $10^6$ | 2,448 | 73.9 | 112 | 4.7 |
| SB-BIC(0) | $10^0$ | 655 | 10.9 | 165 | 2.9 |
| | $10^6$ | 3,498 | 58.2 | 166 | 2.9 |

## 7.3 Large-Scale Computations

A large-scale computation was performed on the simple block model with 784,000 elements and 823,813 nodes (Total DOF= 2,471,439) in Fig. 38.

Linear elastic problem on the geometry was solved by parallel iterative solvers using various types of preconditioning methods with the MPC (multiple point constraint) conditions. Domains are partitioned according to the contact group information described in the previous section. Computations were performed using 16–256 PEs on a Hitachi SR2201 at the University of Tokyo.

Table 4 shows the results for various preconditioners. BIC(1), BIC(2) and SB-BIC(0) provide robust convergence but convergence of BIC(0) is very slow. SB-BIC(0) provides the most efficient performance, although the iteration number for convergence is larger than BIC(1) and BIC(2). Figure 39 and Table 4 show the parallel performance for the same problem solved using 16–256 PEs of Hitachi SR2201. BIC(1) and BIC(2) did not work if the PE number was small due to memory limitation. As shown in Table 4 and Fig. 39 the iteration number for convergence increases according to PE number due to the locality of the preconditioning method, but this increase is very slight (only 14% increase from 16 PEs to 256 PEs for SB-BIC(0)). The speed-up ratio based on elapsed execution time including communication for 256 PEs, is 235 for SB-BIC(0), as extrapolated from the results obtained using 16 PEs.

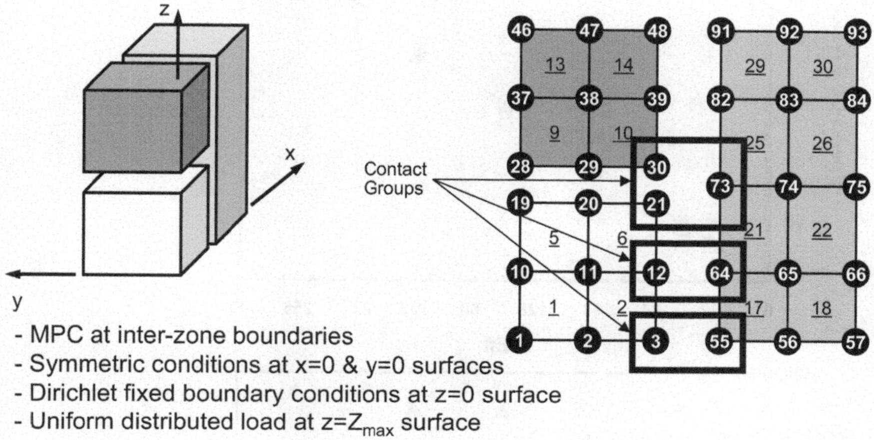


**Fig. 38** Description of the simple block model

**Table 4** Iterations/elapsed execution time (including factorization, communication overhead) for convergence ($\varepsilon=10^{-8}$) on a Hitachi SR2201 with 256 PEs using preconditioned CG for the 3D elastic contact problem for simple block model with MPC condition in Fig. 38 (2,471,439 DOF), domains are partitioned according to the contact group information, **BIC(p)**: Block IC with p-th-order fill-ins, **SB-BIC(0)**: BIC(0) with the selective blocking reordering

| | | PE# | | | | | |
|---|---|---|---|---|---|---|---|
| | | 16 | 48 | 96 | 144 | 192 | 256 |
| BIC(0) | iterations | 14,459 | 15,018 | 15,523 | 15,820 | 16,084 | 16,267 |
| | sec. | 13,500 | 4,810 | 2,410 | 1,630 | 1,270 | 1,230 |
| | speed-up | 16 | 45 | 90 | 133 | 170 | 211 |
| BIC(1) | iterations | | 379 | 402 | 424 | 428 | 452 |
| | sec. | N/A | 236 | 119 | 81 | 62 | 48 |
| | speed-up | | 48 | 95 | 140 | 183 | 236 |
| BIC(2) | iterations | | | 364 | 387 | 398 | 419 |
| | sec. | N/A | N/A | 212 | 140 | 112 | 86 |
| | speed-up | | | 96 | 145 | 182 | 217 |
| SB-BIC(0) | iterations | 511 | 527 | 543 | 567 | 569 | 584 |
| | sec. | 555 | 193 | 96 | 64 | 48 | 38 |
| | speed-up | 16 | 46 | 92 | 139 | 185 | 235 |

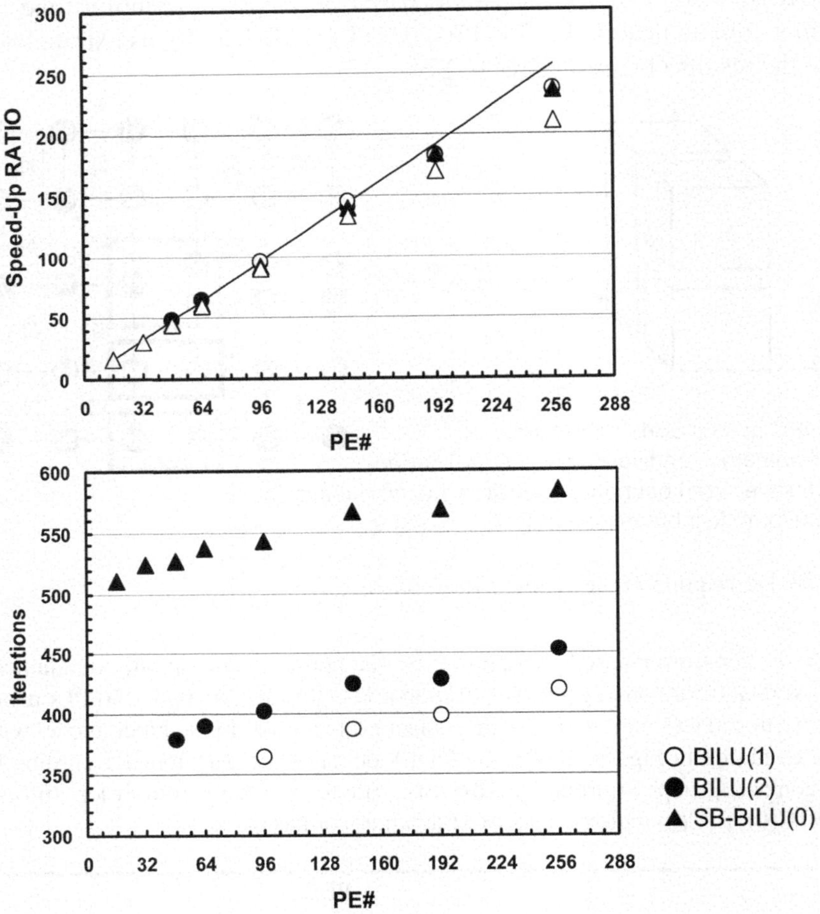


**Fig. 39** Parallel performance based on elapsed execution time including communication and iterations for convergence ($\varepsilon=10^{-8}$) on a Hitachi SR2201 with 16–256 PEs using preconditioned CG for the 3D elastic contact problem with MPC condition ($\lambda=10^{6}$) in Fig. 38 (2,471,439 DOF)

# IV. Algorithms for Optimizing Rheology and Loading Forces in Finite Element Models of Lithospheric Deformation

Youqing Yang and Mian Liu

University of Missouri, Columbia, MO 65211, USA

**Abstract** Lithospheric deformation results from dynamic interplay of tectonic driving forces (loading) and lithospheric properties (rheology and structure). Unlike engineering problems, in lithospheric dynamics both the loading conditions and lithospheric properties are often hard to constrain, forcing many computer models to oversimplification. These simplified models usually cannot take the full advantage of the fast growing observational constraints. In this article, we present algorithms that help to seek optimal loading conditions and rheological parameters in models of lithospheric deformation. In particular, we use genetic algorithms to iterate for the optimal rheological structure, and a regression algorithm for optimizing tectonic loading. We illustrate these algorithms in two models: a plate flexure and a viscous three-dimensional lithospheric deformation. In both cases these algorithms utilize the observational constraints to obtain the optimal driving forces and lithospheric rheology. The results significantly improve over those derived from traditional approaches.

## 1 Introduction

Lithospheric deformation is usually more difficult to simulate than that of engineering structures such as airplanes, automobiles or bridges. In the latter the physical property of the media is usually known, so the deformation can be accurately predicted for various loading conditions. Conversely, studies of lithospheric dynamics often require simultaneous determination of both the rheological structure and the loading forces. The common practice is to assume a simple rheological structure and then to seek the combinations of force balance to fit the observed deformation, or from "known" force balance to determine the lithospheric rheology necessary for fitting the observed deformation (Bird, 1998; Flesch et al.,

H. Xing, *Advances in Geocomputing*, Lecture Notes in Earth Sciences 119,
DOI 10.1007/978-3-540-85879-9_4

2000). The results could vary significantly depending on one's approach and assumptions.

Take, for example, the impact of different rheology on continental deformation in the western United States. Using a viscous thin-sheet model with homogeneous power-law viscosity, Sonder et al. (1986) showed that the shear traction from the Pacific-North American plate boundary is largely limited to regions near the plate boundary, thus contributes little to the Basin and Range extension. Conversely, using a linear viscous model with laterally heterogeneous viscosity, Choi and Gurnis (2003) argued that the plate boundary traction could affect the entire Basin and Range province. The assumed lithospheric viscosity also has major impact on the surface strain rates. For a viscosity of $10^{23}$ Pas, the gravitational force is inadequate to produce the extension rate in the Basin and Range province. But when it is lowered to $10^{22}$ Pas, gravitational spreading could exceed the strain rates measured by the GPS (Liu et al., 2007). Thus, to integrate and interpret the fast growing observational data of crustal deformation, lithosphere dynamic models need better constraints on the rheological structure.

One major goal of geodynamic modeling is to determine the tectonic driving forces. Because of the uncertainty in rheology, the magnitudes of forces and thus their relative roles are often ambiguous. Again use the western US as an example. Different workers has attributed the primary cause of late Cenozoic continental deformation in the western US to gravitational spreading (Jones et al., 1996; Sonder and Jones, 1999), plate boundary force (Atwater, 1970; Zoback et al., 1981), and basal traction (Liu and Bird, 2002). To determine two unknown variables (rheology and force) from only one constraint equation (deformation), the task of numerical modeling of lithospheric dynamics would become easier with algorithms that utilize all available observational constraints to seek optimal values of rheology and loading conditions.

In this paper, we describe some of such algorithms. We illustrate their application in two examples. One is a flexural model for inferring orogenic loading on the edges of the Tarim basin in northwest China; another is a three-dimensional (3D) lithospheric deformation model for simulating lithospheric deformation in the western United States.

## 2 Methodology

Our approach is to start with a simple homogeneous rheologic structure, then modify it if necessary using a genetic algorism. Genetic algorithms (GAs) use adaptive heuristic search, based on the evolutionary ideas of natural selection (Holland, 1973). GAs simulate processes in natural system necessary for evolution, following Charles Darwin's principle of survival of the fittest. As such they represent an intelligent exploitation of a random search within a defined search space. It works well with mixed (continuous and discrete) combinatorial problems.

This approach is better than the traditional try-and-error approach for determining tectonic forces, which is inefficient with serious limitations. Given the rheological heterogeneities in the lithosphere, the tectonic forces obtained through the try-and-error approach could vary significantly depending on subjective selection of the rheological structure. When the numerical model is complex, it is hard to predict how the system will respond to the modified rheology or boundary forces.

The algorithm we developed for searching tectonic forces uses linear regression. For a given rheology, the driving forces on various parts of the model domain can be calculated through least square fitting. The optimal rheology can be found through minimizing the residual errors.

This approach is straightforward in principle, but a number of issues need to be clarified. The first is the non-linear constitutive relation in lithospheric deformation. Lithosphere's response to tectonic forces in geological timescales can be approximated as that of a power-law viscous fluid (Brace and Kohlstedt, 1980; Kirby and Kronenberg, 1987). Because the power-law constitutive relation is nonlinear, the response to two forces is not equal to the sum of two responses to each force. This nonlinear behavior prevents a direct application of least square algorithm. Thus we need to first linearize the power-law stress-strain rate relation, usually written as

$$\tau = B\dot{E}^{1/n-1}\dot{\varepsilon} = \eta\dot{\varepsilon} \qquad (1)$$

where $B$ is stress coefficient, $n$ is power index, $\eta$ is effective viscosity, $\tau$ is deviatory stress vector and $\dot{\varepsilon}$ is strain rate vector, and the effective strain rate is defined as

$$\dot{E} = \sqrt{\frac{2}{3}\dot{\varepsilon}_{ij}\dot{\varepsilon}_{ij}} \qquad (i = 1,2,3 \qquad j = 1,2,3) \qquad (1a)$$

So the effective viscosity can be defined as

$$\eta = B\dot{E}^{1/n-1} \qquad (2)$$

And

$$B = \eta\dot{E}^{1-1/n} \qquad (3)$$

Thus by introducing the effective viscosity, the non-linear rheology is linearized in (1). In practice, the strain rate can be obtained from the GPS (Global Positioning System) or other measurements of crustal deformation.

Once the effective viscosity is obtained, the corresponding power-law constitutive relation can be determined.

The bending and buckling of lithosphere can be simulated in another linear model: Hooke's elasticity. The non-linear effects of faulting and folding and others can be lumped into a simple parameter – the effective elastic thickness of a plate (McNutt, 1984). For a thin plate (in-plane span is about ten times larger than the thickness), the flexure due to tectonic loading around and sedimentary loading within a compensative basin can be expressed as (Turcotte and Schubert, 2002):

$$D\nabla^4 w + p\nabla^2 w + (\rho_m - \rho_s) g w = q_t \tag{4a}$$

or

$$D\nabla^4 w + p\nabla^2 w + \rho_m g w = q_t + \rho_s g w_0 \tag{4b}$$

where $p$ is horizontal load and $q_t$ is vertical tectonic load per unit area on the lithosphere, and the flexural rigidity $D = \frac{Eh^3}{12(1-\nu^2)}$ is defined by elastic modulus $E$, Poisson ratio $\nu$, and plate thickness $h$. The parameters $\rho_m$ and $\rho_s$ are mantle and sedimentary (in some case, it is seawater) density, respectively. Equation (4b) is derived from (4a) by replace the deflection $w$ with the known sedimentary thickness $w_0$. For a thick plate we use three-dimensional elastic theory, which produces a more reliable solution when the thin-plate approximation becomes inadequate.

With the linearized rheology, we apply the least-square algorithm to seek for the optimal loading. For a linear rheology, the displacement (or velocity) caused by force with unit magnitude and direction at specified locations can be expressed as:

$$\vec{v}_i = g(\eta, f_i) \tag{5}$$

where $\eta$ is the linearized rheology, and $f$ is the unit force. Here $i = 0,1,2,...,n$ for segments or regions where uniform force vectors are assumed. When $i = 0$, the force is assumed to be known, which can be vertical sedimentary loading or gravitational force in the model. The synthetic displacement or velocity can be expressed as

$$\vec{v} = (v^x, v^y) = \vec{v}_0 + \sum_{i=1}^{n} \alpha_i \vec{v}_i \tag{6}$$

Here $\alpha_i$ is an unknown coefficient of magnitude of force. Note that Eq. (6) works only for a linear system.

In the model we may have $m$ sites, within each site we have observational constraints, either a horizontal velocity vector or displacement in the form of $\overrightarrow{u_k} = (u_k^x, u_k^y)$ $\quad k = 1, 2, 3, ..., m$,

or flexure deflection $w_k, k = 1, 2, 3, ..., m$.

Our algorithm seeks to minimize the error $E$ defined as:

$$E^2 = \frac{1}{m}\sum_{k=1}^{m}\left[(v_k^x - u_k^x)^2 + (v_k^y - u_k^y)^2\right] \tag{7}$$

or $$E^2 = \frac{1}{m}\sum_{k=1}^{m}(w_k - W_k)^2 \tag{7a}$$

for the plate flexural model, the capitalized $W$ stands for observed deflection.

If the observed deformation has three components, we may use

$$E^2 = \frac{1}{m}\sum_{k=1}^{m}\left[(v_k^x - u_k^x)^2 + (v_k^y - u_k^y)^2 + (v_k^z - u_k^z)^2\right] \tag{7b}$$

Error $E$ is an averaged misfit between the predicted and the observed velocity or displacement. By minimizing $E^2$, we determine the optimal magnitudes of force in each segment using multiple variables regression. The rheology structures are searched using the genetic algorithms; the optimal rheology structure produces the minimum residual error.

## 3 A Plate Flexural Model

Let us start with the problem of bending an elastic lithosphere under the load of mountain ranges. An analytic solution given by Turcotte and Schubert (2002) shows that the effective elastic thickness of the lithosphere can be uniquely determined by the position of the forebulge relative to that of the load. If the amplitude of the forebulge is known, the magnitude of the load can also be estimated. In other words, the mechanic parameters can be uniquely constrained from the deformation.

However, problems of lithospheric flexure are often more complicated. Instead of a single forebulge for constraining the effective elastic thickness, we often have voluminous data of various spatial and time scales that form an overdetermined system. In this case the optimal algorithm described above can be helpful. We used this approach to study orogenic loading around the Tarim Basin in northwestern China (Yang and Liu, 2002). Through much of the Cenozoic, the Tarim basin behaved as an enclosed foreland basin, receiving sediments from the Tibetan Plateau to its south and the Tian Shan mountains to its north. The complete sedimentary records, made available from intensive oil drilling in the basin (Li et al., 1996), provide useful albeit indirect constraints on mountain building in the Tian Shan and the Tibetan Plateau. Figure 1 shows a 2D model, where we applied the sedimentary load based on geological record, and tried to determine the extra orogenic loads near the margins of the basin that are necessary to fit the basement flexure. Through try-and-error, we obtained optimal orogenic load and effective elastic thickness. However, the optimal results derived from one profile are different from another, leading to having different elastic thickness or different optimal loading at the same location where the two profiles intercept.

So we developed a 3D model and divided the margin of the basin into 14 domains (Fig. 2), on each domain the orogenic load is assumed uniform for a given period. Using the regression algorithm described above, we obtained the optimal effective elastic thickness and tectonic loads on each segment of the margins of the basin during various periods. The optimal orogenic load in the paleogene is shown in Fig. 3. By modeling the basement flexure through various periods of the Cenozoic, we also find a thickening trend of the effective thickness of the Tarim plate through the Cenozoic. This thickening trend is consistent with the cooling history of the Tarim basin based on hydrocarbon maturity data (Jia et al., 1996). By dividing the margins of the Tarim basin into segments and seeking of the optimal orogenic loading on each segment for a given period, we were able to infer the spatial and temporal history of Cenozoic mountain building in the northern Tibetan Plateau and the Tian Shan mountains (Yang and Liu, 2002).

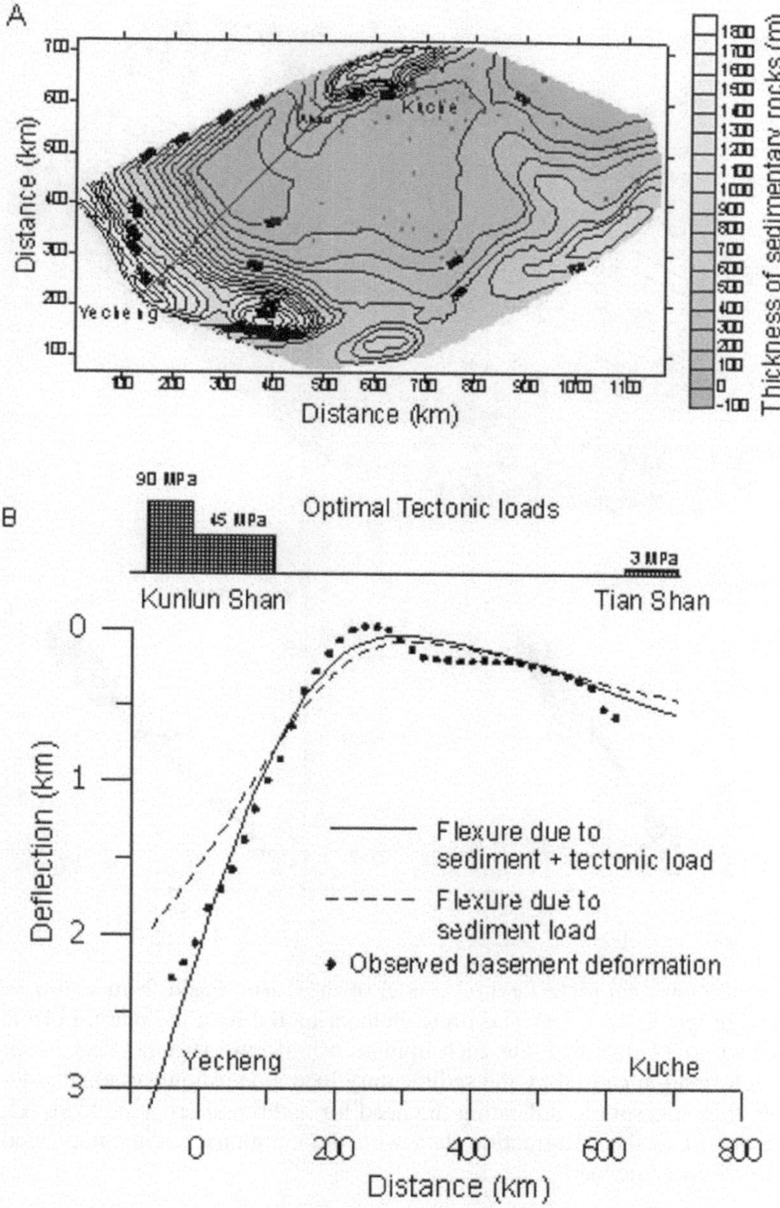


**Fig. 1** Optimal orogenic loading in a two-dimensional elastic plate flexural model of the Tarim basin. (**A**) Isopach of Paleogene sedimentary thickness; the *dots* show part of drill holes that is used for constraining the isopach (Zhao et al., 1997). (**B**) Fitting of the flexural model with different loading conditions to the basement deformation along a profile from Yecheng to Kuche. The profile position is shown in (**A**)

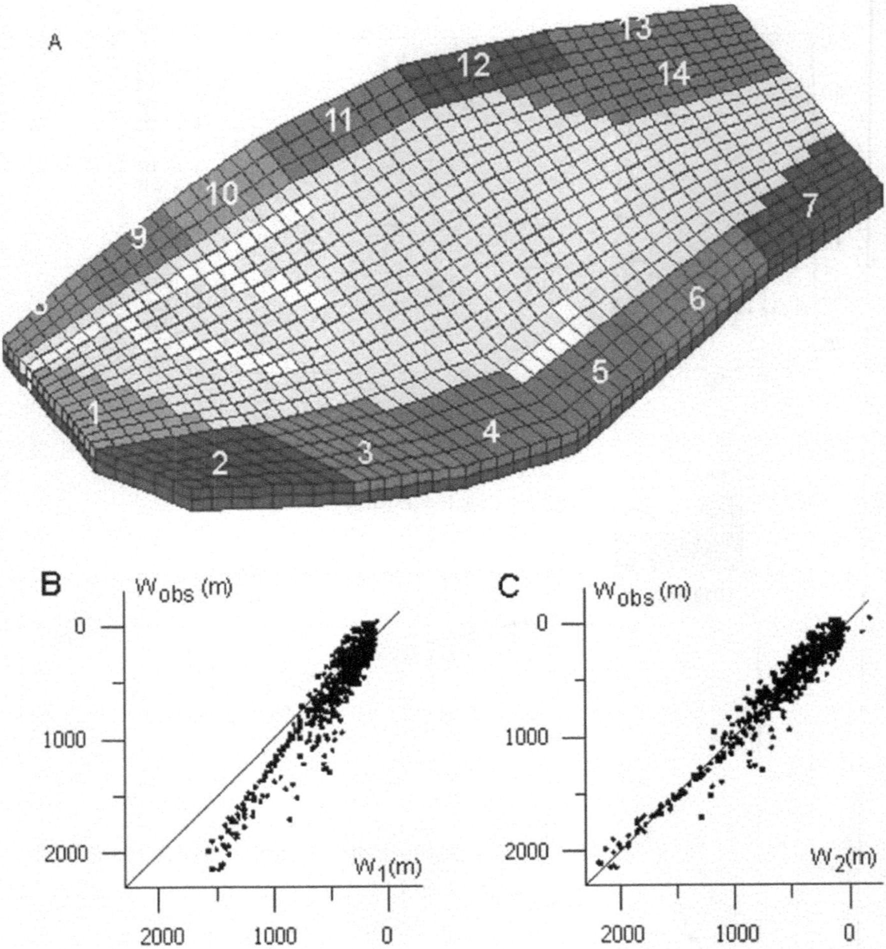


**Fig. 2** Three-dimensional plate flexural model of the Tarim basin from sedimentary load and orogenic load. (**A**) The finite element model with the margin of the basin divided into 14 domains. On each domain a uniform orogenic load is applied. (**B**) Deformation caused by the sedimentary load alone show systematic deviation from the observation, indicating the need for additional orogenic load. (**C**) An satisfactory fit to the deformation data with the combined sedimentary load and the optimal orogenic load

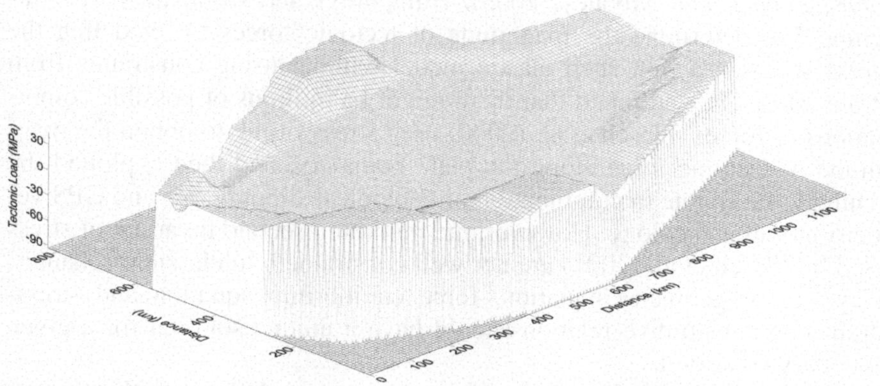


**Fig. 3** The optimal orogenic loading on the margins of the Tarim basin, shown in equivalent average height of mountain ranges. These results suggest that the Tian Shan begin rising in Early Tertiary

## 4 A Three-Dimensional Viscous Model of Lithospheric Deformation

In this example, we apply the optimization algorithm to explore the rheological structure and driving forces responsible for active tectonics in the western United States. The target function consists of mainly surface displacement or strain rates from GPS measurements or geological data. Although the target function can include stresses, stress measurement is difficult and incomplete. In-situ stress measurement varies with lithology and local geological structure. Faults, folds, and dykes may indicate the principle stress direction but not the full stress state. Earthquake mechanism solutions contain no information about the magnitude of stress. So in this study we use only the stress orientation as an additional constraint to our model. The misfit of stress orientation is given by:

$$misfit = \frac{1}{m}\sum_{k=1}^{m}\left|\alpha_k - \beta_k\right| \tag{8}$$

Here $\alpha$ and $\beta$ stand for model predicted and the observed directions of the maximum horizontal principal stress, respectively.

The lithospheric rheology and driving forces responsible for the widespread crustal deformation in the western US have been the subject of intensive studies. Comprehensive reviews and extensive citation lists can be found in many papers (Humphreys and Coblentz, 2007; Jones et al., 1998; Zoback and Mooney, 2003). Humphreys and Coblentz (2007) attempted to determine the magnitude of tectonic forces by modeling the stress states in a thin shell elastic model without using constraints from strain rates. They admitted that there might be millions of possible combinations of forces. Flesch et al. (2000) used stress fitting to obtain the magnitude of tectonic force along the plate boundary, and then exploited the synthetic strain rate (a combination of geological slipping rate and GPS velocity) to infer rheology. However, the minimum second invariant of stress used by Flesch et al. (2001) are not well constrained. In classic mechanics, a system of geometry equation, force equilibrium equation and stress-strain rate constitutive relation should have a unique solution for a given boundary condition.

We solve the driving forces for the western US in a 3D viscous model (Fig. 5A). The model domain is 100 km thick and laterally spans from the Rockies and the Colorado plateau to the San Andreas Fault. The topography (based on the Etopo5 data) and the associated topographic loading are included in the model, and the crustal thickness is based on the Crust2.0 database. In the model we divided southwestern US according to the tectonic provinces; the Basin and Range province is further divided into three subdomains because of their distinct active deformation (Bennett et al., 2003). The surface of the model is free; the bottom is fixed in vertical direction and free traction in horizontal. The model is fixed along the eastern boundary and subject to forces along other boundary. We constrained the magnitudes of the forces along the plate boundary by fitting the observed crustal motion. We also compare the stress states predicted from our model and those from World Stress Map database. Using the genetic algorithm, we iterated the rheological structure. Each new generation of rheological model inherits the good features from the last generation and adds mutation (new factors).

We started our search with a homogeneous rheologic structure in the entire model domain. In the first generation of models, we tested with (1) one strong homogeneous plate, (2) one weak homogeneous plate, (3) a strong plate with moderate faults, (4) and a strong plate with weak faults. The results showed that an average effective viscosity of $5.0\times10^{22}$ Pa s is a good approximation of the western U.S., so it is used as a reference value for adjusting rheology in later heterogeneous rheology models. With a homogeneous effective viscosity $5.0\times10^{22}$ Pa s, we find the averaged shear traction along the San Andreas Fault (SAF) to be 16 MPa by least square regression shown in Eq. (7). Integrating over a 100 km thick lithosphere, the force per unit length is 1.6 TN/m, similar to the value of 1.0~2.0 TN/m estimated by Humphreys and Coblentz (2007). Considering that the central SAF and south SAF segments are weak (Townend and Zoback, 2004), we assumed zero shear traction along these two segments in one case, and predicted shear traction to be 28 MPa along the northern SAF and 37 MPa along the Big Bend segment of the SAF. Our results were not sensitive to the rheology of the fault zone, because the traction force is directly balanced by other forces. Had a velocity boundary condition been applied, the results would be affected by the rheology of the fault zone.

Previous research has shown that the San Andreas fault is weak (Bird and Kong, 1994) and the shallow part is locked during interseismic period. So we kept this feature in the second generation of the rheological model, in which we added lateral variations. In particular, we raised viscosity in the Great Valley-Sierra Nevada, the Colorado Plateau, and the central Basin and Range province, and lowered it in the Eastern California Shear Zone, the Wasatch fault zone, and the Colorado River Extensional Corridor, according to the observed strain rates (Bennett et al., 2003). We explored more than 30 different combinations of rheology parameters. Decreasing viscosity by 30% or more in the weak zones lead to too much gravitational spreading, thus worsen the fit to the GPS velocities. Table 1 lists four cases that show how varying viscosity in the central Basin and Range Province, the Great Valley, and the Colorado Plateau would affect the predicted velocity and stress field. Compared with the model of homogeneous rheology, the improvement is significant. Misfit to GPS velocity is reduced from 3.7 mm/yr to less than 2.9 mm/yr. Comparison of stress direction between model M2b and M2d in Table 1 suggests that a large rheology contrast between the Colorado Plateau (the strongest block in the model) and the Eastern California Shear Zone (the weakest part) is not needed. From models M2c and M2d, we found that weakening of the southern Basin and Range province by 20% did not have a notable impact.

In the third generation, we adjusted the viscosity of the Great Valley and northern California (the triangular area in Fig. 5A). The models M3b and M3d produced better velocity fits. Compared with the second-generation models, the third-generation models included rheological adjustment to more regions, but the improvement is insignificant.

**Table 1** Rheology structure and fitting quality

| Domain | 1 | 2 | 3 | 4 | 5 | 6 | 7 | 8 | 9 | 10 | GPS misfit | WSM misfit |
|---|---|---|---|---|---|---|---|---|---|---|---|---|
| 1 HomoS | 1 | 1 | 1 | 1 | 1 | 1 | 1 | 1 | 1 | 1 | 3.7 | 21 |
| HomoW | 0.4 | 0.4 | 0.4 | 0.4 | 0.4 | 0.4 | 0.4 | 0.4 | 0.4 | 0.4 | 7.5 | 23 |
| StrongF | 1 | 1 | 1 | 1 | 1 | 1 | 1 | 1 | 1 | 0.4 | 3.8 | 21 |
| WeakF | 1 | 1 | 1 | 1 | 1 | 1 | 1 | 1 | 1 | 0.02 | 3.9 | 21 |
| 2 M2a | 0.8 | 4 | 0.8 | 0.8 | 1 | 4 | 4 | 1 | 1 | 1/0.02 | 3.02 | 22.2 |
| M2b | 0.8 | 20 | 0.8 | 0.8 | 1 | 20 | 20 | 1 | 1 | 1/0.02 | 2.87 | 26.8 |
| M2c | 0.8 | 10 | 0.8 | 0.8 | 1 | 10 | 10 | 1 | 1 | 1/0.02 | 2.89 | 24.1 |
| M2d | 0.8 | 10 | 0.8 | 0.8 | 0.8 | 10 | 10 | 1 | 1 | 1/0.02 | 2.87 | 24.3 |
| 3 M3a | 0.8 | 10 | 0.8 | 0.8 | 0.8 | 4 | 10 | 1 | 1 | 1/0.02 | 2.74 | 23.6 |
| M3b | 0.8 | 10 | 0.8 | 0.8 | 0.8 | 2 | 10 | 1 | 1 | 1/0.02 | 2.64 | 23.0 |
| M3c | 0.8 | 10 | 0.8 | 0.8 | 0.8 | 2 | 10 | 2 | 1 | 1/0.02 | 2.72 | 22.9 |
| M3d | 0.8 | 10 | 0.8 | 0.8 | 0.8 | 2 | 10 | 1.4 | 1 | 1/0.02 | 2.68 | 22.9 |
| 4 M4a | 0.8 | 10 | 0.8 | 0.8 | 0.8 | 2 | 10 | 1.4 | 1 | 1/0.02 | 2.72 | 23.0 |
| M4b | 0.8 | 10 | 0.8 | 0.8 | 0.8 | 2 | 10 | 1.4 | 1 | 1/0.02 | 2.64 | 23.0 |
| M4c | 0.8 | 10 | 0.8 | 0.8 | 0.8 | 2 | 10 | 1.4 | 1 | 1/0.02 | 2.67 | 22.9 |
| M4d | 0.8 | 10 | 0.8 | 0.8 | 0.8 | 2 | 10 | 1.4 | 1 | 1/0.02 | 2.64 | 22.9 |
| 5 M5a | 0.8 | 10 | 0.8 | 0.8 | 0.8 | 2 | 10 | 1.4 | 1 | 1/0.2 | 2.68 | 23.3 |
| M5b | 0.8 | 10 | 0.8 | 0.8 | 0.8 | 2 | 10 | 1.4 | 1 | 1/0.002 | 2.80 | 23.8 |
| M5c | 0.8 | 10 | 0.8 | 0.8 | 0.8 | 2 | 10 | 1.4 | 1 | 1/0.06 | 2.66 | 23.1 |
| M5d | 0.8 | 10 | 0.8 | 0.8 | 0.8 | 2 | 10 | 1.4 | 1 | 1/0.006 | 2.66 | 23.1 |
| 6 M6a | 1.2 | 15 | 1.2 | 1.2 | 1.2 | 3 | 15 | 2.1 | 1.5 | 2. | 2.64 | 23.9 |
| (sandwich) | 0.4 | 5 | 0.4 | 0.4 | 0.4 | 1 | 5 | 0.7 | 0.5 | 0.5 | | |
| | 0.8 | 10 | 0.8 | 0.8 | 0.8 | 2 | 10 | 1.4 | 1 | 0.02 | | |
| M6b | 1.6 | 20 | 1.6 | 1.6 | 1.6 | 4 | 20 | 2.8 | 2 | 1 | 3.05 | 24.7 |
| (declining) | 0.8 | 10 | 0.8 | 0.8 | 0.8 | 2 | 10 | 1.4 | 1 | 0.5 | | |
| | 0.6 | 8 | 0.6 | 0.6 | 0.6 | 1.6 | 8 | 1.1 | 0.8 | 0.02 | | |
| | | | | | | | | | | | mm/yr | Deg. |

Note 1: Geological domains codes: 1 – Eastern California shear zone; 2 – Central Basin and Range Province; 3 – Wasatch Zone; 4 – Colorado river Extensional corridor; 5 – South Basin and Range; 6 – Great Valley; 7 – Colorado Plateau; 8 – North California; 9 – South California; 10 – San Andreas Fault.

2: Reference viscosity: $5.0\times10^{22}$ Pa s. The two values for San Andreas fault stand for the crust and mantle viscosity, respectively.

3: In Model M6a and M6b, the parameters in the upper line are for upper crust, the middle for lower crust, and the bottom for mantle lithosphere from the Moho to 100 km depth.

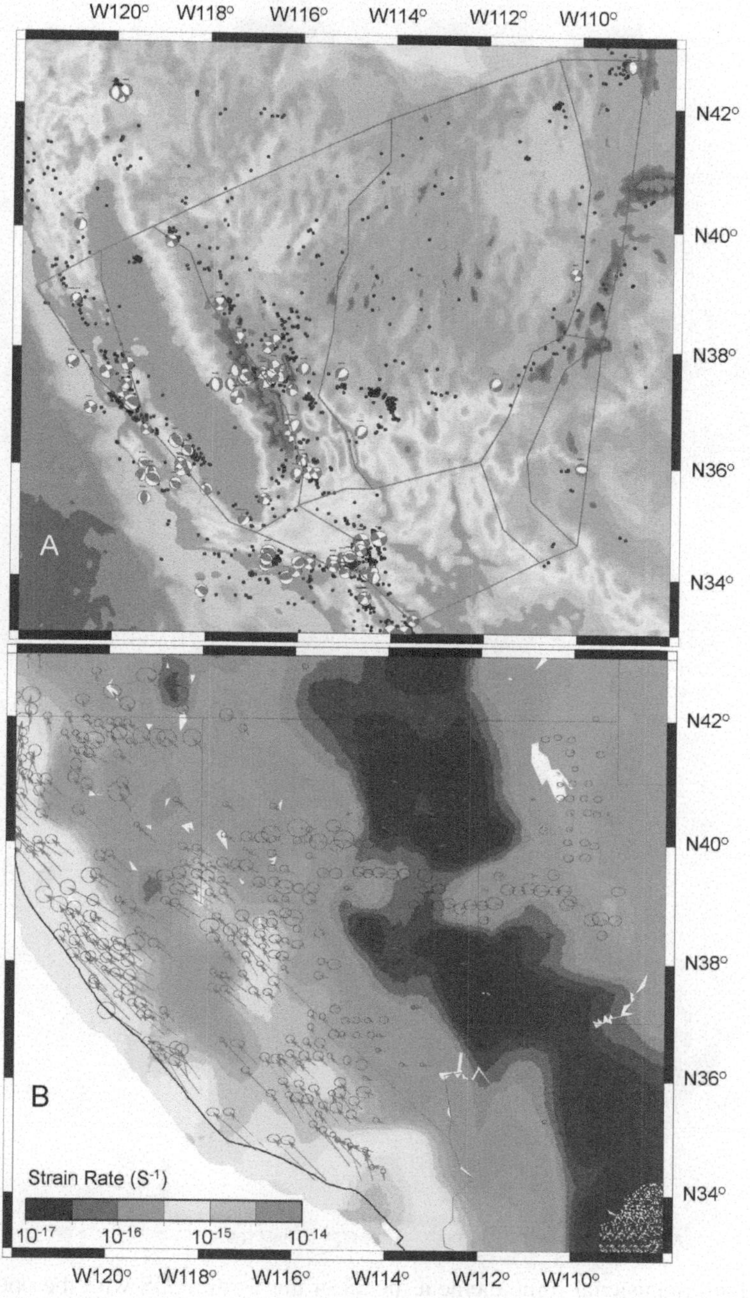


**Fig. 4** Diffusive deformation in the western United States as shown by the seismicity and GPS data. The earthquake focal mechanism solutions show the stress state in the crust. The GPS velocities are interpolated to a regular grid

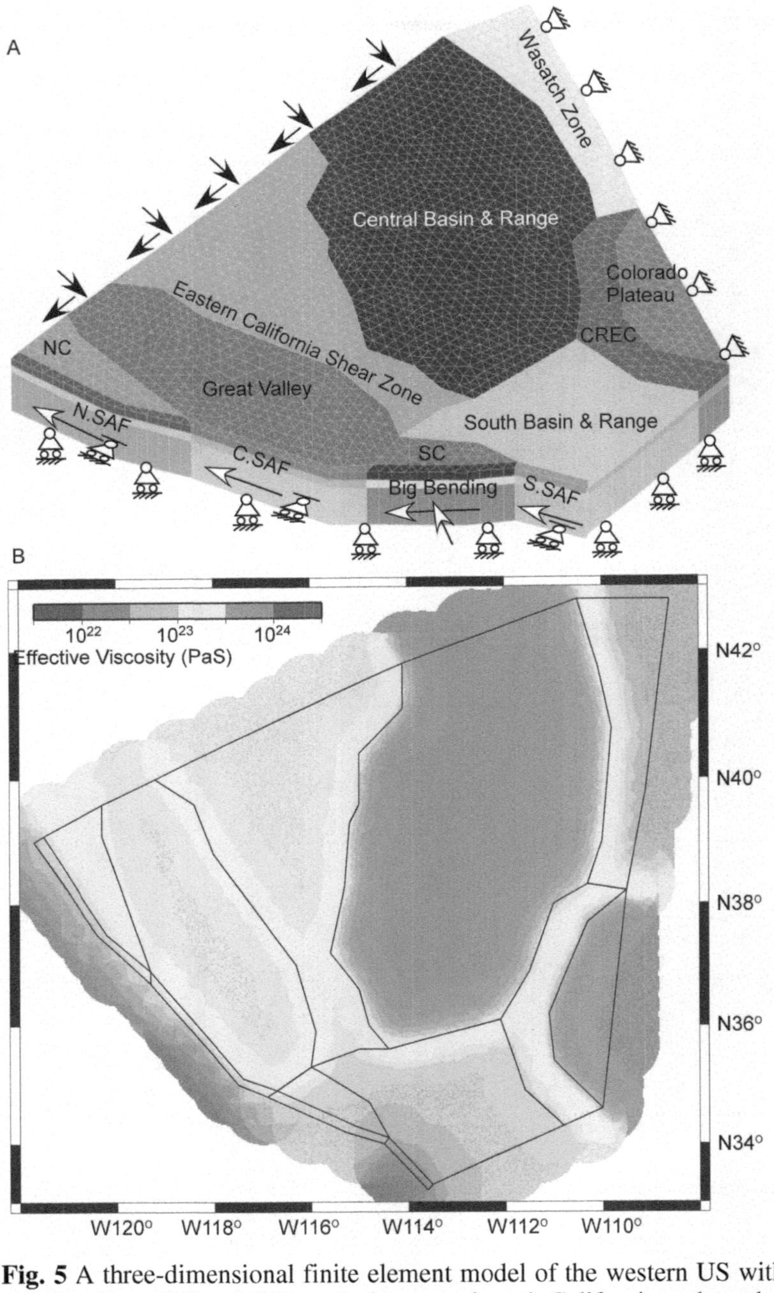


**Fig. 5** A three-dimensional finite element model of the western US with the optimal rheology. NC and SC stands for part of north California and south California in the model

In the fourth generation, we adjusted the force applied to the northern boundary of the model. At the very beginning of modeling, we tried to regress all forces along the plate boundary and other model boundaries. Doing so has produced some unrealistic results, such as the wrong sense of shear on some part of the SAF. The better approach is to impose some "common sense" conditions in the model. Not knowing the exact forces on the northern boundary, we first assigned an estimated force, and then calculated the optimal forces along the SAF through regression. We then adjust the force along northern boundary and repeat the regression. Through try-and-error, we found the best combinations of forces on the model boundaries in model M4d (Table 1), which produced minimum misfit to the GPS velocity and satisfactory fitting to the stress directions.

The merit of M4d is passed to the fifth generation, in which we varied the rheology of San Andreas Fault. All cases (Models M5a, M5b, M5c, and M5d) produce worse results than model M4d. Thus we conclude that the optimal viscosity for the SAF is close to $5.0\times10^{22}$ Pa s for the crust and $1.0\times10^{21}$ Pa s for the mantle. The lateral rheological variations of model M4d is shown in Fig. 5B.

The predicted velocity by model M4d is compared with the GPS velocity in Fig. 6. At most GPS sites, the predicted velocity falls well within the 95% confident level of the GPS measurements. The predicted stress states (Fig. 7A) is consistent with that indicated by earthquake mechanisms (Fig. 4A), and the predicted strain rate (Fig. 7A) is comparable to that derived from GPS data (Fig. 4B). The predicted horizontal directions of the maximum principal stress fit the observations in most places (Fig. 7B). The averaged misfit is 22.9°. Considering stress fluctuates due to strain accumulation and release in seismic cycles, we further reduced the shear traction along the SAF in a series of new models, from 25 MPa to 17 MPa on average. This reduced the misfit to 20.7°.

The 3D model allows us to explore the variation of lithosphere strength in depth. We tried two vertically heterogeneous models: one is a classic sandwich structure with a weak lower crust; another features a weak mantle as suggested from studies of postseismic deformation (Freed and Burgmann, 2004). The model parameters are list in Table 1. In both cases the fitting to the GPS velocity and to the stress direction are not significantly improved.

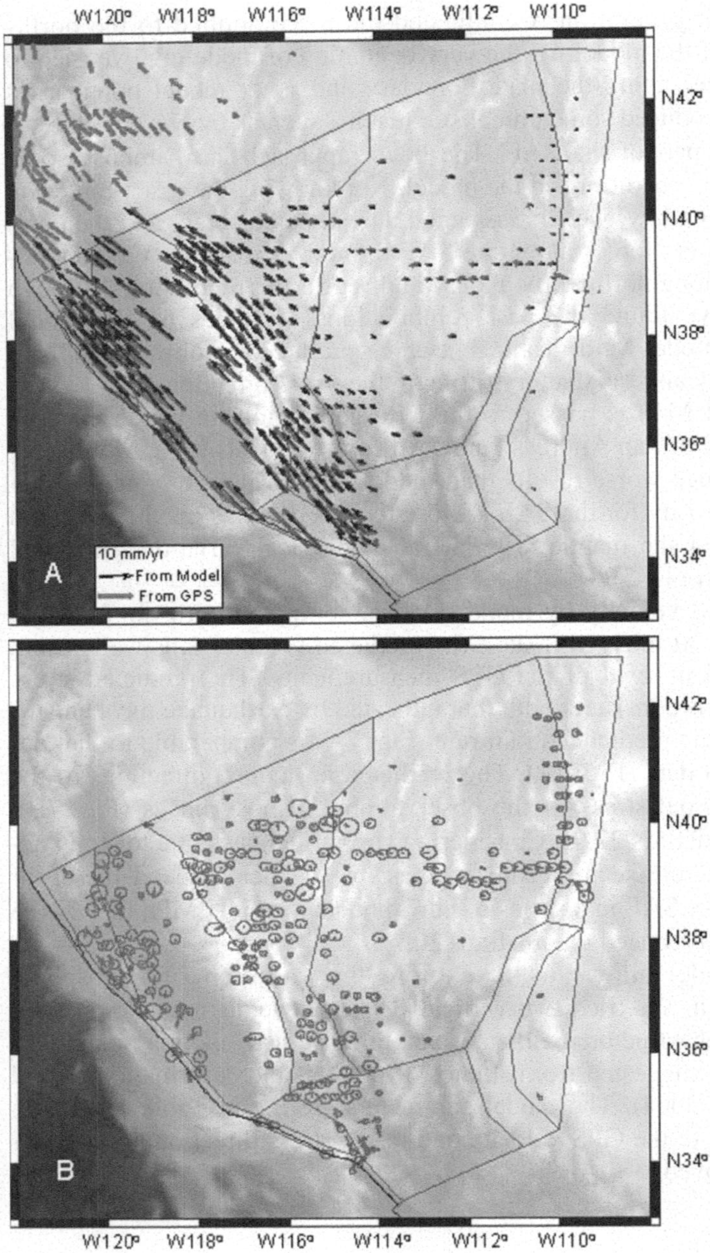


**Fig. 6** Comparison of the predicted surface velocity (*black arrows*) with the GPS velocity (*red arrows*). The *right panel* shows the residual errors. The results show that most of the predicted surface velocities fall within the 95% confidence ellipses of the GPS velocities

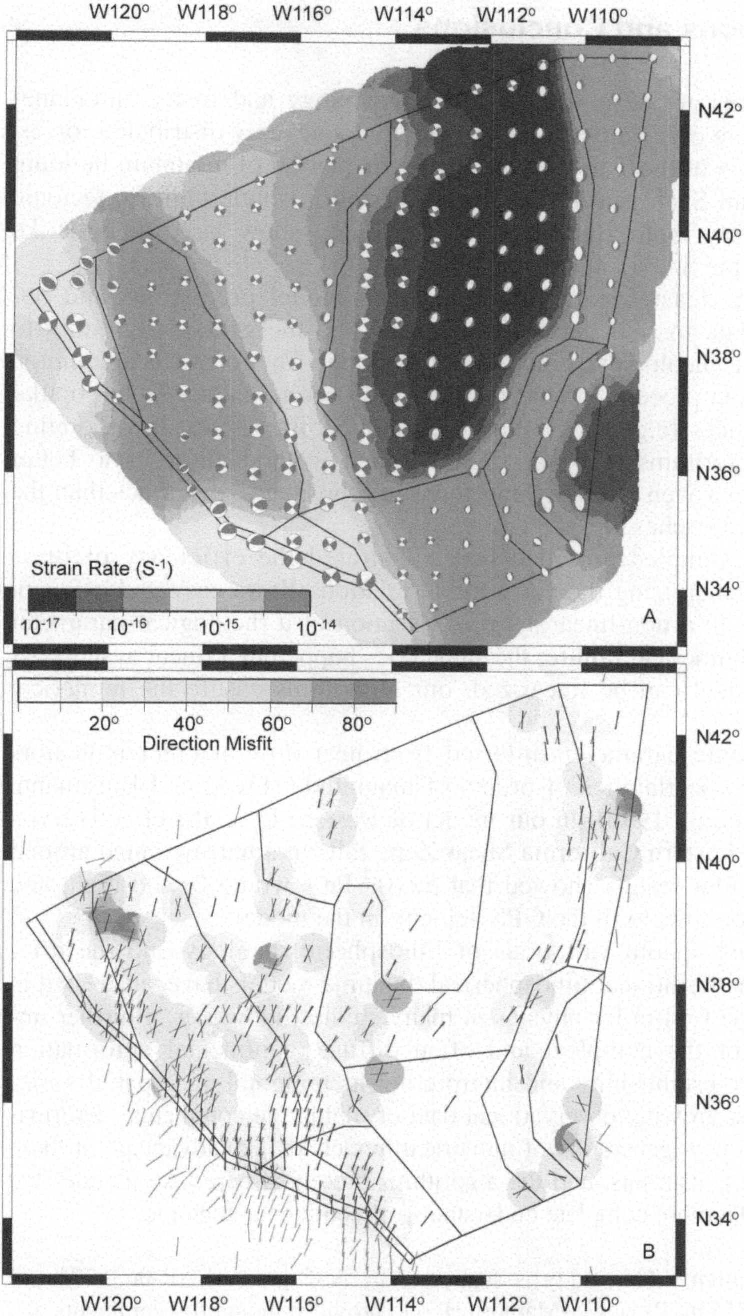


**Fig. 7** (**A**) The model predicted stress states and strain rate in the crust. (**B**) Comparison of the predicted stress direction (*black bars*) with that from the World Stress Map database (*red bars*)

## 5 Discussions and Conclusions

Our algorithms are designed for solving rheology and forces simultaneously. They are especially useful in estimating unevenly distributed forces. In the example of the Tarim basin, the segmentation of mountain building along the Tian Shan can not be simulated with a single uniform tectonic force. Similarly, both slip rates and seismicity show large along-strike variations in the SAF (Li and Liu, 2006).

We use the least square fitting between model predictions and observational data to find the optimal forces, and the genetic algorithm to improving the rheological structure of the model. The former is a common practice in data processing, and the later has become increasingly popular in earth sciences (e.g. King, 1995; Sen and Stoffa, 1992). Incorporating these two algorithms in linear finite element method allow us to better solve the interwoven rheology and force in lithospheric dynamics than the traditional approaches.

The two examples provided here illustrated the efficiency of these algorithms. Linearizing the constitutive relation allows easy and efficient computation. In a non-linear system, variation of a rheological parameter may lead to numerical failure; the same does happen in a linear system. As long as a model can be linearized, our algorithms ensure the numerical convergence.

The rheologic parameters inferred from heat flow or other indicators usually have uncertainties of orders of magnitude (Freed and Burgmann, 2004; Jones et al., 1996). In our model of western U.S., the effective viscosity of the Eastern California Shear Zone falls in a narrow range around $4\times10^{22}$ Pa s. Our results showed that $5\times10^{22}$ Pa s would be too high, and $3\times10^{22}$ Pa s too low, to fit the GPS velocity in the model.

By ignoring spatial variations of lithospheric rheology and the driving forces, the traditional lithospherical dynamic models have succeeded in illustrating the first-order physics in many studies. However, a further understanding of the complex, and often diffuse, continental deformation would require establishing and interpreting its temporal and spatial variations. The fast growing observational data of multiscale continental deformation call for a new generation of numerical models to take advantage of these observational constraints, and the algorithms presented here take us one step closer toward a more complete understating of continental tectonics.

**Acknowledgements** This study is supported by NSF grants EAR-0635574 and NSF EAR 0225546. Reviewer Mamoru Hyodo provided thoughtful comments that lead to a much improved paper.

## References

Atwater, T., 1970, Implications of plate tectonics for the Cenozoic tectonic evolution of western North America: Geol. Soc. Am. Bull., v. 81, pp. 3513–3536.

Bennett, R. A., B. P. Wernicke, N. A. Niemi, A. M. Friedrich, and J. L. Davis, 2003, Contemporary strain rates in the northern Basin and Range province from GPS data: Tectonics, v. 22.

Bird, P., 1998, Testing hypotheses on plate-driving mechanisms with global lithosphere models including topography, thermal structure, and faults: J. Geophys. Res., v. 103, pp. 10115–10129.

Bird, P., and X. Kong, 1994, Computer simulations of California tectonics confirm very low strength of major faults: G. S. Am. Bull., v. 106, pp. 159–174.

Brace, W. F., and D. L. Kohlstedt, 1980, Limits on lithospheric stress imposed by laboratory experiments: J. Geophys. Res., v. 85, pp. 6248–6252.

Choi, E., and M. Gurnis, 2003, Deformation in transcurrent and extensional environments with widely spaced weak zones: Geophys. Res. Lett., v. 30.

Flesch, L. M., A. J. Haines, and W. E. Holt, 2001, Dynamics of the India-Eurasia collision zone: J. Geophys. Res., v. 106, pp. 16435–16460.

Flesch, L. M., W. E. Holt, A. J. Haines, and B. Shen-Tu, 2000, Dynamics of the Pacific-North American plate boundary in the Western United States: Science, v. 287, pp. 834–836.

Freed, A. M., and R. Burgmann, 2004, Evidence of power-law flow in the Mojave desert mantle: Nature, v. 430, pp. 548–551.

Holland, J. H., 1973, Genetic algorithms and the optimal allocation of trials: J. Comput., v. 2, pp. 88–105.

Humphreys, E. D., and D. D. Coblentz, 2007, North American dynamics and Western US tectonics: Rev. Geophys., v. 45.

Jia, C., G. Wei, L. Wang, D. Jia, Z. Guo, Z. Zhang, and D. He, 1996, Structural acteristics of the Tarim Basin: Report of the oil-gas resources of the Tarim Basin (in Chinese): Beijing, Petroleum Industry Press, 207p.

Jones, C. H., L. J. Sonder, and J. R. Unruh, 1998, Lithospheric gravitational potential energy and past orogenesis: Implications for conditions of initial basin and range and Laramide deformation: Geology, v. 26, pp. 639–642.

Jones, C. H., J. R. Unruh, and L. J. Sonder, 1996, The role of gravitational potential energy in active deformation in the southwestern United States: Nature, v. 381, pp. 37–41.

King, S. D., 1995, Radial models of mantle viscosity – results from a genetic algorithm: Geophys. J. Int., v. 122, pp. 725–734.

Kirby, S. H., and A. K. Kronenberg, 1987, Rheology of the lithosphere: Selected topics: Rev. Geophys., v. 25, pp. 1219–1244.

Li, D., D. Liang, C. Jia, G. Wang, Q. Wu, and D. He, 1996, Hydrocarbon accumulations in the Tarim Basin, China: AAPG Bull., v. 80, pp. 1587–1603.

Li, Q., and M. Liu, 2006, Geometrical impact of the San Andreas Fault on stress and seismicity in California: Geophys. Res. Lett., v. 33, L08302, p. doi:10.1029/2005GL025661.

Liu, M., Y. Yang, Q. Li, and H. Zhang, 2007, Parallel computing of multi-scale continental deformation in the Western United States: Preliminary results: Phys. Earth Planet. Inter., v. 163, pp. 35–51.

Liu, Z., and P. Bird, 2002, North America plate is driven westward by lower mantle flow: Geophys. Res. Lett., v. 29.

McNutt, M., 1984, Lithospheric flexure and thermal anomalies: J. Geophys. Res., v. 89, pp. 11180–11194.

Sen, M. K., and P. L. Stoffa, 1992, Rapid sampling of model space using genetic algorithms – examples from seismic wave-form inversion: Geophys. J. Int., v. 108, pp. 281–292.

Sonder, L. J., P. C. England, and G. A. Houseman, 1986, Continuum calculations of continental deformation in transcurrent environments: J. Geophys. Res., v. 91, pp. 4797–4818.

Sonder, L. J., and C. H. Jones, 1999, Western United States extension: How the west was widened: Annu. Rev. Earth. Planet. Sci., v. 27, pp. 417–462.

Townend, J., and M. D. Zoback, 2004, Regional tectonic stress near the San Andreas fault in central and southern California: Geophys. Res. Lett., v. 31.

Turcotte, D. L., and G. Schubert, 2002, Geodynamics: New York, Cambridge University Press.

Yang, Y., and M. Liu, 2002, Cenozoic deformation of the Tarim basin and implications for mountain building in the Tibetan plateau and the Tian Shan: Tectonics, v. 26, p. doi:10.1029/2001TC001300.

Zhao, Z., T. Yong, C. Jia, and Z. Zhang, 1997, Stratigraphy in the Tarim Basin: Beijing, Petroleum Industry Press House.

Zoback, M. D., R. E. Anderson, and G. A. Thompson, 1981, Cenozoic evolution of the state of stress and style of tectonism of the Basin and Ranges Province of the western United States: Phil. Trans. Roy. Soc. London, v. Serial A 300, pp. 407–434.

Zoback, M. L., and W. D. Mooney, 2003, Lithospheric buoyancy and continental intraplate stresses: Int. Geol. Rev., v. 45, pp. 95–118.

# V. Mantle Dynamics – A Case Study

Klaus-D. Gottschaldt,[1] Uwe Walzer,[2] Dave R. Stegman,[3]
John R. Baumgardner[4] and Hans B. Mühlhaus[5]

[1]University of Queensland, ESSCC, PO Box 6067, St Lucia, QLD 4067, Australia; now at: Deutsches Zentrum Für Luft- und Raumfahrt, Institut Für Physik der Atmosphäre, 82234 Oberpfaffenhofen, Germany
[2]Friedrich-Schiller-Universität Jena, Institut für Geowissenschaften, Burgweg 11 07749 Jena, Germany
[3]Earth Sciences, The University of Melbourne, Victoria 3010, Australia
[4]University of California, Department of Earth & Planetary Science, 307 McCone Hall, Berkeley, CA 94720-4767, USA
[5]University of Queensland, ESSCC, PO Box 6067, St Lucia QLD 4067, Australia

E-mail: klausgottschaldt@web.de

**Abstract** Solid state convection in the rocky mantles is a key to understanding the thermochemical evolution and tectonics of terrestrial planets and moons. It is driven by internal heat and can be described by a system of coupled partial differential equations. There are no analytic solutions for realistic configurations and numerical models are an indispensable tool for researching mantle convection. After a brief general introduction, we introduce the basic equations that govern mantle convection and discuss some common approximations. The following case study is a contribution towards a self-consistent thermochemical evolution model of the Earth. A crude approximation for crustal differentiation is coupled to numerical models of global mantle convection, focussing on geometrical effects and the influence of rheology on stirring. We review Earth-specific geochemical and geophysical constraints, proposals for their reconciliation, and discuss the implications of our models for scenarios of the Earth's evolution. Specific aspects of this study include the use of passive Lagrangian tracers, highly variable viscosity in 3-d spherical geometry, phase boundaries in the mantle and a parameterised model of the core as boundary condition at the bottom of the mantle.

H. Xing, *Advances in Geocomputing*, Lecture Notes in Earth Sciences 119,
DOI 10.1007/978-3-540-85879-9_5

# 1 Introduction

Terrestrial planets like the Earth, Mars and Venus consist of a metallic core and a silicic mantle. The rocky mantles can be partially molten in small regions, but the vast majority of the mantle material is solid. In response to short-term force (e.g. seismic waves) the mantle behaves like an elastic solid body. However, in geological time scales the rock behaves like a viscous fluid. Thermal and thermo-chemical convection is possible on a range of scales. Mantle convection provides a framework to reconcile observations of planetary magnetic and gravity fields, heat flux, distribution of volcanoes and tectonic structures, geo- & cosmochemistry and mineral physics. Numerical models are an indispensable tool for researching mantle convection.

## 1.1 Energy Budget of the Mantle

Thermal convection in terrestrial bodies is driven by internal heat sources. This heat stems from accretion, the decay of radioactive elements, tidal dissipation, and the gravitational energy which is released during the differentiation into core, mantle and crust. The relative importance of each of these heat sources is a function of time and can vary from one terrestrial body to another. Please note: although solar irradiation can be orders of magnitude bigger than internal heat, it has no effects on internal dynamics. Solar irradiation penetrates less than a few 100 m, driving only processes close to the surface. In contrast, phenomena like volcanism, planetary magnetism, tectonics and seismicity are controlled by internal heat.

## 1.2 Physics of Mantle Convection in a Nutshell

Mantle dynamics can be described by the equations for the conservation of energy, momentum and mass, equation(s) of state, initial and boundary conditions. Solid state creep in the mantle is very slow, in the order of centimetres per year. Therefore inertia is negligible, but frictional terms and constitutive relations are crucial for the fluid dynamics of the mantle. Mantle rheology strongly depends on temperature and pressure and is not very well constrained.

## 1.3 Surface Tectonics

Tectonics is the manifestation of internal dynamics at the outer boundary layer of the mantle. The simplest tectonic regime is a freely deformable surface, a so-called mobile lid. Fluids with low viscosity contrasts – like oceans – display this behaviour, but it is not known from any solid mantle. In contrast, a steep viscosity increase towards the surface produces a stagnant lid. Convection just takes place in the mantle below the lid. Mars and the Moon are examples for this tectonic style. The Earth is the only known planet currently operating plate tectonics. Here the outer lid is broken into several plates that show little internal deformation, but change their shapes and relative positions. Plate material is generated at divergent margins and recycled into the mantle at convergent zones. The tectonic style of a planet may change and the controlling parameters are a current research topic.

## 1.4 Volcanism

Apart from the tectonic style, the distribution of volcanoes is another characteristic feature of each planet. On Earth, most volcanic activity occurs along the rims of tectonic plates, sampling the shallow mantle. Additionally there are at least 32 rising plumes that sample the deep mantle (Montelli et al. 2004; Davies 2005). Plumes move much slower laterally than tectonic plates and therefore appear to be stationary relative to each other. Strongly reduced viscosity in those 'hot spots' permits rising velocities in the order of 10 cm/yr, in narrow conduits of a few 100 km in diameter. On Earth, plume locations seem to be related to large upwellings in the lower mantle, under Africa and the Pacific (McNamara and Zhong 2004). Plumes and plate tectonics are modes of heat transport within and out of the mantle, but their relative importance is not entirely clear (Nolet et al. 2006).

There are only two narrow volcanic centres on Mars and the distribution of volcanoes on Venus seems to be random. The differences are hard to explain, particularly since narrow, hot plumes cannot be resolved satisfactorily in global mantle convection models yet.

## 1.5 Core and Magnetism

The evolution of the metallic core is an important boundary condition and constraint for mantle models. If the heat flux across the core mantle boundary is high enough, convection in a molten core can generate a magnetic field. This field may be preserved in rock just cooling below the

Curie point, thus providing an observational record of the magnetic field history. Core convection acts on much smaller time- and length-scales, but parameterised core models can be coupled to mantle evolution models.

### 1.6 Composition

Planetary mantles are a mixture of minerals and the main chemical components may be inhomogeneous on different scales. However, the resulting variation of parameters relevant to mantle convection (e.g. density, thermal conductivity) is small. Here we consider only one major component: the bulk composition of the mantle.

A more comprehensive presentation of the various aspects of mantle convection can be found in text books like Schubert et al. (2001).

## 2 Physics of Mantle Convection: Basic Equations

We distinguish among several types of computational models. A scaling law describes the efficiency of heat transport in parameterised models. Only a global energy budget is considered in this case, which is computationally inexpensive. These models are useful for investigating a wide parameter space. If one is interested in further details, e.g. convection structure and temperature fields, the full system of equations must be solved. Usually this is done in 2-d or 3-d, Cartesian or spherical geometry. Of those, 2-d Cartesian models generally consume the least computational resources, but 3-d spherical models are most realistic. The choice depends on the specific problem. In the following we start from general formulations of the equations to show the approximations used for mantle dynamics.

### 2.1 Conservation of Mass

Mass conservation is given by

$$\frac{\partial \rho}{\partial t} + \nabla(\rho \vec{v}) = 0 \tag{1}$$

with density $\rho$, time $t$, velocity $\vec{v}$ and the operator $\nabla = \frac{\partial}{\partial x_i}\vec{e}_i$.

Mantle convection is slow and density does not change rapidly. Therefore the first term can be neglected, giving the so-called anelastic liquid approximation,

$$\nabla(\rho\vec{v}) = 0 \tag{2}$$

Mantle rock is compressible. This is relevant for big terrestrial planets like the Earth, but not for the limited pressure range in smaller bodies (e.g. Moon). Equation (2) simplifies to

$$\nabla\vec{v} = 0 \tag{3}$$

for incompressible calculations.

## 2.2 Conservation of Momentum

The change of momentum of a small mantle element is balanced by forces acting on the element surface (pressure gradients, viscous forces) and body forces (e.g. gravity – incl. centrifugal force, Coriolis force)

$$\rho\frac{D\vec{v}}{Dt} = -\nabla p + \nabla\tau + \rho\vec{g} - 2\rho\vec{\omega}\times\vec{v} \tag{4}$$

where $p$ is pressure, $\tau$ is the deviatoric stress tensor, $\vec{g}$ is gravity, $\vec{\omega}$ angular velocity and the material derivative $\frac{D}{Dt} = \frac{\partial}{\partial t} + \vec{v}\cdot\nabla$ (Euler system). Small mantle creep velocities allow the neglection of inertia, leading to

$$0 = -\nabla p + \nabla\tau + \rho\vec{g} \tag{5}$$

## 2.3 Conservation of Energy

Introducing specific heat at constant volume $c_V$, thermal expansivity $\alpha = -\frac{1}{\rho}\left(\frac{\partial\rho}{\partial T}\right)_p$, bulk modulus $K_T$, thermal conductivity $k$ and heat generation rate per unit volume $H$, the energy equation is

$$\rho c_V \frac{DT}{Dt} - \frac{\alpha K_T}{\rho} T \frac{D\rho}{Dt} = \nabla(k\nabla T) + \tau\nabla\vec{v} + \rho H \tag{6}$$

The source terms on the right hand side are thermal conduction, frictional heating and volumetric heating, respectively. The second term on the left hand side vanishes for incompressible calculations.

## 2.4 Equation of State

Furthermore we need an equation of state to reduce the number of independent thermodynamic variables (here: $p$, $\rho$, $T$). Those variables vary much more radially than laterally in planetary mantles. Therefore it is convenient to consider lateral variations as a perturbation to radial reference profiles ($p_r$, $\rho_r$, $T_r$). The basic – but widely used – equation of state for this approach is

$$\rho = \rho_r \cdot \left[ 1 - \alpha (T - T_r) + \frac{p - p_r}{K_T} \right] \tag{7}$$

Only a reference point ($\rho_0$, $T_0$) is needed for the incompressible case and Eq. (7) simplifies to

$$\rho = \rho_0 \cdot [1 - \alpha (T - T_0)] \tag{8}$$

## 2.5 Constitutive Relations

Constitutive equations relate stress $\tau$ and deformation $\varepsilon$ or deformation rate $\dot{\varepsilon}$. This so-called rheology strongly depends on the actual mantle conditions, may be complicated and uncertain. We show just some simple or often used examples.

Elastic behaviour

$$\tau_{ij} = E \cdot \varepsilon_{ij} \tag{9}$$

may be important for short and medium term processes near the surface, but is secondary for long term processes in the deep mantle.

Models of mantle convection often use linear viscous rheology

$$\rho \cdot H_{cool} = \rho \cdot c_v \cdot \frac{d(T - T_m)}{dt} \tag{10}$$

Here $\dot{\varepsilon}_{ij} = \frac{1}{2}\left( \frac{\partial v_i}{\partial x_j} + \frac{\partial v_j}{\partial x_i} \right)$ is the strain rate tensor and $\eta$ is the shear viscosity. We have neglected volume viscosity, but $\eta$ may vary, e.g. according to an Arrhenius-type equation,

$$\eta = \eta_0 \exp\left[\xi \frac{T_m}{T}\right] \tag{11}$$

Based on high-pressure experiments it has been suggested (Boehler 2000) that this viscous rheology (with mantle specific constitutive parameters combined to $\xi \approx 17$, melting temperature $T_m$ and a reference viscosity $\eta_0$) is a good approximation for diffusion creep, the most important deformation mechanism in the Earth's lower mantle. Dislocation climb dominates in the Earth's upper mantle, resulting in a power law rheology of the form

$$\tau \sim \dot{\varepsilon}^n \tag{12}$$

According to Eq. (11), the low temperatures near planetary surfaces would result in an unrealistically rock strength. In reality rock yields above a certain threshold stress $\tau_y$. Combining this plastic stress limiter with the viscous rheology (10) gives the effective visco-plastic viscosity

$$\eta_{eff} = \min\left[\eta(p,T), \frac{\tau_y}{2\sqrt{\dot{\varepsilon}_{II}}}\right] \tag{13}$$

Such a rheology allows strain localisation, which is a prerequisite for plate tectonics.

The six scalar Eqs. (2) or (3), (5), (6) and (7) or (8) are used to determine $T$, $p$, $\rho$ and the three components of $\bar{v}$ for mantle parameters. Heating mode, boundary conditions, initial conditions, rheology, radial reference profiles and other specific parameters are discussed with the case study in the following section.

## 3 Case Study: Stirring in Global Models of the Earth's Mantle

This case study is a contribution towards a self-consistent thermochemical evolution model of the Earth. An approximation for crustal differentiation is coupled to numerical models of global mantle convection, focussing on geometrical effects and the influence of rheology on stirring. This section is based on Gottschaldt et al. (2006).

First we review some Earth-specific geochemical and geophysical constraints, as well as proposals for their reconciliation. Thereafter we describe extensions to the base methodology of the previous section and the numerical approach. The results of selected models will be discussed. We highlight results that could be relevant for the Earth in the final section.

## 3.1 Background

### *3.1.1 Mantle Composition and Crustal Segregation*

The following five Oxides make up 98.5% of the *bulk composition* of the Earth's mantle and crust (O'Neill et al. 1998): MgO (36.3%), $Al_2O_3$ (4.7%), $Si_2O_3$ (45.6%), CaO (3.7%), FeO (8.2%). Depending on pressure, temperature and differentiation history, these oxides form mixtures of different minerals.

Partial melt is extracted from the mantle and upon solidification forms the crust. Hence the crust is an end product of mantle differentiation. On Earth there are two chemically distinct sorts of crust. Thin (0–7 km), dense and mafic oceanic crust (OC) is continuously generated at mid-ocean ridges and recycled almost entirely back into the mantle at subduction zones. By contrast, andesitic continental crust (CC) is less dense and thicker (~40 km). It is buoyant and recycled into the mantle only to a small degree, as sediments or by delamination. While the lifetime of OC is about 100 Ma, that of CC is at least 2 Ga (Hofmann 1997). Suggestions for the segregation history of CC range from rapid early net growth to episodic growth of juvenile CC to continuous net growth (Arndt 2004). Today CC is produced mainly by andesitic volcanism related to subduction and the release of water from the slab. Continents themselves also grow through intraplate volcanism and the accretion of sediments and basaltic terranes. Such terranes could be the product of extensive melting caused by plume heads reaching the surface (Hofmann 1997). Tectonic settings for the formation of Archean cratons may have been different and include rifts (Trendall 2002), (van Thienen 2003), but are still represent on the extraction of partial melt from the shallow (<200 km: (Presnall et al. 2002)) mantle.

### *3.1.2 Phase Transitions in the Mantle*

Olivine transforms into β-spinel at a mean depth of 410 km, γ-spinel to perovskite plus magnesiowüstite at 660 km depth. Olivine and spinel make up 58% of the upper mantle in the respective depth ranges. The other two major phases in the upper mantle undergo phase changes in the same depth range: pyroxenes and garnet change relative proportions throughout the upper mantle and gradually transform into perovskites between 600 and

700 km depth. The mantle below 700 km consists of 80% perovskite and 20% magnesiowüstite, with chemical heterogeneities of a few percent (Trampert et al. 2004). There is only one major phase transition in the lower mantle: perovskite transforms into postperovskite (ppv) near the core-mantle boundary (CMB).

Of all phase changes in the mantle (Table 1), the γ-spinel transition at 660 km depth has the biggest impact on the physical properties and dynamics of the mantle. It defines the boundary between upper (UM) and lower mantle (LM). The ppv transition has a minor effect on the dynamics and mantle temperature, mildly destabilizing the lower boundary layer (Tackley et al. 2007). However, the topology and dynamics of the seismically defined D" layer at the bottom of the mantle are controlled by the ppv transition (Monnereau and Yuen 2007). When compositional effects on the stability of ppv are taken into account, a large potential variety of complex behaviour could occur, generating structures such as discontinuities, gaps or holes and multiple crossings. The different contributions to seismic heterogeneity have different spectral slopes: temperature is long-wavelength, composition is 'white' and ppv is intermediate (Tackley et al. 2007).

**Table 1** Mean parameters of the main phase transitions in the Earth's mantle, based on Schubert et al. (2001) for the first two transitions and on Hirose (2006) for the postperovskite transition. The affected mantle proportion, $c$, is based on a pyrolite composition. $\Gamma = \mathrm{D}p/\mathrm{D}T$ is the Clapeyron slope, ol – olivine, pv – perovskite, ppv – postperovskite

| | Mean depth [km] | Width of the two-phase zone [km] | $\Gamma$ [MPa/K] | $\Delta\rho/\rho$ | c |
|---|---|---|---|---|---|
| ol → $\beta$-sp | 410 | 10–20 | 1.6 | 0.070 | 0.58 |
| $\gamma$-sp → pv | 660 | 4–7 | –2.5 | 0.100 | 0.58 |
| pv → ppv | 2600 | 30–35 | 8.0 | 0.011 | 0.72 |

The above transition depths are for mean mantle temperatures. In fact they occur over some depth range, depending on temperature variations due to convection. However, e.g. topography of the phase boundary around 660 km depth is less than 50 km. Hence it cannot be resolved directly in large scale models. We use a parameterization for the three dynamic effects of all considered phase boundaries.

The first effect is additional buoyancy in the momentum Eq. (5), due to the phase boundary distortion by advection of thermal anomalies. This buoyancy is,

$$F_B = \Gamma \cdot \frac{\Delta\rho}{\rho} \cdot A \cdot \Delta T \tag{14}$$

over a phase boundary area $A$, which has a temperature difference $\Delta T$ with respect to the mean temperature at that depth. $\Delta\rho$ is the density difference of the advected anomaly with respect to the surrounding density $\rho$. We neglect the possible temperature dependence of $\Gamma$ (Hirose 2002).

The second effect is the release or absorption of latent heat, $E_{latent}$, which distorts the phase boundary by changing the temperature. We calculate the latent heat from the Clausius-Clapeyron formula,

$$E_{latent} = \Gamma \cdot \Delta V \cdot T \tag{15}$$

with $\Delta V \approx V \cdot \Delta\rho / \rho$ and the processed volume $V = v_r \cdot \Delta t \cdot A \cdot c$, during time step $\Delta t$, over area $A$, having a mean vertical velocity $v_r$, affected mantle proportion, $c$. The rate of latent heat production,

$$H_{latent} = E_{latent} / \rho V \Delta t \tag{16}$$

is added to the energy Eq. (6), but only at the grid layers next to the phase boundaries, with appropriate weighting. The work due to volume change at the phase transition, $p\Delta V$ is reflected in the density profile and therefore automatically accounted for in the energy Eq. (6).

The third effect is the expansion or contraction due to the release or absorption of latent heat. It is automatically accounted for via the temperature field, like any other thermal buoyancy.

### 3.1.3 Geochemistry – a Primer

Radiogenic *trace elements* in the mantle record differentiation events that change the parent/daughter ratio. The subsequent decay in isolated reservoirs allows dating of the differentiation event, if the decay time constant is in a matching order of magnitude.

The existence of geochemically distinct reservoirs in the Earth's mantle is inferred from the observation of worldwide rather homogeneous mid-ocean ridge basalts (MORB) on the one hand and heterogeneous ocean island basalts (OIB) on the other (Hofmann 1997). Of course, what we observe today is the result of the interplay between chemical differentiation and convective stirring that started with the formation of the Earth and is still going on.

Seismic tomography provides a snapshot of the modern mantle. Wave speed anomalies are interpreted as evidence for subducting slabs that extend from the surface to the lower or lowermost mantle (Grand et al. 1997), (Trampert et al. 2004) and for superplume upwellings from the core mantle boundary (CMB) region (Su et al. 1994; Ritsema et al. 1999; Ritsema and van Heijst 2000). Together with the prevailing surface plate velocities these seismic observations lead most investigators to conclude there has been considerable mass exchange between upper and lower layers of the mantle for at least the last 100 Ma.

The reconciliation of geochemical and geophysical evidence has long been an unresolved problem in geodynamics. How can different geochemical reservoirs be maintained in the presence of large-scale convection for a long time? Why is one part of the mantle more homogeneous on a global scale than other parts?

### *3.1.4 Geochemical Heterogeneities*

The size of observed geochemical heterogeneities ranges from cm-scale structures in high-temperature peridotites (Allègre and Turcotte 1986) to the DUPAL anomaly (Hart 1984). The latter seems to have a different origin in the Pacific than in the southern Atlantic and Indian Ocean (Hanan et al. 2004), so it is not global. Yet its existence for at least 115 Ma (Weiss et al. 1989) indicates that there is limited large-scale lateral stirring somewhere in the mantle.

The OIB isotopic compositions are far more diverse than MORB compositions (Allègre 2002), with the global variance of isotopic ratios for OIB being three times larger than the corresponding global variance of MORB (Allègre et al. 1987). The degree to which mantle heterogeneities are reflected in the resulting basalt depends on the degree of partial melting, magma mixing and extraction, the size of the volume that is sampled by partial melt, and the dimension of geochemical heterogeneity itself. Differences in composition and heterogeneity between OIB and MORB could be due to different sampling processes (Meibom and Anderson 2003), sampling of different geochemical reservoirs (Hofmann 1997) or a combination of these (Kellogg et al. 2002). The concept of a reservoir is here used to reflect a scale of systematic variation of mantle geochemistry that is too large to be erased by the sampling process. Sampling at mid-ocean ridges, for instance, acts on a scale of 30–200 km in depth and several 100 km in width (Presnall et al. 2002) and hence the scale of the implied reservoir is larger.

### *3.1.5 Mantle Degassing*

From the amount of $^{40}Ar$ in the atmosphere it has been estimated, that only about 50% of this isotope have been degassed from the mantle (Allègre et al. 1996). The low concentration in MORB suggests there is another reservoir containing the missing $^{40}Ar$. This is a strong argument against simple whole-mantle convection (Hofmann 1997). However, it is subject to challenge, because the amount of $^{40}K$ (exclusively producing $^{40}Ar$) in the Earth (Coltice and Ricard 2002), the efficiency of Ar degassing (Watson et al. 2007) and the role of Ar recycling (Rüpke et al. 2003) are poorly constrained. Seawater recycling might actually control the argon chemistry of the mantle (Holland and Ballentine 2006).

Nearly all the $^{3}He$ coming from the mantle is likely to be primordial, supporting the conclusion that the Earth has never been completely degassed (Gonnermann and Mukhopadhyay 2007). The ratio $^{3}He/^{4}He$ is very uniform in MORB, but varies in OIB. There could be a reservoir of absolutely high $^{3}He$ concentration, possibly 3.5 times higher than currently estimated from He flux (Ballentine et al. 2002), possibly poorly degassed or primordial mantle. High $^{3}He/^{4}He$ signatures might also originate in recycled material, if He degassing near the surface is less efficient than extraction of $^{4}He$ producing U and Th (Watson et al. 2007), or if recycled material is convectively isolated for long enough (Ferrachat and Ricard 2001).

### *3.1.6 Interpretation of Reservoirs*

The MORB source region is sampled by the network of divergent margins all over the world and therefore is thought to occupy the shallow mantle (Hofmann 1997). The fixity of OIB-producing hot spots relative to surface plate movements is a compelling indicator that these plumes originate in and sample deeper regions of the mantle that are more or less decoupled from the surface motion. This is independently supported by tomographic images of deep-rooted plumes under OIB hot spots (Montelli et al. 2004), but hard to reconcile with a proposed (Meibom and Anderson 2003) shallow origin for all geochemical heterogeneity.

The roughly complementary geochemical signatures of CC and MORB are interpreted to be the result of primary extraction of CC from the original, primitive mantle (PM) (Hofmann 1988) or from an early depleted reservoir (EDR) that had already lost incompatible elements to a sunken enriched reservoir (early enriched reservoir – EER) (Boyet and Carlson 2005).

Mixing of all OIB would not give MORB. Other reservoirs proposed to contribute to OIB in differing proportions are EM1, EM2 (enriched mantle) and HIMU (high $\mu = {}^{238}U/{}^{204}Pb$). HIMU could be subducted oceanic crust, EM1 delaminated lower CC (Hofmann 1997) or subducted oceanic plateaus (Albarède 2001) and EM2 subducted continental sediments. The MORB source (depleted MORB mantle, DMM) is variably polluted on a regional scale by other components (Hanan et al. 2004). However, the absence of PM samples does not preclude the existence of a primitive reservoir (Kellogg et al. 2002).

### 3.1.7 Age of Reservoirs

The mean age of CC is 2–2.5 Ga, and that of oceanic basalts 1–1.3 Ga (Hofmann 1997). However, that is not necessarily the age of original differentiation (Albarède 2001). The depleted signature of MORB may (partly) reflect differentiation in a terrestrial magma ocean, 4.53 Ga ago (Boyet and Carlson 2005). The oldest HIMU signature is about 2 Ga (Hofmann 1997). Not introducing HIMU into the mantle prior to 2–2.5 Ga before present, due to a change in the surface oxidization environment or to subduction zone processes, could explain that age (Xie and Tackley 2004). The coincidence with the assumed end of primary CC segregation is striking.

### 3.1.8 Size of Reservoirs

How much of the mantle must have been depleted in incompatible elements to form the present volume of CC? Estimates depend on the geochemical models used and range from 25 to 90% (Hofmann 1997), but more likely 40–50% (Allègre 2002) – if CC was extracted from a primitive reservoir. If there was an early differentiation event (into EDR and EER) or if the bulk composition of the earth is different to primitive chondritic meteorites, then roughly 96% of the mantle must be as incompatible element depleted as the MORB source (Boyet and Carlson 2005) (Carlson et al. 2007). There must be one or more reservoir(s) containing the missing elements, in particular, the heat-producing nuclides (${}^{235}U$, ${}^{238}U$, ${}^{40}K$, ${}^{232}Th$) and ${}^{40}Ar$ (Albarède and van der Hilst 2002).

### 3.1.9 Reconciliation of Geophysical and Geochemical Constraints

There are several proposals how the depleted, well-mixed DMM could be separated from the other reservoirs and evolve independently for billions of years.

*Phase boundaries*: The γ-spinel-to-perovskite-plus-magnesiowüstite phase transition at 660 km depth hinders convection. It is appealing from a geochemical point of view to assume that it forces the upper mantle to convect separately from the lower mantle, but results of seismic tomography (e.g. van der Hilst et al. (1997), Trampert et al. (2004)) show penetration of the boundary today. On the other hand, temporal layering has been proposed on geological (e.g. (Condie 1997), geophysical (Breuer and Spohn 1995) and geochemical (Hofmann 1997; Allègre 2002) grounds. The necessary change in the effect of the phase transition could be due to the temperature-dependence of the Clapeyron slope (Hirose 2002) or the decreasing Rayleigh number (Ra) of the Earth (e.g. Tackley (1996)). However, at the Ra range assumed for the Earth, dynamic models commonly show avalanches through the phase boundary rather than long-term layering (Tackley 1996).

*Small-scale heterogeneities plus D"*: The assumption that large-scale convection implies very efficient stirring leads to a cartoon with nearly the entire mantle being homogeneous DMM. Only D" remains as a possible repository of enriched reservoirs and hence as the OIB source region (e.g. Stegman et al. (2002)). OIB plumes originating at shallower depths (Montelli et al. 2004) are an argument against this model. Additionally, D" is a rather small volume in which to fit all the heterogeneities (Hofmann 1997; Albarède and van der Hilst 2002). On the other hand, by assuming that the 'homogeneous' mantle is riddled with small scale (~8 km: (Helffrich and Wood 2001)) up to regional scale (~100 km: (Meibom and Anderson 2003)) blobs of recycled material, the model becomes geochemically more plausible. Coltice and Ricard (2002) suggest a marble-cake mantle with recycled OC and peridotites containing primitive veins. In this case the mantle is assumed merely to be better mixed within the MORB source region than below it. Differing geochemical signatures are formed by mixing different amounts of these components, and because of segregation of dense OC, possibly in D". Heterogeneities smaller or comparable to the scale of sampling are supported by observations, but larger scale heterogeneities also exist (Trampert et al. 2004). None of above models, however, conclusively explains the coexistence of a shallow, more homogeneous reservoir and a deeper, less mixed one.

*Different intrinsic densities*: The nature of the D" layer is not entirely clear yet. It might be dominated by the effects of the perovskite to post-perovskite phase transition (Monnereau and Yuen 2007), or higher internal density (Nakagawa and Tackley 2004). If the compositional part of D" is due to subducted OC (eclogite), it could be the HIMU reservoir (Hofmann 1997). If it is foundered early crust (Tolstikhin and Hofmann 2005), D" could serve as a source for primordial gases and missing incompatible elements.

A similar proposal of early differentiation suggests that incompatible elements became enriched in a late crystallising layer of the terrestrial magma ocean – near the surface of the Earth. The residual of a magma ocean solidifies into a heavy layer that may sink, dragging the early enriched reservoir to the bottom of the mantle. There it could remain unsampled and undetected until today. Such a scenario might explain that primitive meteorites – believed to be the building blocks of the Earth – have a slightly skewed isotopic composition with respect to mantle samples (Boyet and Carlson 2005; Carlson et al. 2007). However, the solar nebula – from which the Earth and meteorites formed – was isotopically heterogeneous in the first place (Andreasen and Sharma 2006; Ranen and Jacobsen 2006). This could also explain the differences between chondritic meteorites and the Earth, supported by the fact that the Moon has an isotopic signature similar to the Earth's mantle. The Moon likely formed from a giant impact into the Earth's mantle, but simulations suggest that at least two-thirds of the material that makes up the Moon derived from the impactor, not from Earth (Canup 2004). Either the impactor had a similar history of silicate differentiation as the early Earth or both formed from material that had a different isotopic signature to chondritic meteorites.

Alternatively, a stable layer of dense melt may have formed at the base of the mantle early in the Earth's history (Labrosse et al. 2007), possibly having a thickness of about 1000 km. Such an initial, basal magma ocean would have undergone slow fractional crystallization, and it would be an ideal candidate for an unsampled geochemical reservoir. It could host a variety of incompatible species (most notably the missing budget of heat-producing elements) and might also explain the geochemical observations of Boyet and Carlson (2005).

A stable dense layer at 1500–2000 km depth was put forward by Kellogg et al. (1999) and could represent a leftover feature from the mantle's early evolution, e.g., magma ocean crystallisation (Hofmann 1997). Its seismic invisibility combined with fluid-dynamical constraints, however, weighs against the presence of such a layer today (Oldham and Davies 2004).

Dense material near the CMB could also be swept into 'piles' beneath upwellings and thinned, possibly to zero, under downwellings (Tackley 2000). This model is consistent with seismic tomography (Tackley 2002; Trampert et al. 2004), but the volume of the piles is still rather small to satisfy geochemical constraints (Hofmann 1997). It is not known whether dense piles can be formed and maintained by modern tectonic processes at the surface, or by core-mantle interactions, or represent a leftover expression of magma ocean fractional crystallisation, or a combination of these van der Hilst (2004).

*Viscosity contrasts*: Blobs having a 10–100 times higher viscosity resist stretching and mixing by the surrounding mantle (Manga 1996). It is conceivable, that 35–65% of the mantle could consist of such PM blobs and be concentrated in the LM (Becker et al. 1999). The lack of sampling of any such PM components at mid-ocean ridges and their need for persisting negative or neutral buoyancy despite substantial heating internal to the blobs are arguments against the model.

Other studies focus on radial viscosity stratification. Neither sufficient layering in the stirring efficiency (Stegman et al. 2002) nor significant convective isolation (van Keken and Ballentine 1998), (van Keken and Zhong 1999), (Ferrachat and Ricard 2001) were found for moderate viscosity contrasts. A model with more extreme contrasts (> factor of 1000) developed small-scale dominated whole mantle convection and separate reservoirs for DMM and PM, but no deep subduction (Walzer and Hendel 1999).

Allègre (2002) surmises that vigorous convection in the asthenosphere and sluggish stirring in the high viscosity transition layer both contribute to maintaining the MORB source region.

*Filtering in the transition zone*: The solubility of water above the β-to-γ-spinel transition is smaller than in the transition zone beneath it. Water lost from rising material lowers the melting temperature, possibly leading to a thin layer of partial melt at 410 km depth. Melt at that depth is denser than the matrix and will not rise. Incompatible elements concentrate in the melt and therefore only depleted material would reach the shallow mantle (Bercovici and Karato 2003). This hypothised filtering mechanism must be less efficient in hotter environments, allowing OIB forming plumes to retain enriched signatures. The temperature dependence would also allow CC to be formed from a volume larger than that above the transition zone.

### 3.2 Model Setup

Our convection calculations in this case study are restricted to uniform chemical composition and only consider the phase boundaries at 410 and 660 km depth. Thermal conductivity is constant in all models: $k = 12\ \mathrm{Wm^{-1}K^{-1}}$. Radial profiles for $p_r$, $\rho_r$, $g$, $K_T$ were taken from PREM (Dziewonski and Anderson 1981), for $c_V$ (Fig. 1a) and $\gamma$ (Fig. 1b) were derived by (Walzer et al. 2003; 2004a) based on PREM. $\alpha$ is shown in Fig. 1d, $T_r$ in Fig. 3a. Lateral variations of $\rho$ are allowed according to Eq. (7).

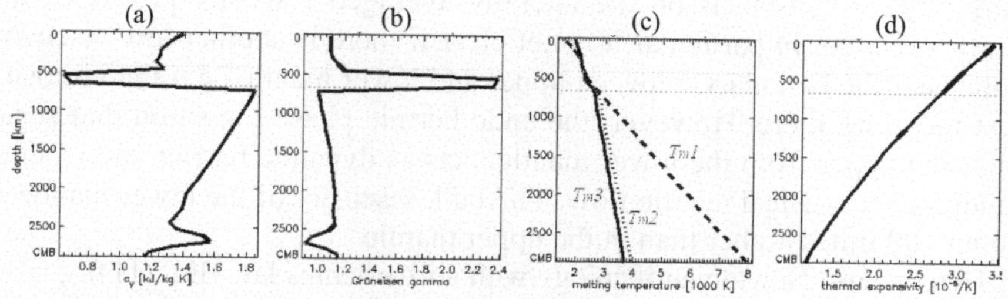


**Fig. 1** All models feature these radial profiles of (**a**) specific heat at constant volume, (**b**) Grüneisen parameter, (**c**) melting temperature $T_m1$, $T_m2$, $T_m3$ and (**d**) thermal expansivity

We use three different profiles for the melting temperature $T_m$ ($T_m1$, $T_m2$, $T_m3$) (Fig. 1c). In $T_m1$ (Walzer et al. 2003) the melting temperature is linearly interpolated between 660 km depth and CMB, according to experimental results of (Zerr and Boehler 1993; 1994). In $T_m2$ the melting temperature at the CMB was adjusted to results of (Boehler 2000). $T_m3$ is taken from Walzer et al. (2004a).

The heat generation rate per unit volume due to the decay of $^{40}K$, $^{242}Th$, $^{235}U$, $^{238}U$ in the models is spatially homogeneous, but decays exponentially with time. Parameters are based on Walzer et al. (2004a). The observed scale of plumes (Montelli et al. 2004) is on the order of our grid resolution and their existence depends on the strong lateral temperature dependence of viscosity. Therefore, a self-consistent formation of small-scale plumes is not possible in our models. On the other hand, major partial melting is expected only near the surface, possibly above the CMB and near 410 km depth. In order to prevent unrealistically high temperatures, we considered cooling due to volcanism and small-scale plumes in the energy equation of some models. It is active only where $T > T_m$ and can be restricted to certain depths and basically means cutting $T > T_m$ to $T = T_m$ each time step:

$$\rho \cdot H_{cool} = \rho \cdot c_v \cdot \frac{d(T - T_m)}{dt} \tag{17}$$

### 3.2.1 Rheology

There is a low viscosity asthenosphere beneath the lithosphere and there are layers of higher viscosity beneath the asthenosphere. Otherwise there is currently no consensus on the laterally averaged viscosity profile of the Earth's mantle. In particular it is not clear if there is another low-viscosity zone near the boundary between upper and lower mantle or if the viscosity just increases there. However, the endothermic phase transition that separates the upper from the lower mantle, acts as dynamic barrier and is a net hindrance to vertical mantle flow. The bulk viscosity of the lower mantle is about 100 times higher than in the upper mantle.

Here we use Newtonian rheology with an Arrhenius law (Eq. (11)), $\xi = 17$ (Karato et al. 2001) and $\eta_0 = 10^{21}$ Pa s. The radial variation of $\eta(T)$ has been included in a profile $\eta_r(r)$, which also accounts for pressure dependence and compositional effects in the real Earth. For numerical reasons the lateral variability had to be damped by a factor $c_t$, which does depend on radius (Fig. 2a) in some models and is constant in others.

$$\eta(r,\theta,\varphi,t) = \eta_r(r) \cdot \exp\left[c_t(r) \cdot T_m(r) \cdot \left(\frac{1}{T(r,\theta,\varphi,t)} - \frac{1}{T_{av}(r,t)}\right)\right] \tag{18}$$

Viscoplastic yielding (Eq. (13)) is considered in the uppermost 285 km of the mantle. This weakening is not affected by $c_t$, but limited to $\eta_{eff} \geq 0.002 \cdot \eta$.

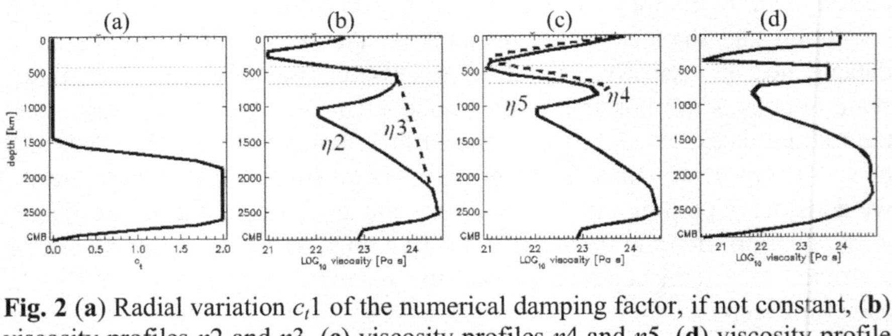


**Fig. 2** (**a**) Radial variation $c_t1$ of the numerical damping factor, if not constant, (**b**) viscosity profiles $\eta 2$ and $\eta 3$, (**c**) viscosity profiles $\eta 4$ and $\eta 5$, (**d**) viscosity profile $\eta 6$. (**a**) and (**b**) are based on Walzer et al. (2003) and (**d**) on Walzer et al. (2004a)

We use different radial profiles $\eta_r(r) = \eta\#$ (Figs. 2b, c, d). The simplest case, $\eta 1$, is a constant viscosity of $10^{23}$ Pa s throughout the mantle. $\eta 6$ and $T_m 3$ are favoured for the modern mantle and were derived (Walzer et al. 2004b) from PREM by using solid-state physics considerations, thermodynamic relations, the Grüneisen parameter and Lindemann's law. The profile is anchored at $10^{21}$ Pas in the asthenosphere. That value is higher than assumed for the Earth (Dixon et al. 2004), but reflects our present numerical capabilities. $\eta 2$ is an earlier version (Walzer et al. 2003) of $\eta 6$, with a weaker lithosphere and smoother gradients. Physically more desirable steep gradients were taken out there for numerical reasons. A non-conventional feature of $\eta 2$ is a second asthenosphere below the transition zone. This weak layer is omitted in $\eta 3$ in order to test its influence on the convection. When applied to models using $\eta 2$, the module based on (Eq. (13)) had virtually no effect. Therefore $\eta 4$ and $\eta 5$ were introduced, featuring stiffer lithospheres but still keeping gradients smooth. Both are generally weaker than $\eta 6$ and can therefore be interpreted to show effects of a supposedly hotter, early mantle. However, no attempt has been made here to address shape and amplitude of the viscosity profile for the Archean mantle. We stress that profiles based on PREM are valid for the modern mantle only and that our present models do not include time dependence in the assumed radial viscosity profile.

### 3.2.2 Boundary Conditions

The computing domain is a thick spherical shell, mimicking the silicate part of the Earth. The inner boundary at $r_{CMB}$ = 3480 km and the outer boundary at $r_E$ = 6371 km are free of tangential stresses, because the viscosities of liquid outer core and atmosphere/hydrosphere are negligible compared to mantle rocks.

A constant temperature of 288 K is assumed for the outer boundary, because it is the mean surface temperature today and there has been liquid water on the surface for at least 3.8 Ga.

Different thermal boundary conditions at the CMB reflect the continuous development of our models rather than being a focus of this paper. Condition *CMB*1 means a heat flux of 28.9 mW m$^{-2}$ (Anderson 1998), temporally (Schubert et al. 2001) and laterally (Walzer et al. 2003) constant. This was changed to spatially constant temperature, which is adjusted every time step to ensure a mean heat flux constant in time (*CMB*2). The physically most appealing approach is to couple a parameterised model of the evolution of the core to mantle convection. The heat flux across the CMB is equalled by secular cooling of inner and outer core, latent heat from freezing or melting of the inner core, release of gravitational potential energy due to the preferred segregation of heavy elements at the inner core boundary and the radiogenic heat production of the core. This energy balance was calculated every time step to obtain heat flux and spatially constant temperature at the CMB as well as the radius of the inner core. The core model is based on Labrosse (2003) and the implementation is described in detail by Gottschaldt (2003). This kind of boundary condition is characterised in the following by the concentration of $^{40}K$, which is thought to be the major radiogenic heat source in the core.

### 3.2.3 Initial Conditions

Because of the assumption of infinite Prandtl number and homogeneous composition, we need to consider only the initial temperature field. The thermal state of the early Earth is highly speculative, and we have explored several different scenarios. Frequent impacts may have determined the heat structure of the outer layers (Arrhenius and Lepland 2000), leading to an early thermally stable stratification. A global magma ocean (Solomatov 2000) or several large scale melting events (Kleine et al. 2004) are also conceivable. Fractional crystallisation and subsequent overturn has the potential to result in compositionally or thermally stable layering, too (Elkins-Tanton et al. 2003; Zaranek and Parmentier 2004). In this context we used a starting profile of constant temperature $T_{av}(r)$, which is stable in a compressible mantle. Only in the upper boundary layer did we introduce a smooth transition to the surface temperature (Fig. 3b). Together with a small lateral perturbation of the form

$$T(r,\theta,\varphi) = T_{av}(r) + T(\theta,\varphi) \cdot \cos\left[\frac{\pi}{r_E - r_{CMB}} \cdot \left(r - \frac{r_E + r_{CMB}}{2}\right)\right] \tag{19}$$

with

$$T(\theta,\varphi) = \sum_{l=2}^{15} \sum_{m=1}^{l} 0{,}002 \cdot (-1)^m \cdot P_l^m(\cos\theta) \cdot \left[\cos(m\varphi) - \sin(m\varphi)\right] \tag{20}$$

this initial condition is called *IC*1. $P_l^m$ is a Legendre function. The above perturbation is used for all models of this study. In *IC*2 a lower thermal boundary layer was added (Fig. 3c). An adiabatic profile plus boundary layers (Stacey 1992) and the solidus $T_m3$ are more obvious starting profiles, used in *IC*3 (Fig. 3d) and *IC*4 respectively.

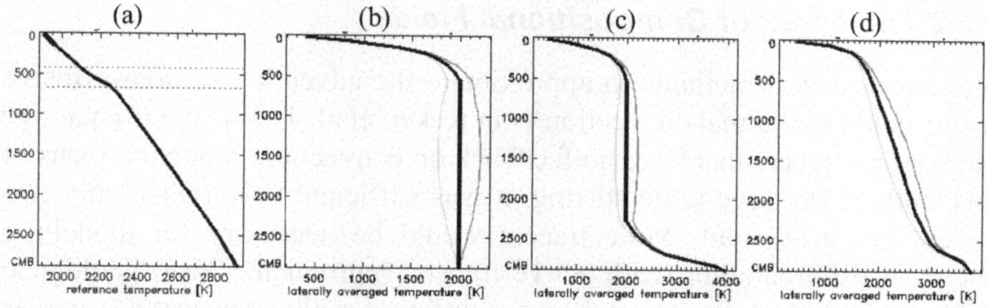


**Fig. 3** Radial profiles of (**a**) reference temperature $T_r$, (**b**) initial temperature (*black*) and maximum perturbation (*grey*) for *IC1*, (**c**) *IC2* and (**d**) *IC3*. *Dotted lines* in figure (**a**) mark the phase boundaries, which are considered in the code

## 3.3 Numerics

### *3.3.1 Mantle Convection Code: TERRA*

The coupled system of equations is solved numerically with the well-established Fortran code TERRA (Baumgardner 1983; Yang 1997) for the whole evolution of the entire mantle. The computational grid is based on a projection of the regular icosahedron onto a sphere and successive dyadic refinements (Baumgardner and Frederickson 1985). Concentric copies of such spherical layers of nodes build the domain in radial direction. All models have been run on a grid with 1394250 grid points, corresponding to a spatial resolution on the order of 100 km. Equations (2) and (4) are discretised in an Eulerian approach by finite elements with linear basis functions and solved simultaneously (modified after Ramage and Walthan (1992)) using a multigrid method. The energy equation is discretised by the MPDAT algorithm (Smolarkiewicz 1984) and integrated with a Runge-Kutta scheme using explicit time stepping. Domain decomposition is used for the parallelisation (Bunge and Baumgardner 1995) with MPI. TERRA was benchmarked for constant viscosity convection by Bunge et al. (1997) with numerical results of Glatzmaier (1988) for Nusselt numbers, peak temperatures, and peak velocities. A good agreement ($\leq 1.5\%$) was found. However, strong viscosity variations challenge the stability of the code and results should be interpreted as approximations. No detailed error analysis (DeVolder et al. 2002) was done.

### 3.3.2 Treatment of Compositional Fields

There are several methods to approximate the advection of non-diffusive scalar fields in thermal convection (van Keken et al. 1997). We use passive Lagrangian tracers that have no feedback on convection. Since the focus of this study is on large scale stirring, it was sufficient to initialise only two tracers per grid point. More tracers would be necessary for modelling chemical differentiation with active tracers, reproducing local geological features or characterise mixing across different scales. For testing, case H was repeated, starting with eight tracers per grid point. Without fine tuning nearly every second tracer was lost during the run, still giving twice the standard resolution at the last time step. A spherical harmonic analysis of the chemical field reveals not much structure above degree 30, so we cut it off there. The relevant results (Figs. 4, 5) are basically identical to the low resolution run.

Each tracer represents half of the mass of its initial cell and keeps this attribute throughout the entire run. Trajectories of the tracers are calculated synchronous to the integration of the heat equation. They move independently of the TERRA grid, but are continuously re-indexed with respect to the nearest grid point. This allows efficient interpolation between the Lagrangian particles and the local neighbourhood of Eulerian grid points. Memory requirements limit the number of tracers that can be hosted by a single grid point. If a grid point becomes overcrowded, surplus tracers are deleted randomly. With ~2% lost tracers and ~5% empty cells after a typical evolution run coverage is reasonably good. The accuracy of tracer trajectories was tested with prescribed velocity fields. Deviations from the analytic solution were less than 0.006%, which means ~10 km for a typical run.

### 3.3.3 Definition of Two Components

Here we want to track the dispersion of the material that comes close to the surface ('degasses') and characterise its subsequent distribution in the mantle at discrete time steps. In this framework we shall define two components, degassed tracers and residuum, and apply two constraints in our definition:

1. Each component will comprise 50% of the mantle at the end of the convection calculation. This ensures the comparability of different models. The half-half distribution is the most sensitive for measuring the state of stirring between two components.
2. It must represent a reasonable approximation for chemical differentiation near the surface. This corresponds to the fact that major differentiation

processes like the extraction of CC and degassing are related to partial melting near the Earth's surface. Hence the term 'degassing' is used here for a range of chemical modifications near the surface.

The following algorithm is used to satisfy both constraints: No compositional information is assigned to tracers during the convection calculation, but each tracer remembers its closest approach to the surface. This information is output together with the Cartesian coordinates and the mass of each tracer. In postprocessing, the tracers are ordered according to that distance, starting with the smallest. Then their masses are added successively until the sum has reached 50% of the mantle mass. The distance attribute of the last tracer incrementing this sum is taken as degassing depth ($d_d$). In other words, the degassing depth is an output, not an input. All tracers which have been closer to the surface than the degassing depth are assigned the concentration $c = 0$, while the residuum retains a concentration $c = 1$. This is identical to ongoing differentiation throughout the run with complete degassing above and none below the degassing depth. The algorithm describes modification of pristine mantle near the surface and degassing of primordial, non-radiogenic gases. It is not necessary to prescribe degassing locations such as mid-ocean ridges or volcanoes, because mantle rock can reach the surface only at these locations. The concentration of undegassed material in each grid cell is the mean of the concentrations of all the tracers it contains, weighted by their masses.

## 3.4 Model Results

The ultimate goal of convection-differentiation models is to reproduce observations. For geochemistry this would be plots of isotopic concentrations in surface samples. Since no isotopic systems are modelled, this is not possible here. We focus on large-scale geometrical characteristics of the convective flow instead. For easy comparison, results are grouped together here and will be discussed in the next section. The parameter space covered by this study and some output values are summarized in Table 2. Maximum and root mean square (rms) surface velocity may be used to compare the convective vigour of our models to the Earth.

**Table 2** Summary of models contributing to this study. Rheology is determined by the viscosity profile, denoted in column $\eta$, the numerical damping factor $c_t$ and the yield stress $\tau_y$. The melting temperature profile, volcanism model, boundary condition at the CMB and initial condition are given in columns $T_m$, *vol*, *CMB*, and *IC* respectively. $\tau_y$ is given in [$10^8$ Pa] and the concentration of $^{40}K$ in the core in column *CMB* in [ppm]. The results for root mean square velocity ($v_{rms}$) and maximum ($v_{max}$) surface velocity are in [cm/year], the degassing depth $d_d$ is in [km]. All are for the last time-step

| Model | $\eta$ | $c_t$ | $\tau_y$ | $T_m$ | vol | CMB | IC | $v_{rms}$ | $v_{max}$ | $d_d$ |
|---|---|---|---|---|---|---|---|---|---|---|
| A | $\eta$1 | $c_t$1 | – | $T_m$1 | – | CMB1 | IC1 | 0.2 | 0.7 | 144 |
| B | $\eta$2 | $c_t$1 | – | $T_m$1 | – | | | 1.5 | 2.7 | 64 |
| C | $\eta$2 | $c_t$1 | – | $T_m$1 | – | | IC3 | 1.4 | 2.5 | 62 |
| D | $\eta$2 | $c_t$1 | – | $T_m$1 | – | 100 | IC2 | 1.6 | 2.6 | 67 |
| E | $\eta$2 | $c_t$1 | – | $T_m$1 | – | 200 | IC2 | 1.7 | 2.9 | 66 |
| F | $\eta$3 | $c_t$1 | – | $T_m$1 | – | CMB1 | IC1 | 1.7 | 2.6 | 69 |
| G | $\eta$4 | 1.0 | 1.8 | $T_m$1 | vol1 | CMB1 | IC1 | 1.2 | 1.8 | 99 |
| H | $\eta$5 | 1.0 | 1.8 | $T_m$1 | vol1 | CMB1 | IC1 | 0.5 | 1.4 | 163 |
| I | $\eta$5 | 1.0 | 1.4 | $T_m$1 | vol1 | CMB1 | IC1 | 0.6 | 1.7 | 152 |
| J | $\eta$5 | 1.5 | 1.35 | $T_m$3 | vol2 | | IC4 | 0.7 | 1.4 | 148 |
| K | $\eta$6 | 1.0 | 1.4 | $T_m$1 | vol1 | | IC1 | 1.2 | 2.8 | 223 |
| L | $\eta$6 | 1.0 | 1.35 | $T_m$1 | – | 200 | IC2 | 0.8 | 1.9 | 176 |
| M | $\eta$6 | 1.0 | 1.35 | $T_m$1 | – | 200 | IC3 | 0.9 | 2.0 | 155 |
| N | $\eta$6 | 1.0 | 1.35 | $T_m$1 | – | CMB2 | IC3 | 0.6 | 2.2 | 150 |
| O | $\eta$6 | 1.0 | 1.35 | $T_m$2 | vol2 | CMB2 | IC3 | 0.8 | 1.7 | 151 |
| P | $\eta$6 | 1.0 | 1.35 | $T_m$2 | vol2 | 200 | IC3 | 0.8 | 2.4 | 160 |
| Q | $\eta$6 | 1.75 | 1.35 | $T_m$2 | vol2 | 200 | IC3 | 0.7 | 2.1 | 144 |
| R | $\eta$6 | 1.5 | 1.35 | $T_m$3 | vol2 | CMB2 | IC4 | 0.7 | 1.8 | 158 |

Besides of Table 2, results are presented graphically in Figs. 4, 5 and in the Supplementary Material (available on accompanying DVD). Temperature evolution plots display the laterally averaged temperature profile versus time in a contour plot. Accordingly, degassing evolution plots are based on laterally averaged profiles of the concentration of pristine material. The scale of heterogeneity in the distribution of degassed material is determined by a spherical harmonic analysis. We applied the method used by Yang (1997) to the field of concentration minus the mean concentration at each radial level. The resulting spectral heterogeneity map is a contour plot of the rms amplitude of the scalar field at different depths for spherical harmonic degrees 0–30. The colour scale of each plot is given as percentage of the maximum value occurring in that plot for all degrees and depths. A spherical harmonic analysis is only done for the last time step, corresponding to the modern Earth. The planform of the radial velocity component at 65 and 2825 km depths is given for three time steps. Only each group of three pictures shares a colour scheme. Cutaway views of the mantle show the degassing field at the same time steps as the velocity planform pictures. All cutaway views use the same colour scheme (for more information Please see the accompanying DVD).

Since some parameter variations have only minor influences on the convective flow, pictures are just given for selected models.

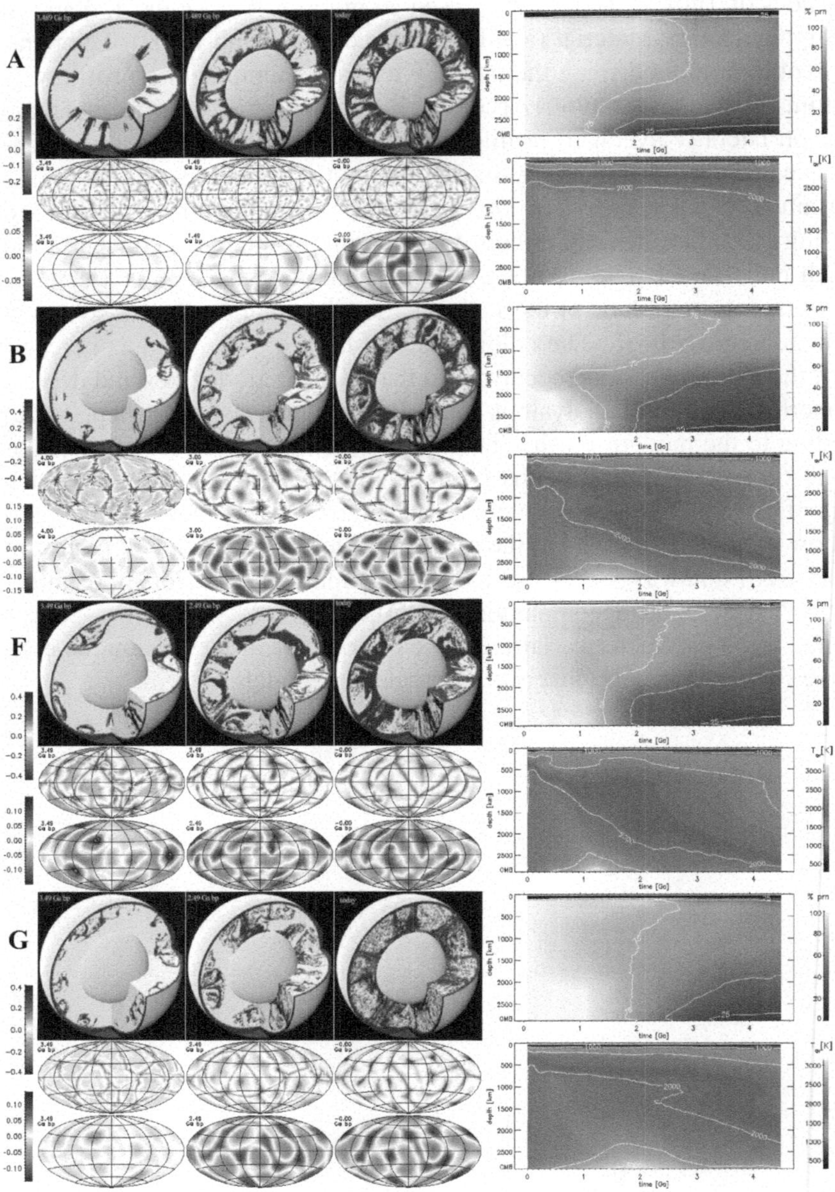


**Fig. 4** Results of selected models. Three cutaway views of the mantle are placed to the *right* of the letter that is denoting the model. They show the degassing field at the same time steps as the planform of the radial velocity component is given below. The *upper row* always is for 65 km depth and the *lower row* is for 2825 km. Velocity is given in [cm $a^{-1}$] and each legend is valid for a *whole row*. Depicted

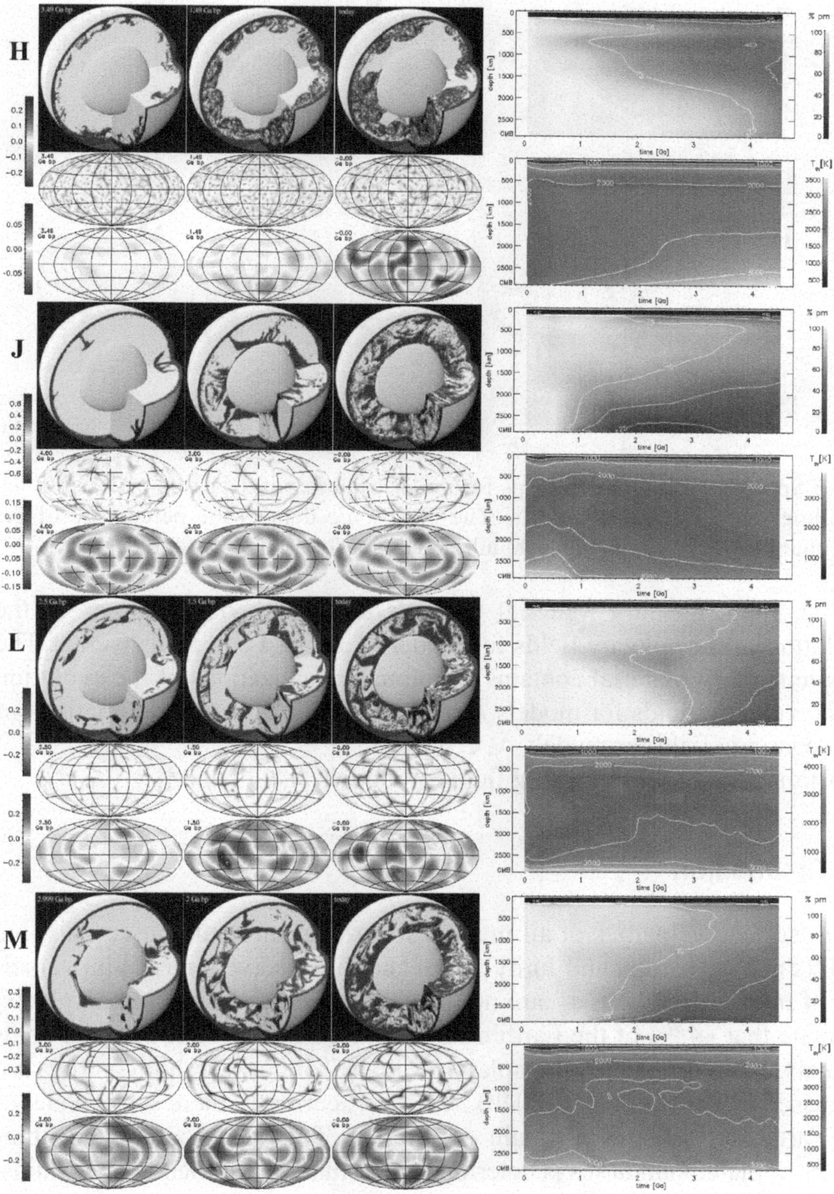


time steps were chosen separately for each model. Degassing evolution plot and temperature evolution plot are placed in the *right column*. *Light colours* in the degassing evolution plot indicate undegassed material (white = 100% pm). The legend is also valid for the cutaway views. Pictures of not displayed models (in *brackets*) are similar to: B ~ (C, D, E), H ~ (I), L ~ (K), M ~ (N, O, P, Q, R)

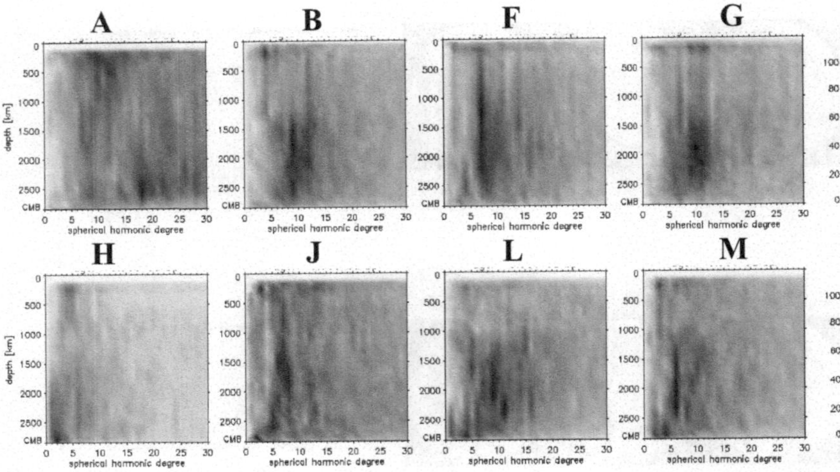


**Fig. 5** Spectral heterogeneity maps for the distribution of degassed material in the last time step of selected models. The *grey scale* depicts the percentage of the maximum value occurring in each individual plot for all degrees and depths

The Supplementary Material contains additional visualizations of the evolution of the velocity fields for models A, F, C, I and K, as well as The Supplementary Material contains additional visualizations of the evolution of the velocity fields for models A, F, C, I and K, as well as animations of the degassing field for models A, C, I and K.

Animations of the degassing field for models A, C, I and K.

## 3.5 Discussion

Mean surface velocities in all models are smaller than the observed value of 3.9 cm $a^{-1}$ (Gordon and Jurdy 1986) and the observed maximum velocities of more than 10 cm $a^{-1}$ are not reached either. O'Connell et al. (1991) estimate that ~45% of the Earth's surface velocity field is in toroidal components, but our models all display less than 5%. Further shortcomings are discussed in Walzer et al. (2004b). However, currently there is no model of the Earth that satisfies all main constraints. The behaviour of the lithosphere at plate boundaries is determined by small-scale phenomena (fluids, melt, cracking and elasticity) that are not resolved in our global models. Besides a more realistic lateral variability of rheology, coupling between different scales is desirable in future models. Especially the outer boundary needs improvement but the deep mantle could still be adequately represented here. We do not attempt to relate the timescale of our models to the real Earth. Despite the rather low convective vigour (measured by surface velocity), we find degassing depths between 62 and 258 km, which are in the same order of magnitude as concluded by other studies: 60–115 km (Hirth and Kohlstedt 1996), 65 km (Regenauer-Lieb and Kohl 2003) and the depth of partial melting (<200 km: (Presnall et al. 2002)).

### 3.5.1 Influence of Geometry

In model A, no radial viscosity variation is assumed. This is certainly unrealistic for the Earth, but most clearly demonstrates an effect of 3-d spherical geometry. Thermal convection in general is organised in cell-like structures, which may vary in their shape, symmetry, size and time dependence. Let's consider a single roll of constant angular velocity in an incompressible, isoviscous cube. There is a line of pure rotation through the centre of the roll, which lies at mid-depth in the cube. Conservation of mass requires that flux through a vertical plane below this stagnation line equals the flux through a plane above the line. Now let's shrink the bottom of the cube, making it a frustum of an upside down pyramid. The stagnation line will move upwards until the flux through the vertical plane above the line is the same as below. This is analogous to what happens in 3-d spherical geometry. The assumption of constant angular velocity is certainly not realistic and asymmetries in the velocity distribution of a cell will overlay the illustrated geometric effect. For example, the planform area of upwellings may be smaller or larger than that of downwellings, depending on the model and even depth. A detailed investigation of the implications and relative importance of these aspects is beyond the scope of this paper. In A degassed material from the top sinks right to the bottom of the mantle. There is only little hindrance at the depth of nonzero $c_t$, because viscosity is increased locally in cold downwellings. In lateral average degassed material fills the mantle from the bottom. The upper half of the mantle keeps its pristine signature for the longest time. This is due to the combined effects of cell geometry, flux asymmetry and lateral averaging. The latter just means that there is less volume in the LM for the same amount of degassed material than in the UM, giving LM a more degassed signature on average.

### 3.5.2 Influence of Rheology

Model B is identical to A except for the viscosity profile. Downwellings are hindered and partially deflected, when they hit the high viscosity zone

(HVZ) in the mid LM. This is evident in the first cutaway view and the high concentration of degassed material in about 1500 km depth during the first billion years of evolution. Furthermore the long-term preservation of pristine material in the upper half of the mantle is more pronounced than in A, with the lowest concentration of degassed material even closer to the surface. Velocities are lower in the HVZ of the lower mantle and consequently stagnation points of whole-mantle cells move further upwards. This effect is an additional reason for the more degassed signature of the LM. It has been observed in axisymmetrical models (van Keken et al. 2001) and even in Cartesian geometry (Ferrachat and Ricard 2001). A distinct concentration gradient at 1500 km depth is maintained after ~1.5 Ga. This suggests that there is some degree of decoupling. Cold material sinks efficiently into the HVZ and produces long-wavelength heterogeneity there due to sluggish stirring. On the other hand, ambient upwelling from the HVZ delivers an average, less degassed signature upwards. The planform of radial velocity shows that the area of up- and downwellings is about equal in the HVZ but in the UM downwellings are narrow.

This is also true for the planform of model F, where the lower low viscosity zone (LVZ) has been omitted. However, $l \sim 7$ heterogeneity is dominant throughout the mantle. This is a notable difference to B, where heterogeneity is pronounced in the HVZ. During its way down degassed material is stirred more efficiently, allowing only smooth concentration gradients in the lateral average. We conclude that the second asthenosphere is crucial for the decoupling of the stirring behaviour in the mid mantle.

Support for that conclusion comes from model G, which features the lower LVZ but a stiffer lithosphere plus yielding. That variation hinders convection as indicated by lower surface velocities. In the first billion years degassed material is dispersed in the upper half of the mantle, similar to F. When the stable initial temperature profile is overcome, large cells fill the mantle from the bottom with degassed material. At the end, heterogeneity is pronounced in the HVZ. Hence the resulting layering of stirring behaviour is not an artefact of the early mode of dispersion in the model.

Compared to G, the viscosity profile for H and I has a slightly thicker and stiffer lithosphere. This minor change triggers a rather different evolution. During the first billion years convection is hindered by the phase boundary and viscosity increase in 660 km depth, combined with the effects of the stable initial temperature profile and the stiffer lithosphere. Small-scale cells develop which efficiently disperse degassed material in the upper mantle. After about 1.5 Ga the boundary between the vigorous convecting zone and pristine mantle moves down through the second LVZ until it reaches the HVZ of the lower mantle. While hindered there, degassed material is effectively stirred in the upper half of the mantle. However, the band of less degassed material in the degassing evolution plot at 500 km depth extends to at least 3 Ga. A less visible band develops at 1000 km depth. This indicates, that there are still some cells confined to the UM. Other cells are operating between the two asthenospheres, with their stagnation points in the stiffer transition zone. Between 3 and 4 Ga the lower half of the mantle starts to convect and the system switches to long wavelength convection. Regions with persisting small-scale convection are dragged into the mantle by this catastrophic overturn. No stability analysis (e.g. like Turcotte and Schubert (2002), p. 270) has been carried out, but obviously that behaviour may be explained as follows. Small-scale convection efficiently cools the upper half of the mantle, but it is not able to penetrate into the HVZ, where longer wavelength perturbations are unstable. Further cooling of the upper half and heating of the lower half increases the Rayleigh number of the whole system, finally leading to the rise of the lower layer. The long wavelength of that rising layer is thereby overprinted on the entire mantle. Walzer and Hendel (1999) found a similar behaviour in a 2-d convection-differentiation model without stable initial stratification.

The different yield stresses in H and I have only a gradual influence.

Profile $\eta 6$ in the otherwise identical model K initially leads to medium scale cells in the upper half of the mantle. There is no significant layering at the high viscosity transition zone or the phase boundary at 660 km depth, but instead at the lower HVZ. Large-scale overturn takes over before the upper half of the mantle is well stirred. This difference from models H and I is possibly due to the longer wavelength of the initial convection in the upper half of the mantle.

Profiles $\eta 5$ and $\eta 6$ are compared again in models J and R, but with an unstable initial temperature profile. There is insignificant or no layering of convection now. In both models, degassed material sinks from the top to the CMB in large-scale cells from the beginning. As in B, stagnation points occur in the upper 1000 km of the mantle (pronounced in J) and heterogeneity is expressed mostly in the lower HVZ (pronounced in R).

### 3.5.3 Influence of Initial Conditions

The comparisons of H versus J and K versus R indicate, that stable initial layering is necessary to confine early convection to the upper layers of the mantle. This conclusion is supported by L versus M, which differ only in their initial temperature profiles. Layering on top of the lower HVZ is much weaker in M, which has an unstable initial profile. As demonstrated by A and B, stable initial layering is not sufficient to trigger layered convection. Walzer et al. (2004b) find the small-scale convective regime also in 2-d models with unstable initial temperature profiles, but with larger viscosity differences in the UM.

### 3.5.4 Minor Influences

According to Eq. (18) a higher melting temperature acts like an increased lateral temperature dependence of viscosity ($c_t$) at a given depth, and the volcanism module (*vol**) simply removes unrealistic peak temperatures. It follows from N versus O, that this difference affects degassing and global thermal history only slightly.

The thermal boundary at the CMB could affect the temperature profile and therefore has the potential to alter the whole evolution. However, comparison of models with core evolution versus constant heat flux (M versus N and P versus O) shows no significant difference in their degassing histories. The same is also true for D versus E, which feature different concentrations of $^{40}$K in the core. All models with core evolution develop higher CMB heat fluxes than assumed for the other models. This leads to steeper thermal boundaries at the bottom. A wider parameter space ought to be explored, especially for the influence of the CMB heat flow in models with stable initial layering.

## 3.6 Conclusions

Despite the deficiencies of our models the following conclusions are considered to be reasonably robust for Earth-like parameters:

1. During large-scale convection in 3-d spherical geometry with a highly viscous LM, the upper 1000 km of the mantle are least affected by material that was subject to differentiation near the surface.
2. Realistic radial viscosity variations – with two low viscosity zones in the upper half of the mantle and a highly viscous LM – result in a layering of the stirring behaviour.
3. Small-scale convection confined to the upper parts of the mantle is possible under certain conditions.
4. The convective regime may change from small-scale to large-scale convection.

The timescale in our models is unlikely to correspond to the real Earth, because surface velocities are smaller than observed, the viscosity profile does not evolve and we only speculate about the initial conditions.

### 3.6.1 Relevance for the Earth

We promote the following cartoon (Fig. 6) to be considered in the list of possible scenarios for the convection-differentiation history of the Earth.

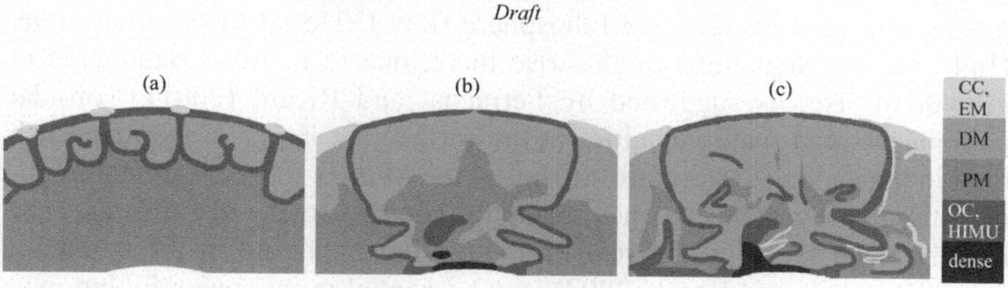


**Fig. 6** During an episode of layered convection CC is extracted near the surface, leaving behind a depleted, well stirred residue above deeper pristine material (**a**) Later large-scale convection with deep subduction starts to work on this configuration (**b**) Further differentiation and segregation of CC takes place near the surface, but the shallow mantle is likely to retain its depleted, homogeneous signature due to 3-d spherical geometry and high viscosity in the mid lower mantle. Additionally, stirring is least effective in the mid lower mantle, allowing primitive and recycled material to coexist for the longest time in the convecting mantle. High density components may segregate in D". The modern MORB source would be a leftover from the episode of small-scale convection plus variable contamination from the lower half of the mantle. Small-scale plumes sampling different regions of the heterogeneous lower mantle produce the observed scatter in OIB geochemistry (**c**)

There may have been layered convection in the Archean mantle. CC is extracted near the surface and the residuum is well stirred by small-scale cells in the uppermost convecting layer, as in models H and I. A layer of homogeneous depleted material may grow on top of pristine mantle, as found in a convection-differentiation model of Walzer and Hendel (1999). The depleted layer is not necessarily confined to the UM and the boundary between both regions is irregular.

At some stage the system must have switched to the style of convection, which is favoured for the Earth today. There are downwellings plunging from the surface into the lower mantle or even to the CMB. Because of 3-d spherical geometry and high viscosity in the lower mantle the stagnation points of the cell-like convective structures lie at rather shallow depths. Additionally, broad up- and small downwellings would fill the mantle from the bottom with degassed material and rise old material that has been there before. Laterally averaged, material just below the depth of partial melting can preserve its original signature for the longest time. The original signature may be the depleted signature that is left from the episode of small-scale convection. Stirring is less effective in the HVZ of the lower mantle and heterogeneities can persist there longer than in other parts of the mantle. Primordial material could have survived there, or, if stabilized by higher density, in D". Inefficient stirring in the HVZ also provides a mechanism to store recycled lithosphere (low U/He ratio) for some time. High $^3He/^4He$ signatures might arise there, due to retarded radiogenic in growth of $^4He$, as suggested by Ferrachat and Ricard (2001). Toroidal components and plate velocities at the surface are larger in the real Earth than in our models. Therefore stirring of the MORB source is probably underestimated here and the heterogeneity contrast to the HVZ in the LM may be even larger in the real Earth. A filter operating in the transition zone (Bercovici and Karato 2003) is not essential in this scenario, but may help to maintain a homogeneous, depleted MORB reservoir.

Reasons for a change in the convective regime are speculative. The overcoming of initially stable layering – thermally or compositionally – might be a possibility. An impermeable boundary in the mid mantle is unlikely today, but compositional layering may have contributed to convective isolation of lower parts of the mantle in the early Earth. Layering is more sustainable with higher convective vigour (Davaille 1999; Oldham and Davies 2004). Therefore it is also likely that the phase boundaries (Tackley 1996) and steep viscosity gradients (Walzer and Hendel 1999) had a stronger layering effect in a more vigorous regime. Almost certainly the viscosity was generally lower in the Archean. Radiogenic heating has been steadily decreasing. Both effects support higher convective vigour in the early Earth and likely had the potential to change the convective style of the mantle. In our models the viscosity profile is fixed for each run, gradients are smoothed for numerical reasons and composition is uniform. Therefore layered convection in the Earth could have lasted longer than in the models – opposite to the effect of too small convective vigour on the model timescale. In the light of these competing shortcomings we cannot determine the duration of the layered period or the percentage of mantle processed from our models.

Additional feedback could come from spatially inhomogeneous radiogenic heating, with the pristine lower layer being heated more than the depleted layer. The proposed change in the style of subduction (Xie and Tackley 2004) could be contributing to or resulting from a change of the convective mode.

Two extreme scenarios are conceivable for the transition from small- to large-scale convection. One is the catastrophic overturn, captured in its initial stage by models H and I. The overturn is driven by the globally unstable layering and could result in a large-scale inversion of geochemical signatures, with pristine on top of degassed material. This is the opposite of what is being concluded from OIB and MORB signatures (Ballentine et al. 2002). They also argue that this kind of overturn results in rising upper mantle temperatures, but this is not observed in the laterally averaged temperature profiles of H and I. The other scenario is more like that occurring in model G. Large-scale cells appear early, but are accompanied by small-scale cells. The small cells die out, leaving behind a well-stirred, depleted upper part of the mantle. Then large-scale convection takes over without catastrophic overturn. At some stage the existing long wavelength cells just start to penetrate the lowermost mantle. Local instabilities rather than unstable global layering control this onset of deep subduction. The second scenario is favoured here, but both are only a small change in lithospheric rheology apart (see discussion G versus H) and hence aspects of both may have been operating in the Earth. Neither mechanism has been investigated in detail as yet. Models G, H, and I show only certain aspects of each.

### 3.6.2 Other Hints for a Change of Convective Mode

A consequence of the scenario outlined above is that the upper parts of the mantle are heterogeneous in the early stages of small-scale convection and more homogeneous before the onset of large-scale convection. Large-scale convection increases heterogeneity again, leading to the modern geochemical scatter in mantle-derived rocks. Rare earth signatures in picrites and komatiites from the Late Archean are indeed more homogeneous than in samples from the Early Archean, Proterozoic and Phanerozoic respectively (Campbell 1998).

Breuer and Spohn (1995) propose a flush instability to explain geological and climate changes at the Archean-Proterozoic transition. There are several indications for rapid growth of CC, continents and ocean water mass during that time. CC growth is attributed to increasing UM temperatures, convective vigour and the lower mantle reservoir becoming available for CC extraction. Increased degassing of the mantle could lead to increased ocean water mass. This would be consistent with the catastrophic overturn scenario. Archean island-arc magmas originated in relatively warm, basaltic crust and at lower pressures than at present. This could be explained by a smaller plate scale (Taylor and McLennan 1995).

Based on supercontinent cycles and greenstone ages, Condie (1997), favours layered convection before 2.8 Ga bp, episodicity of catastrophic overturn and hindering of deep subduction until 1.3 Ga bp and large-scale convection afterwards. Allègre (2002) reviews geochemical constraints and concludes that they are consistent with layered convection earlier and large-scale convection today. Breuer and Spohn (1995), Condie (1997) and Allègre (2002) assume the phase boundary at 660 km depth leads to temporal layering. However, their general arguments are also valid for the types of layering discussed here.

### 3.6.3 Outlook

Future models aiming at the reconciliation of geophysical and geochemical constraints may consider the following points:
(1) 3-d spherical geometry is important for global stirring.
(2) One or more changes of the convective mode during the evolution of the mantle are a possibility.
(3) Rheology has a strong influence on the stirring properties of the mantle. Shortcomings of our models are static viscosity profiles and an over-simplified lithospheric rheology that yields unrealistic (toroidal) velocities at the surface. Viscosity profiles should evolve according to the mantle temperature and more realistic lateral viscosity variations need to be included.
(4) Neither trace element nor major element compositional differences have been modelled here. A more direct comparison of models and observations would be possible by coupling the evolution of isotopic systems (like Xie and Tackley (2004)) or reservoirs (like Walzer and Hendel (1999), (2008)) to models of mantle convection. Inhomogeneous internal heating, different densities and viscosities could provide a feedback on thermal convection.
(5) We need a rational vision of Precambrian scenarios. Was there stable compositional or thermal layering? What was the tectonic style of the early Earth and how did chemical differentiation work, especially the segregation of CC?

As a working assumption we propose a marble cake like structure for the modern mantle, with small-scale geochemical structure dominating in the MORB source and heterogeneities larger than the sampling volume of deep-rooted plumes occurring in the LM. This includes segregation and temporal or permanent storage of dense recycled material above the CMB. The more homogeneous structure of the shallow mantle could be a leftover from an episode of small-scale convection and differentiation near the surface, combined with the effects of 3-d spherical geometry and the viscosity profile. Primitive material is most likely to have survived in the poorly mixed HVZ in the mid LM. It may reach the UM only as small-scale veins that are homogenised during sampling. A pure recycling model is also conceivable. The relative importance and time-dependence of additional mechanisms, like zoning by variable depth of subduction, a filter contributing to the homogeneity of DMM, differences during melt extraction and the statistical nature of sampling and the possibility of a doming regime with dense piles in upwellings may be investigated in more sophisticated future models.

**Acknowledgements** The calculations for this paper were done on the supercomputers of the John-von-Neumann Institute for Computing at Forschungszentrum Jülich (Germany) and the Earth Systems Science Computational Centre at the University of Queensland (Australia). The authors thank D. Breuer, T. Burghardt, N. Coltice, S. Ferrachat, R. Hendel, W. Jacoby, G. Jentzsch, and L. Viereck-Götte for helpful discussions. Comments from D. Murphy and an anonymous reviewer improved the manuscript and were highly appreciated.

## References

Albarède F (2001) Radiogenic in growth in systems with multiple reservoirs: applications to the differentiation of the mantle-crust system. Earth and Planetary Science Letters 189: 59–73

Albarède F, van der Hilst RD (2002) Zoned mantle convection. Philosophical Transactions of the Royal Society of London: A 360: 2569–2592

Allègre CJ (2002) The evolution of mantle mixing. Philosophical Transactions: Mathematical, Physical and Engineering Sciences 360(1800): 2411–2431

Allègre CJ, Hamelin B, Provost A, Dupre B (1987) Topology in isotopic multispace and origin of mantle chemical heterogeneities. Earth and Planetary Science Letters 81(4): 319–337

Allègre CJ, Hofmann AW, O'Nions K (1996) The argon constraints on mantle structure. Geophysical Research Letters 23: 3555–3557

Allègre CJ, Turcotte DL (1986) Implications of a two-component marble-cake mantle. Nature 323: 123–127

Anderson OL (1998) The Grüneisen parameter for iron at outer core conditions and the resulting conductive heat and power in the core. Physics of the Earth and Planetary Interiors 109: 179–197

Andreasen R, Sharma M (2006) Solar nebula heterogeneity in p-process Samarium and Neodymium isotopes. Science 314(5800): 806–809

Arndt NT (2004) The Precambrian Earth: Tempos and events. In: Eriksson PG, Altermann W, Nelson DR, Mueller WU, Catuneanu O Developments in Precambrian Geology, 12, Elsevier, 155–158

Arrhenius G, Lepland A (2000) Accretion of Moon and Earth and the emergence of life. Chemical Geology 169: 69–82

Ballentine CJ, van Keken P, Porcelli D, Hauri EH (2002) Numerical models, geochemistry and the zero-paradox noble-gas mantle. Philosophical Transactions of the Royal Society of London: A 360: 2611–2631

Baumgardner JR (1983) A three-dimensional finite element model for mantle convection, Los Angeles: University of California

Baumgardner JR, Frederickson PO (1985) Icosahedral discretization of the 2-sphere. Siam Journal on Numerical Analysis 22(6): 1107–1115

Becker TW, Kellogg JB, O'Connell RJ (1999) Thermal constraints on the survival of primitive blobs in the lower mantle. Earth and Planetary Science Letters 171: 351–365

Bercovici D, Karato S-I (2003) Whole-mantle convection and the transition-zone water filter. Nature 425: 39–44

Boehler R (2000) High-pressure experiments and the phase diagram of lower mantle and core materials. Reviews of Geophysics 38(2): 221–245

Boyet M, Carlson RW (2005) $^{142}$Nd evidence for early (>4.53 Ga) global differentiation of the silicate Earth. Science 309: 576–581

Breuer D, Spohn T (1995) Possible flush instability in mantle convection at the Archaean–Proterozoic transition. Nature 378: 608–610

Bunge H-P, Baumgardner JR (1995) Mantle convection modeling on parallel virtual machines. Computers in Physics 9(2): 207–215

Bunge H-P, Richards MA, Baumgardner JR (1997) A sensitivity study of three-dimensional spherical mantle convection at $10^8$ Rayleigh number: effects of depth-dependent viscosity, heating mode, and an endothermic phase change. Journal of Geophysical Research 102(B6): 11991–12007

Campbell IH (1998) The Earth's Mantle: Composition, structure and evolution. In: Jackson I, Cambridge University Press, Cambridge, 259–310

Canup RM (2004) Simulations of a late lunar-forming impact. Icarus 168(2): 433–456

Carlson RW, Boyet M, Horan M (2007) Chondrite Barium, Neodymium, and Samarium isotopic heterogeneity and early Earth differentiation. Science 316(5828): 1175–1178

Coltice N, Ricard Y (2002) On the origin of noble gases in mantle plumes. Philosophical Transactions of the Royal Society of London: A 360: 2633–2648

Condie KC (1997) Plate Tectonics and Crustal Evolution. Butterworth-Heinemann

Davaille A (1999) Simultaneous generation of hotspots and superswells by convection in a heterogeneous planetary mantle. Nature 402: 756–760

Davies GF (2005) A case for mantle plumes. Chinese Science Bulletin 50(1): 1–14

DeVolder B, Glimm J, Grove J, Kang Y, Lee Y, Pao K, Sharp DH, Ye K (2002) Uncertainty quantification for multiscale simulations. Journal of Fluids and Engineering 124: 29–41

Dixon JE, Dixon TH, Bell DR, Malservisi R (2004) Lateral variation in upper mantle viscosity: role of water. Earth and Planetary Science Letters 222(2): 451–467

Dziewonski AM, Anderson DL (1981) Preliminary reference Earth model. Physics of the Earth and Planetary Interiors 25: 297–356

Elkins-Tanton LT, Parmentier EM, Hess PC (2003) Magma ocean fractional crystallization and cumulate overturn in terrestrial planets: Implications for Mars. Meteoritics and Planetary Science 38(12): 1711–1875

Ferrachat S, Ricard Y (2001) Mixing properties in the Earth's mantle: Effects of the viscosity stratification and of oceanic crust segregation. Geochemistry Geophysics Geosystems 2: 1013, doi: 10.1029/2000GC000092

Glatzmaier GA (1988) Numerical simulations of mantle convection: Time-dependent, three-dimensional, compressible, spherical shell. Geophysical and Astrophysical Fluid Dynamics 43: 223–264

Gonnermann HM, Mukhopadhyay S (2007) Non-equilibrium degassing and a primordial source for helium in ocean-island volcanism. Nature 449: 1037–1040

Gordon RG, Jurdy DM (1986) Cenozoic global plate motions. Journal of Geophysical Research 91: 12389–12406

Gottschaldt K-D (2003) Vermischung in 3D sphärischen Konvektionsmodellen des Erdmantels, Jena: Friedrich-Schiller-Universität

Gottschaldt K-D, Walzer U, Hendel RF, Stegman DR, Baumgardner JR, Mühlhaus H-B (2006) Stirring in 3-d spherical models of convection in the Earth's mantle. Philosophical Magazine 86(21–22): 3175–3204

Grand SP, van der Hilst RD, Widiyantoro S (1997) Global seismic tomography: a snapshot of convection in the Earth. GSA Today 7: 1–7

Hanan BB, Blichert-Toft J, Pyle DG, Christie DM (2004) Contrasting origins of the upper mantle revealed by hafnium and lead isotopes from the Southeast Indian Ridge. Nature 432: 91–94

Hart S (1984) A large-scale isotope anomaly in the southern hemisphere mantle. Nature 309: 753–757

Helffrich GR, Wood BJ (2001) The Earth's mantle. Nature 412: 501–507

Hirose K (2002) Phase transitions in pyrolitic mantle around 670 km depth: Implications for upwelling of plumes from the lower mantle. Journal of Geophysical Research 107(B4): 2078, doi:10.1029/2001JB000597

Hirose K (2006) Postperovskite phase transition and its geophysical implications. Reviews of Geophysics 44(2005RG000186): RG3001

Hirth G, Kohlstedt DL (1996) Water in the oceanic upper mantle: implications for rheology, melt extraction and the evolution of the lithosphere. Earth and Planetary Science Letters 144: 93–108

Hofmann AW (1988) Chemical differentiation of the earth – The relationship between mantle, continental crust, and oceanic crust. Earth and Planetary Science Letters 90(3): 297–314

Hofmann AW (1997) Mantle geochemistry: the message from oceanic volcanism. Nature 385: 219–229

Holland G, Ballentine CJ (2006) Seawater subduction controls the heavy noble gas composition of the mantle. Nature 441: 186–191

Karato S-I, Riedel MR, Yuen DA (2001) Rheological structure and deformation of subducted slabs in the mantle transition zone: implications for mantle circulation and deep earthquakes. Physics of the Earth and Planetary Interiors 127: 83–108

Kellogg JB, Jacobsen SB, O'Connell RJ (2002) Modeling the distribution of isotopic ratios in geochemical reservoirs. Earth and Planetary Science Letters 204: 183–202

Kellogg LH, Hager BH, van der Hilst RD (1999) Compositional stratification in the deep mantle. Science 283: 1881–1884

Kleine T, Mezger K, Palme H, Münker C (2004) The W isotope evolution of the bulk silicate Earth: constraints on the timing and mechanisms of core formation and accretion. Earth and Planetary Science Letters 228(1–2): 109–123

Labrosse S (2003) Thermal and magnetic evolution of the Earth's core. Physics of the Earth and Planetary Interiors 140(1): 127–143

Labrosse S, Hernlund JW, Coltice N (2007) A crystallizing dense magma ocean at the base of the Earth's mantle. Nature 450: 866–869

Manga M (1996) Mixing of heterogeneities in the mantle: Effect of viscosity differences. Geophysical Research Letters 23(4): 403–406

McNamara AK, Zhong SJ (2004) Thermochemical structures within a spherical mantle: Superplumes or piles? Journal of Geophysical Research 109: B07402

Meibom A, Anderson DL (2003) The statistical upper mantle assemblage. Earth and Planetary Science Letters 217: 123–139

Monnereau M, Yuen D (2007) Topology of the postperovskite phase transition and mantle dynamics. Proceedings of the National Academy of Sciences 104:9156–9161, doi:10.1073/pnas.0608480104

Monnereau M, Yuen D (2007) Topology of the postperovskite phase transition and mantle dynamics. PNAS

Montelli R, Nolet G, Dahlen FA, Masters G, Engdahl ER, Hung S-H (2004) Finite-frequency tomography reveals a variety of plumes in the mantle. Science 303: 338–343

Nakagawa T, Tackley PJ (2004) Effects of a perovskite-post perovskite phase change near core-mantle boundary in compressible mantle convection. Geophysical Research Letters 31(L16611)

Nolet G, Karato S-I, Montelli R (2006) Plume fluxes from seismic tomography. Earth and Planetary Science Letters 248: 685–699

O'Connell RJ, Gable CW, Hager BH (1991) Toroidal-poloidal partitioning of lithospheric plate motion. In: Sabadini K, Lambeck K, Boschi E Glacial Isostasy, Sea Level, and Mantle Rheology, Kluwer Academic Publishers, Dordrecht, 535–551

O'Neill HSC, Palme H, Jackson I (1998) The Earth's mantle: Composition, structure and evolution. Cambridge University Press, Cambridge, UK

Oldham D, Davies HW (2004) Numerical investigation of layered convection in a three-dimensional shell with application to planetary mantles. Geochemistry Geophysics Geosystems 5(12): Q12C04

Presnall DC, Gudfinnsson GH, Walter MJ (2002) Generation of mid-ocean ridge basalts at pressures from 1 to 7 GPa. Geochimica et Cosmochimica Acta 66(12): 2073–2090

Ramage A, Walthan AJ (1992) Iterative solution techniques for finite element discretizations of fluid flow problems. Copper Mountain Conference on Iterative Methods, Copper Mountain, Colorado

Ranen MC, Jacobsen SB (2006) Barium isotopes in chondritic meteorites: implications for planetary reservoir models. Science 314(5800): 809–812
Regenauer-Lieb K, Kohl T (2003) Water solubility and diffusivity in olivine: its role in planetary tectonics. Mineralogical Magazine 67(4): 697–715
Ritsema J, van Heijst HJ (2000) Seismic imaging of structural heterogeneity in Earth's mantle: evidence for large-scale mantle flow. Science Progress 83(3): 243–259
Ritsema J, van Heijst HJ, Woodhouse JH (1999) Complex shear wave velocity structure imaged beneath Africa and Iceland. Science 286(5546): 1925–1928
Rüpke L, Phipps-Morgan J, Hort M, Connolly J, Ranero C (2003) Serpentine and the chemical evolution of the earth's mantle. Geophysical Research Abstracts 5: 09637 See: http://www.cosis.net/abstracts/EAE03/09637/EAE03-J-09637.pdf
Schubert G, Turcotte DL, Olson P (2001) Mantle convection in the Earth and Planets. Cambridge University Press, Cambridge
Smolarkiewicz PK (1984) A fully multidimensional positive definite advection transport algorithm with small implicit diffusion. Journal of Computational Physics 54(2): 325–362
Solomatov VS (2000) Fluid dynamics of a terrestrial magma ocean. In: Canup RM, Righter K Origin of the Earth and Moon, University of Arizona Press, Tucson, 323–338
Stacey FD (1992) Physics of the Earth. Brookfield Press, Brisbane
Stegman DR, Richards MA, Baumgardner JR (2002) Effects of depth-dependent viscosity and plate motions on maintaining a relatively uniform mid-ocean ridge basalt reservoir in whole mantle flow. Journal of Geophysical Research 107(B6): 10.1029/2001JB000192
Su W-J, Woodward RL, Dziewonski AM (1994) Degree 12 model of shear velocity heterogeneity in the mantle. Journal of Geophysical Research 99(B4): 6945–6980
Tackley PJ (1996) Effects of strongly variable viscosity on three-dimensional compressible convection in planetary mantles. Journal of Geophysical Research 101(B2): 3311–3332
Tackley PJ (2000) Mantle convection and plate tectonics: Toward an integrated physical and chemical theory. Science 288: 2002–2007
Tackley PJ (2002) Strong heterogeneity caused by deep mantle layering. Geochemistry Geophysics Geosystems 3(4)
Tackley PJ, Nakagawa T, Hernlund JW (2007) Post-Perovskite: The last mantle phase transition. Geophysical Monograph Series 174: 229–247
Taylor SR, McLennan SM (1995) The geochemical evolution of the continental crust. Reviews of Geophysics 33(2): 241–265
Tolstikhin I, Hofmann AW (2005) Early crust on top of the Earth's core. Physics of the Earth and Planetary Interiors 148: 109–130
Trampert J, Deschamps F, Resovsky J, Yuen D (2004) Probabilistic tomography maps chemical heterogeneities throughout the lower mantle. Science 306: 853–856

Trendall AF (2002) Precambrian sedimentary environments: A modern approach to depositional systems. In: Altermann W, Corcoran PL IAS spec. publ., 44, Blackwell, 33–66

Turcotte DL, Schubert G (2002) Geodynamics. Cambridge University Press, Cambridge

van der Hilst RD (2004) Changing views on Earth's deep mantle. Science 306(5697): 817–818

van der Hilst RD, Widiyantoro S, Engdahl ER (1997) Evidence for deep mantle circulation from global tomography. Nature 386: 578–584

van Keken P, Ballentine CJ, Porcelli D (2001) A dynamical investigation of the heat and helium imbalance. Earth and Planetary Science Letters 171: 533–547

van Keken P, Zhong SJ (1999) Mixing in a 3D spherical model of present-day mantle convection. Earth and Planetary Science Letters 171: 533–547

van Keken PE, Ballentine CJ (1998) Whole-mantle versus layered mantle convection and the role of a high-viscosity lower mantle in terrestrial volatile evolution. Earth and Planetary Science Letters 156(1–2): 19–32

van Keken PE, King SD, Schmeling H, Christensen UR, Neumeister D, Doin MP (1997) A comparison of methods for the modeling of thermochemical convection. Journal of Geophysical Research 102(B10): 22477–22495

van Thienen P (2003) Evolving dynamical regimes during secular cooling of terrestrial planets: insights and inferences from numerical models, Universiteit Utrecht, Utrecht

Walzer U, Hendel R (2008) Mantle convection and evolution of growing continents. Journal of Geophysical Research 113: B09405, doi: 10.1029/2007JB005459

Walzer U, Hendel RF (1999) A new convection-fractionation model for the evolution of the principal geochemical reservoirs of the Earth's mantle. Physics of the Earth and Planetary Interiors 112: 211–256

Walzer U, Hendel RF, Baumgardner JR (2003) Viscosity stratification and a 3D compressible spherical shell model of mantle evolution. High Performance Computing in Science and Engineering 2003: 419–428

Walzer U, Hendel RF, Baumgardner JR (2004a) The effects of a variation of the radial viscosity profile on mantle evolution. Tectonophysics 384: 55–90

Walzer U, Hendel RF, Baumgardner JR (2004b) Toward a thermochemical model of the evolution of the Earth's mantle. High Performance Computing in Science and Engineering 2004: 395–454

Watson EB, Thomas JB, Cherniak DJ (2007) 40Ar retention in the terrestrial planets. Nature 449: 299–304

Weiss D, Bassias Y, Gautier I, Mennesier J-P (1989) Dupal anomaly in existence 115 ma ago: Evidence from isotopic study of the Kerguelen plateau (South Indian Ocean). Geochimica et Cosmochimica Acta 53: 2125–2131

Xie S, Tackley PJ (2004) Evolution of Helium and Argon Isotopes in a convecting mantle. Physics of the Earth and Planetary Interiors 146(3–4): 417–439

Yang W-S (1997) Variable viscosity thermal convection at infinite Prandtl number in a thick spherical shell, University of Illinois, Urbana-Champaign

Zaranek SE, Parmentier EM (2004) Convective cooling of an initially stably stratified fluid with temperature-dependent viscosity – Implications for the role of solid-state convection in planetary evolution. Journal of Geophysical Research 109(B3): B03409

Zerr A, Boehler R (1993) Melting of (Mg,Fe)SiO3-perovskite to 625 kilobars: Indication of a high melting temperature in the lower mantle. Science 262: 553–555

Zerr A, Boehler R (1994) Constraints on the melting temperature of the lower mantle from high-pressure experiments on MgO and magnesiowüstite. Nature 371: 506–508

# VI. The ESyS_Particle: A New 3-D Discrete Element Model with Single Particle Rotation

Yucang Wang and Peter Mora

ESSCC, The University of Queensland, St. Lucia QLD 4072, Brisbane, Australia

**Abstract** In this paper, the Discrete Element Model (DEM) is reviewed, and the ESyS_Particle, our new version of DEM, is introduced. We particularly highlight some of the major physical concerns about DEMs and major differences between our model and most current DEMs. In the new model, single particle rotation is introduced and represented by a unit quaternion. For each 3-D particle, six degrees of freedom are employed: three for translational motion, and three for orientation. Six kinds of relative motions are permitted between two neighboring particles, and six interactions are transferred, i.e., radial, two shearing forces, twisting and two bending torques. The relative rotation between two particles is decomposed into two sequence-independent rotations such that all interactions due to the relative motions between interactive rigid bodies can be uniquely determined. This algorithm can give more accurate results because physical principles are obeyed. A theoretical analysis about how to choose the model parameters is presented. Several numerical tests have been carried out, the results indicate that most laboratory tests can be well reproduced using our model.

## 1 Introduction: A Review of the Discrete Element Method

The Discrete Element Method (DEM), pioneered by Cundall (Cundall et al., 1979), has been a powerful numerical tool in many scientific and engineering applications. The basic idea behind DEM is to treat the sample to be modeled as an assemblage of discrete particles interacting with one another. At each time step, the calculations performed in DEM alternate between integrating equations of motion for each particle, and applying the force-displacement law at each contact.

H. Xing, *Advances in Geocomputing*, Lecture Notes in Earth Sciences 119,
DOI 10.1007/978-3-540-85879-9_6

One of the major advantages of DEM is that highly complex systems can be modeled using basic methodologies without any assumptions on the constitutive behaviors of the materials and any predisposition about where and how cracks may occur and propagate. This advantage allows DEM to study with great simplicity many problems which are highly dynamic with large deformations and a large number of frequently changing contacts. These simulations include blast and impact of brittle materials at high strain rates (Donze et al., 1997; Magnier et al., 1998; Hentz et al., 2004a; 2004b), rock fractures (Donze et al., 1997; Young et al., 2000; Hazzard et al., 2000b; 2000c; Wang et al., 2000; place et al., 2001; 2002; Boutt et al., 2002; Chang et al., 2002; Matsuda et al., 2002; Hunt et al., 2003; Potyondy et al., 1996; 2004), soil mechanics and shear bands (Bardet et al., 1991; Anandarajah, 1994; Ng et al., 1994; Iwashita et al., 1998; 2000; Oda et al., 1998; 2000; Thornton et al., 2000; 2003; 2006; McDowell et al., 2002; Cheng et al., 2003; Tordesillas et al., 2004; Jiang et al., 2006), earthquake faults and gouges (Mora et al., 1993; 1994; 1998; 1999; 2000; 2002a; 2002b; Scott 1996; Morgan, 1999; Morgan 2004; Place et al., 1999; 2000; 2001; Wang et al., 2000; 2004; Hazzard et al., 2000a; 2002; 2004; Latham et al., 2006; Hu et al., 2004), frictional instability (Mora et al., 1994; Guo et al., 2004), granular or powder flow (Schwarz et al., 1998; Langston et al., 2004; Sheng et al., 2004), landslides and slope instability (Chang, 1992; Cleary et al., 1993; Campbell et al., 1995; Eberhardt et al., 2005) and mining and other industrial applications (Holst et al., 1999; Huang et al., 1999; Cleary, 2000; Cleary et al., 2002; Morrison et al., 2004; 2007; Prochazka, 2004).

In the past three decades, DEMs have been greatly developed and many different types of DEMs have appeared. Although the basic concepts are similar, these DEMs vary in the following aspects, all of which are the major physical concerns in DEM simulations and affect the simulated dynamics in various ways.

## 1.1 Dimensionality: 2-D or 3-D

Much of the early work on DEM was restricted to 2-D mainly because of computational restrictions and the geometric simplicity of the 2-D case. In some cases however 3-D simulations differ from their 2-D counterpart not only quantitatively but also qualitatively. Numerical and laboratory tests suggest that there are some significant differences regarding macroscopic friction and wing crack extensions between the 2-D and 3-D cases (Hazzard *et al.*, 2003; Dyskin *et al.*, 2003). As will be demonstrated later in this paper, it is more difficult to theoretically deal with finite particle rotation in 3-D than in 2-D. In recent years 3-D DEM simulations have become more common thanks to an increase in computer speeds.

## 1.2 Contact Laws: Linear or Non-Linear

The simplest contact law is the linear contact law, in which contact stiffnesses are constants and independent of the relative displacement at contacts. It is found that this law can not reproduce the realistic phenomena in certain circumstances (such as cycling loading (Dorby et al., 1992)). Non-linear contact laws, mainly based on Hertz and Mindlin theory (Mindlin et al., 1953; Johnson, 1987) provide more successful results in such circumstances (Morgan et al., 1999; Yang *et al.*, 2000; Guo *et al.*, 2004; Li *et al.*, 2005).

## 1.3 Particle Shapes: Disks/Spheres or Polygons/Polyhedrons

Most DEMs employed disks (in 2-D) and spheres (in 3-D) due to their simplicity. It has been noted that disks and spheres tend to roll or rotate more easily. For this reason more complex shapes such as ellipse (Ting et al., 1993), ellipsoid (Lin et al., 1997), polygons (Issa et al., 1992; Matuttis et al., 2000; D'Addetta et al., 2002; Feng et al., 2004b), polyhedra (Hart et al., 1988; Cundall, 1988; Ghaboussi et al., 1990) and superquadric (Mustoe, 1992; Hogue, 1998) are used. However polygons and polyhedrons introduce some disadvantages, these include difficulties in detecting contacts and calculating forces and torque in cases of edge–edge, edge–corner, corner–corner contact, and difficulties in extending from 2-D to 3-D and bonding particles together. Another alternate may be aggregates or clumps of disks and spheres bonded together or cluster (Jensen et al., 1999), agglomerates (Ning et al., 1997; Cheng et al., 2003; Lu et al., 2007). All of the disadvantages discussed above can be overcome without too much effort and crushing and fracture of aggregates can be easily modeled. The disadvantage is that this method requires a larger number of particles. However with developing computing power and paralleled techniques, this will not be a major problem in the future.

## 1.4 Single Particle Rotation: With or Without

In some DEMs, single particle rotations are prohibited. However, recent studies suggest that in some cases unrealistic results are generated if single particle rotation is ignored (Wang *et al.*, 2008a). The reason behind this is that the rotation of a particle provides the mechanisms for local transmission of shear forces and moments that are not present in non-rotational DEM simulations.

Particle rotation in 2-D can be represented easily using a scalar orientation and single angular velocity. In 3-D orientations require three independent parameters and **explicit** representations for such orientations are needed. By **explicit**, we mean that once the explicit variable is given, the orientation of a 3-D particle is uniquely determined. Candidates of such variants include matrix, quaternion, vector, Euler Angles.

In some existing DEMs, particle orientations are ***implicitly*** represented using three angular velocities around three orthogonal axes. As will be discussed in Sect. 6, this method has its shortcomings, because in 3-D one can not extract the orientation of a particle by simply integrating three angular velocities, limited by the physical principle. In the ESyS_Particle, unit quaternion is used as an explicit representation of particle orientation (Wang et al., 2006). To our knowledge, the few DEMs have employed quaternion (Johnson et al., 2007; Fleissner et al., 2007).

## 1.5 Algorithm for Integrating the Equations of Motion

Typical integration methods used in DEMs are the Molecular Dynamic algorithm, such as Velocity Verlet and leapfrog (Allen et al., 1987). If rotation is involved, integration of rotational equations should also be considered, including updating angular velocities and updating the orientation degree of freedom (Dullweber et al., 1997; Kol et al., 1997; Omelyan, 1998a; 1998b; Buss, 2000; Miller et al., 2002; Munjiza et al., 2003; Krysl et al., 2005; Wang, 2008b).

## 1.6 Bonded or Not Bonded

In the bonded (cohesive) DEMs, particles are bonded together so that tensile forces can be transmitted, but in the un-bonded (cohesionless) DEMs, only repulsive forces can be transmitted between particles. The former is often used to model wave propagation and fracture of intact materials such as rocks (Mora et al., 1993; 1994; 1998; Place et al., 1999; 2002; Chang et al., 2002; Hazzard et al., 2000a; 2000b; Potyondy et al., 1996; 2004; Donze et al., 1997; Hentz et al., 2004a; 2004b). While the latter, or "real discrete" model, is used to model motions of power and behaviors of particulate materials (Morgan et al., 1999; Sheng et al., 2004; Sitharam, 2000).

## 1.7 Interactions Between Particles: Complete or Simplified

Theoretically, for 2-D bonded models, three interactions (normal force, shear force and a bending moment) between two bonded particles should be permitted; whereas in 3-D case, six interactions should be permitted (i.e. normal force, two shear forces, two bending moments and a twisting moment, see detail in Sect. 3.1.1). Under certain circumstances it is a good approximation to use a simple force-displacement law and ignore certain interactions, such as models in which only normal forces exist (Mora et al., 1993; 1994; 1998; Toomey, et al., 2000), or models in which only normal and shear forces are transmitted (Scott, 1996; Chang et al., 2002; Hazzard et al., 2000a; 2000b). However studies from experiments and numerical simulations suggest that rolling resistance at contacts has a significant influence on the behaviors of granular media (Oda et al., 1982; 1998; 2000; Schlangen et al., 1997; Iwashita et al., 1998; 2000; Tordesillas et al., 2002; Kuhn et al., 2004). When only radial forces are transmitted between bonded particles, or rolling resistance is absent, the widely-observed laboratory tests of wing-crack extension can not reproduced (Wang et al., 2008a). Based on these analyses, DEM software, the ESyS_Particle, has been extended to include the complete set of interactions in 3-D (Wang et al., 2006; Wang, 2008b). We point out that the calculation of these interactions is important to the accuracy of simulations.

## 1.8 Criterion for Bond Breakage

For bonded DEMs, a criterion is required to judge the breakage of bonds. The widely used criterion is that a bond breaks either when the tensile strength or the shear strength is reached (either $f_r \geq F_{r0}$ or $|f_s| \geq F_{s0}$) (Potyondy et al., 1996; Donze et al., 1997; Magnier et al., 1998). A combined criterion, which takes into account the effect of normal force on shear failure, is used in the ESyS_Particle (Eq. (21)). A similar but parabolic-type criterion is employed by Pelenne et al. (Pelenne et al., 2004). Several different criteria are presented by Davie et al. (Davie et al., 2003). More work is required to compare the different types of criteria.

## 1.9 Frictional Forces

Frictional interactions play important roles in DEM simulations, particularly for granular flow. DEMs commonly employ a simple

Coulomb friction model in which a shear stiffness is introduced for touching particles that resists tangential movement between the particles until shear force exceeds a threshold (Matuttis et al., 2000; Hazzard et al., 2000c; Matsuda et al., 2002; Cheng et al., 2003; Hentz et al., 2004a; 2004b; Delenne et al., 2004; Potyondy et al., 2004).

In the earlier version of ESyS_Particle, a rate- and state- dependent fictional law (motivated by laboratory studies of frictional sliding) was implemented (Abe et al., 2002) but found to be too computationally expensive. It is also unclear whether this law is the basic friction mechanism at particle scale, or it is an emergent phenomenon which should be reproduced at macroscopic scale. In the current version, a simple stick-slip frictional algorithm is used which accurately captures the transition between static and sliding frictional states. Besides, rolling friction is considered by some modelers (Zhang et al., 1999; Zhou et al., 1999; Feng et al., 2004a).

## 1.10 Parameter Calibration

Values for the micro-physical elastic stiffness parameters in DEM simulations are typically chosen via empirical or trial-and-error methods (Chang et al., 2002; Hazzard et al., 2000a; 2000b; Boutt et al., 2002; Matsuda et al., 2002). Empirical calibration involves simulation of uni-axial or tri-axial tests on a particle assembly and measurement of the macroscopic elastic properties (Young's modulus and Poisson's ration) in a numerical analogue for the laboratory studies with real materials. Sometimes researchers outside DEM community have the wrong impression that DEM parameters are chosen arbitrarily and DEM does not generate the realistic macroscopic elasticity, as continuum models such as the Finite Element Method do. For regular lattices of equal-sized spheres it is possible to derive relations between the microscopic parameters (particle scale stiffnesses) and macroscopic elastic properties (Wang et al., 2008c). Choice of micro-physical fracture parameters is more difficult, requiring empirical calibration against laboratory results.

Other major challenges in DEMs include coupling between different physical processes such as heat, pore flow (Abe et al., 2000; Sakaguchi et al., 2000) and coupling between DEM and Finite Element Method (Munjiza et al., 1995; Owen et al., 2001).

In this paper, we will give a detailed description of the ESyS_Particle, our DEM. The paper is arranged as follows: the basic features of the model, and a detailed description, including equations and algorithm to integrate the equations (rotational) are presented in Sect. 2; Sect. 3

introduces three kinds of contacts and the ways to calculate the forces and torques. In Sect. 4 model parameters are discussed; Sect. 5 gives some of the simulations and Sect. 6 highlights the major differences between our model and the most existing DEMs, followed by conclusions in Sect. 7.

## 2 The Model, Equations and Numerical Algorithms to Integrate These Equations

### 2.1 A Brief Introduction to the ESyS_Particle

The ESyS_Particle, previously called Lattice Solid Model or LSMearth, is similar to the Discrete Element Model (Cundall et al., 1979) and Molecular Dynamics (MD, Allen et al., 1987; Rapaport, 1995) but involves a different computational approach (Place et al., 1999). It has been applied to the study of physical process such as rock fracture (Mora et al., 1993; Place et al., 2002), stick-slip friction behavior and friction (Mora et al., 1994; Place et al., 1999), granular dynamics (Mora et al., 1998; 1999), heat-flow paradox (Mora et al., 1998; 1999), localization phenomena (Place et al., 2000), Load-Unload Response Ratio theory (Mora et al., 2002b; Wang et al., 2004) and Critical Point systems (Mora et al., 2002a).

In the current ESyS_Particle model, particles can be disks (2-D) or spheres (3-D). Single particle rotations have been introduced recently and are represented by unit quaternion (Wang et al., 2006), therefore each particle brings three degrees of freedom in 2-D, and six in 3-D. There are three kinds of basic contacts when two particles touch: bonded, Solely normal repulsive and cohesionless frictional contact. In bonded contacts, the full set of interactions are involved, that is, all three interactions (normal, shearing forces and bending moment) are transmitted in 2-D and six (normal, shearing forces, bending and twisting moment) in 3-D are transmitted between each bonded particle pair. A new technique has been developed to decompose the relative rotations between two rigid bodies in such a way that the torques and forces caused by such relative rotations can be uniquely determined (Wang, 2008b). In the case of the solely normal repulsive contacts, only normal forces exist when two particle contact, and for cohesionless frictional contacts, frictional forces appear in tangential directions.

The source code is written in C++ with a Python script interface. Before running the code, the initial conditions, physics parameters, integration steps, types of particles (simple or rotational particle), types of loading walls, the contact properties (elastic, frictional, bonded contacts), artificial viscosity, ways of loading (force controlled or displacement controlled) and output fields are specified in the script.

Pre-processing includes a particle generation package, which can generate regular or random sized particles (initial position, radius and a unique identifier for each particle) and bond information (if needed). Aggregates (grains), gouges and faults can also be made. Post-processing includes Povray and VTK visualization packages, which can visualize the article and fields (velocity, displacements). The code is Parallelized using MPI, which significantly increases the computational capabilities.

## 2.2 Equations

Particle motion can be decomposed into two completely independent parts, translational motion of the center of mass and rotation about the center of mass. The former is governed by the Newtonian equation

$$\ddot{\boldsymbol{r}}(t) = \boldsymbol{f}(t)/M \tag{1}$$

where $\boldsymbol{r}(t)$, $\boldsymbol{f}(t)$ and $M$ are position of the particle, total forces acting on the particle and the particle mass respectively. A velocity Verlat scheme has been used to integrate the equation above (Allen et al., 1987; Mora et al., 1994; Place et al., 1999).

The particle rotation depends on the total applied torque and usually involves two coordinate frames, one is fixed in space, called space-fixed frame, in which Eq. (1) is applied. The other is attached to the principal axes of the rotation body, referred to as body-fixed frame. In the body-fixed frame, the dynamic equations are (Goldstein, 1980)

$$\begin{aligned}
\dot{\omega}_x^b &= \frac{\tau_x^b}{I_{xx}} + \left(\frac{I_{yy} - I_{zz}}{I_{xx}}\right)\omega_y^b\omega_z^b \\
\dot{\omega}_y^b &= \frac{\tau_y^b}{I_{yy}} + \left(\frac{I_{zz} - I_{xx}}{I_{yy}}\right)\omega_z^b\omega_x^b \\
\dot{\omega}_z^b &= \frac{\tau_z^b}{I_{zz}} + \left(\frac{I_{xx} - I_{yy}}{I_{zz}}\right)\omega_x^b\omega_y^b
\end{aligned} \tag{2}$$

where $\tau_x^b, \tau_y^b$ and $\tau_z^b$ are components of the total torque $\boldsymbol{\tau}^b$ expressed in body-fixed frame, $\omega_x^b$, $\omega_y^b$ and $\omega_z^b$ are components of the angular velocities $\boldsymbol{\omega}^b$ measured in the body-fixed frame, and $I_{xx}$, $I_{yy}$ and $I_{zz}$ are the three principle moments of inertia in body-fixed frame in which the inertia tensor is diagonal. For 3-D spheres, $I = I_{xx} = I_{yy} = I_{zz}$.

Due to the singularity caused by Euler angles, the unit quaternion $q = q_0 + q_1 i + q_2 j + q_3 k$ is usually used to describe the orientation in numerical simulations (Evans, 1977; Evans et al., 1977), where *i, j, k* satisfy $i^2 = j^2 = k^2 = ijk = -1$, $ij = -ji = k$, $jk = -kj = i, ki = -ik = j$ and $\sum_{i=0}^{3} q_i^2 = 1$.

The physical meaning of a quaternion is that it represents a one-step rotation around the vector $q_1\hat{i} + q_2\hat{j} + q_3\hat{k}$ with a rotation angle of $2\arccos(q_0)$ (Kuipers, 1998). It is a very convenient tool in that sequences of rotations can be represented as a quaternion product (Kuipers, 1998).

A quaternion for each particle satisfies the following equations of motion (Evans, 1977; Evans et al., 1977),

$$\dot{\boldsymbol{Q}} = \frac{1}{2}\boldsymbol{Q_0}(q)\boldsymbol{\Omega}, \tag{3}$$

where $\dot{\boldsymbol{Q}} = \begin{pmatrix} \dot{q}_0 \\ \dot{q}_1 \\ \dot{q}_2 \\ \dot{q}_3 \end{pmatrix}$, $\boldsymbol{Q_0}(q) = \begin{pmatrix} q_0 & -q_1 & -q_2 & -q_3 \\ q_1 & q_0 & -q_3 & q_2 \\ q_2 & q_3 & q_0 & -q_1 \\ q_3 & -q_2 & q_1 & q_0 \end{pmatrix}$, $\boldsymbol{\Omega} = \begin{pmatrix} 0 \\ \omega_z^b \\ \omega_z^b \\ \omega_z^b \end{pmatrix}$.

## 2.3 Algorithms to Integrate the Equations of Rotation

Now Eqs. (2) and (3) can be solved numerically and the numerical method we use is outlined below (Fincham, 1992).

In the numerical time integration scheme we obtain the quaternion $q(t+dt)$ at the next time step using:

$$\begin{aligned} q(t+dt) &= q(t) + dt\,\dot{q}(t) + \frac{dt^2}{2}\ddot{q}(t) + O(dt^3) \\ &= q(t) + dt\,\dot{q}(t+dt/2) + O(dt^3) \end{aligned} \tag{4}$$

Hence, the quaternion derivative at mid-step ($\dot{q}(t+dt/2)$) is required. Equation (3) indicates that $q(t+dt/2)$ and $\boldsymbol{\omega}^b(t+dt/2)$ are also required, where the former can be easily calculated using

$$q(t+dt/2) = q(t) + \dot{q}(t)dt/2 \tag{5}$$

where $\dot{q}(t)$ again is given by Eq. (3), and $\boldsymbol{\omega}^b(t)$ can be calculated using

$$\boldsymbol{\omega}^b(t) = \boldsymbol{\omega}^b(t-dt/2) + I^{-1}\boldsymbol{\tau}^b(t)dt/2 \tag{6}$$

and $\boldsymbol{\omega}^b(t+dt/2)$ can be obtained using

$$\boldsymbol{\omega}^b(t+dt/2) = \boldsymbol{\omega}^b(t-dt/2) + I^{-1}\boldsymbol{\tau}^b(t)dt \tag{7}$$

In this algorithm, only $\boldsymbol{\omega}^b(t-dt/2)$ and $q(t)$ need to be stored, whereas the other quantities, such as $q(t+dt/2)$, $\boldsymbol{\omega}^b(t)$, $\dot{q}(t)$, $\dot{q}(t+dt/2)$ etc. are treated as temporary and auxiliary values.

To avoid build-up errors, it is a common practice to renormalize quaternion at frequent intervals (usually done at every time-step). The whole algorithm proceeds as follows:

Step 1: calculate the force $f$ and torque $\boldsymbol{\tau}^b(t)$ at time t according to Eqs. (11), (12), (13), (14), (15), (16), (17) and (18).

Step 2: update $\boldsymbol{\omega}^b(t)$ using the stored $\boldsymbol{\omega}^b(t-dt/2)$ according to Eq. (6).

Step 3: obtain $\dot{q}(t)$ using Eq. (3).

Step 4: calculate $\boldsymbol{\omega}^b(t+dt/2)$ using the stored $\boldsymbol{\omega}^b(t-dt/2)$ according to Eq. (7).

Step 5: compute $q(t+dt/2)$ using Eq. (5).

Step 6: evaluate $\dot{q}(t+dt/2)$ using Eq. (3).

Step 7: calculate $q(t+dt)$ using Eq. (4).

Step 8: renormalize quaternion $q(t+dt)$.

# 3 Contact Laws, Particle Interactions and Calculation of Forces and Torques

When particles come into contact, three kinds of interactions can exist in the current ESyS_Particle model: bonded, solely normal repulsive and cohesionless frictional interaction. Bonded interaction permits tensile forces to be transmitted between particles and can be used to model behaviors of continuum or intact materials. The breakage of bonds provides an explicit mechanism for microscopic fracture. While the solely normal repulsive and cohesionless frictional interaction do not allow tensile forces to be transmitted between particles, and therefore suitable for modeling the motions and behaviors of "real discrete" or granular materials (such as powders, sands and earthquake gouges).

## 3.1 Bonded Interaction

For bonded interactions, the three important issues that need to be specified are types of interactions being transmitted between each particle pair, the algorithm to calculate the interactions between bonded particles due to the relative motion and the criterion for a bond to break.

### 3.1.1 The Bonded Model

In the ESyS_Particle, three parameters are used to represent the centre position of each 3-D particle, and three parameters to represent rotations about the center, the complete set of interactions requires six kinds of interactions being transmitted between each particle pair in the 3-D case, and three in the 2-D case.

Under small deformations, the relationship between interactions and relative displacements between two bonded particles can be written in the linear form (Fig. 1)

$$
\begin{aligned}
&\boldsymbol{f}_r = K_r \Delta \boldsymbol{r}, \\
&\boldsymbol{f}_{s1} = K_{s1} \Delta \boldsymbol{s}_1, \\
&\boldsymbol{f}_{s2} = K_{s2} \Delta \boldsymbol{s}_2, \\
&\boldsymbol{\tau}_t = K_t \Delta \boldsymbol{\alpha}_t, \\
&\boldsymbol{\tau}_{b1} = K_{b1} \Delta \boldsymbol{\alpha}_{b1}, \\
&\boldsymbol{\tau}_{b2} = K_{b2} \Delta \boldsymbol{\alpha}_{b2},
\end{aligned}
\tag{8}
$$

where $\Delta \boldsymbol{r}$, $\Delta \boldsymbol{s}_1(\Delta \boldsymbol{s}_2)$ are the relative displacements in normal and tangent directions. $\Delta \boldsymbol{\alpha}_t$ and $\Delta \boldsymbol{\alpha}_{b1}(\Delta \boldsymbol{\alpha}_{b2})$ are the relative angular displacements caused by twisting and bending. $\boldsymbol{f}_r, \boldsymbol{f}_{s1}, \boldsymbol{f}_{s2}, \boldsymbol{\tau}_t, \boldsymbol{\tau}_{b1}$ and $\boldsymbol{\tau}_{b2}$ are forces and torques, $K_r$, $K_{s1}$, $K_{s2}$, $K_t$, $K_{b1}$ and $K_{b2}$ are relevant stiffnesses. If the bond is identical in every direction, $K_s = K_{s1} = K_{s2}$ and $K_b = K_{b1} = K_{b2}$.

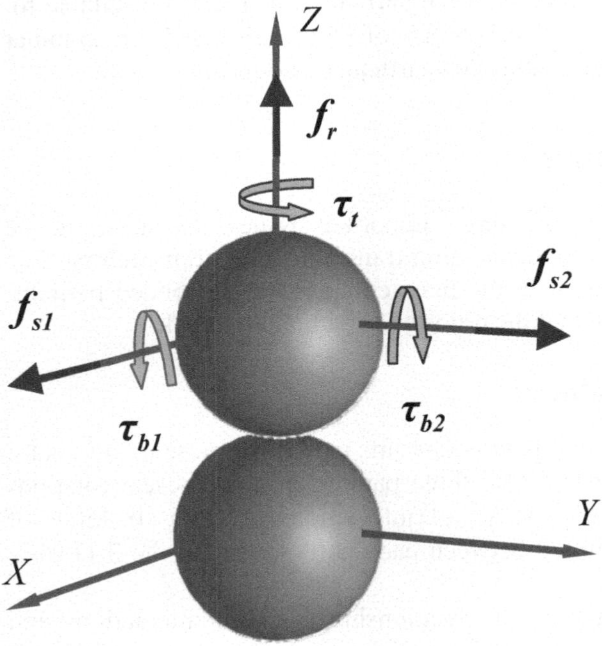


**Fig. 1** Six kinds of interactions between bonded particles. $f_r$ is normal force, $f_{s1}$ and $f_{s2}$ are shear forces , $\tau_t$ is twisting torque, $\tau_{b1}$ and $\tau_{b2}$ are bending torque

### 3.1.2 Calculation of Interactions due to Relative Motion

The calculation of the interactions between bonded particles due to relative motion requires decomposition of the relative motions. Suppose in the space-fixed frame, the initial positions of particle 1 and 2 at time $t = 0$ are $\boldsymbol{r}_{10} = x_{10}\boldsymbol{i} + y_{10}\boldsymbol{j} + z_{10}\boldsymbol{k}$ and $\boldsymbol{r}_{20} = x_{20}\boldsymbol{i} + y_{20}\boldsymbol{j} + z_{20}\boldsymbol{k}$, the initial orientations of particle 1 and 2 are $p^0 = 1 + 0i + 0j + 0k$ and $q^0 = 1 + 0i + 0j + 0k$. At time *t*, the current positions are $\boldsymbol{r}_1 = x_1\boldsymbol{i} + y_1\boldsymbol{j} + z_1\boldsymbol{k}$ and $\boldsymbol{r}_2 = x_2\boldsymbol{i} + y_2\boldsymbol{j} + z_2\boldsymbol{k}$, the current orientations are $p = p_0 + p_1 i + p_2 j + p_3 k$ and $q = q_0 + q_1 i + q_2 j + q_3 k$. Let $X_1Y_1Z_1$ and $X_2Y_2Z_2$ represent the body-fixed frame of particle 1 and 2, the relative translational displacements of particle 1 with respect to $X_2Y_2Z_2$ is (Fig. 2)

$$\Delta\boldsymbol{r} = \boldsymbol{r}_b - \boldsymbol{r}_0 = q^{-1}\boldsymbol{r}q - \boldsymbol{r}_0, \tag{9}$$

where $\boldsymbol{r}_0 = \boldsymbol{r}_{10} - \boldsymbol{r}_{20}$ is initial relative position of particle 1 with respect to $X_2Y_2Z_2$, and

$$\boldsymbol{r}_b = q^{-1}\boldsymbol{r}q = x_{12}\boldsymbol{i} + y_{12}\boldsymbol{j} + z_{12}\boldsymbol{k} \tag{10}$$

is the current position of particle 1 with respect to $X_2Y_2Z_2$, in which the possible rotation of particle 2 has been taken into account, and $\boldsymbol{r} = \boldsymbol{r}_1 - \boldsymbol{r}_2$.

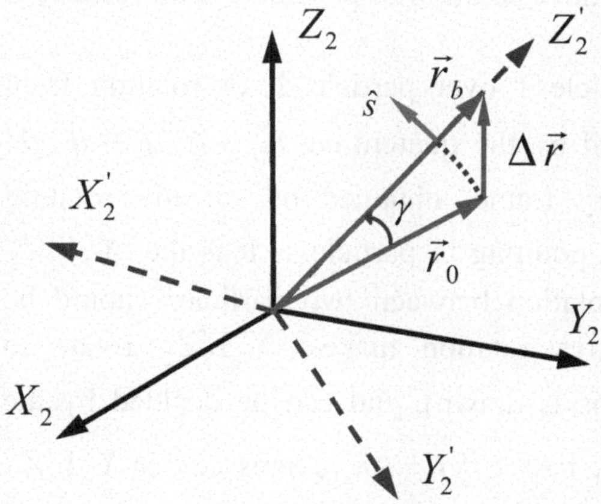


**Fig. 2** The relative translational displacements of particle 1 relative to particle 2 with respect to $X_2Y_2Z_2$ co-ordinates is $\Delta\boldsymbol{r}$. $\boldsymbol{r}_0$ and $\boldsymbol{r}_b$ are initial and current relative position of particle 1 in $X_2Y_2Z_2$.

In case of $|\Delta\boldsymbol{r}| << |\boldsymbol{r}_b|$ and $|\Delta\boldsymbol{r}| << |\boldsymbol{r}_0|$, we have normal force

$$\boldsymbol{f}_r = K_r(|\boldsymbol{r}_b| - |\boldsymbol{r}_0|)\boldsymbol{r}_b / |\boldsymbol{r}_b| \tag{11}$$

Shearing force has two contributors

$$\boldsymbol{f}_s = \boldsymbol{f}_s^t + \boldsymbol{f}_s^r \quad (12)$$

where $\boldsymbol{f}_s^t$, caused by translational displacement relative to $X_2Y_2Z_2$ without relative rotation, can be expressed as

$$\boldsymbol{f}_s^t = K_s |\boldsymbol{r}_b| \gamma \boldsymbol{s} \quad (13)$$

where $\cos\gamma = \frac{\boldsymbol{r}_0 \cdot \boldsymbol{r}_b}{|\boldsymbol{r}_0||\boldsymbol{r}_b|}$ and unit vector $\boldsymbol{s} = \frac{\boldsymbol{r}_b \times (\boldsymbol{r}_b \times \boldsymbol{r}_0)}{|\boldsymbol{r}_b \times (\boldsymbol{r}_b \times \boldsymbol{r}_0)|}$.

Since $\boldsymbol{f}_s^t$ is exerted on surface, rather than the center of mass, it generates a torque

$$\boldsymbol{\tau}_s^t = \frac{1}{2}|\boldsymbol{r}_b||\boldsymbol{f}_s^t|\boldsymbol{t} \quad (14)$$

where unit vector $\boldsymbol{t} = \frac{\boldsymbol{r}_0 \times \boldsymbol{r}_b}{|\boldsymbol{r}_0 \times \boldsymbol{r}_b|}$. Note that Eqs. (11), (13) and (14) are evaluated in the second particle's body-fixed frame $X_2Y_2Z_2$.

$\boldsymbol{f}_s^r$ in Eq. (12), caused by relative rotation of particle 1 with respect to $X_2Y_2Z_2$, is given later.

The relative rotation of particle 1 over particle 2, or rotation from $X_2Y_2Z_2$ to $X_1Y_1Z_1$, is represented by the quaternion $r_{21}^o = q^{-1}p = q^* p$. Suppose $X_2'Y_2'Z_2'$ is an auxiliary frame, obtained by directly rotating $X_2Y_2Z_2$ such that its $Z_2'$-axis is pointing to particle 1. It is the $X_2'Y_2'Z_2'$ system in which the relative rotation between two particles should be evaluated (Fig. 1). Such relative rotation makes $X_2'Y_2'Z_2'$ rotate to $X_1'Y_1'Z_1'$ (Fig. 3, only the $Z_1'$-axis is drawn), and can be decided by the quaternion $r_{21} = m^{-1} r_{21}^o m = r_0 + r_1 i + r_2 j + r_3 k$ (expressed in $X_2'Y_2'Z_2'$ system), where the quaternion $m$ specifies the rotation from $X_2Y_2Z_2$ to $X_2'Y_2'Z_2'$, and is given by

$$m_0 = \frac{\sqrt{2}}{2}\sqrt{\frac{\sqrt{x_{12}^2 + y_{12}^2 + z_{12}^2} + z_{12}}{\sqrt{x_{12}^2 + y_{12}^2 + z_{12}^2}}},$$

$$m_1 = -\frac{\sqrt{2}}{2}\sqrt{\frac{\sqrt{x_{12}^2 + y_{12}^2 + z_{12}^2} - z_{12}}{\sqrt{x_{12}^2 + y_{12}^2 + z_{12}^2}}}\frac{y_{12}}{\sqrt{x_{12}^2 + y_{12}^2}},$$

$$m_2 = \frac{\sqrt{2}}{2}\sqrt{\frac{\sqrt{x_{12}^2 + y_{12}^2 + z_{12}^2} - z_{12}}{\sqrt{x_{12}^2 + y_{12}^2 + z_{12}^2}}}\frac{x_{12}}{\sqrt{x_{12}^2 + y_{12}^2}},$$

$$m_3 = 0 \qquad (15)$$

By using quaternion algebra, it has been proved that (Wang et al., 2008a) an arbitrary rotation between two rigid bodies or two coordinate systems cannot be decomposed into three mutually independent rotations around three orthogonal axes. However it can be decomposed into two rotations, one pure axial rotation of angle $\psi$ around its $Z_2'$-axis, and one rotation of $Z_2'$-axis over $\theta$ on certain plane controlled by another parameter $\varphi$. Figure 3 describes such a decomposition. These two rotations, corresponding to the relative twisting and bending between two bodies in our model, are sequence-independent. Such a two-step rotation is controlled by three independent parameters $\psi$, $\theta$ and $\varphi$, which can be decided as follows,

$$cos\frac{\psi}{2} = \frac{r_0}{\sqrt{r_0^2 + r_3^2}},$$

$$sin\frac{\psi}{2} = \frac{r_3}{\sqrt{r_0^2 + r_3^2}},$$

$$cos\,\theta = r_0^2 - r_1^2 - r_2^2 + r_3^2,$$

$$cos\,\varphi = \frac{r_1 r_3 + r_0 r_2}{\sqrt{(r_0^2 + r_3^2)(r_1^2 + r_2^2)}},$$

$$sin\,\varphi = \frac{r_2 r_3 - r_0 r_1}{\sqrt{(r_0^2 + r_3^2)(r_1^2 + r_2^2)}}. \qquad (16)$$

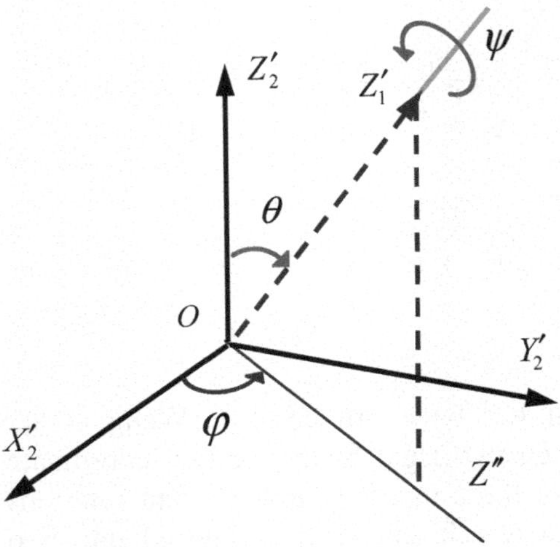


**Fig. 3** An arbitrary rotation between two rigid bodies or two coordinate systems can be decomposed into two-step rotations, one pure axial rotation of angle $\psi$ around its $Z_2'$-axis, and one rotation of $Z_2'$-axis over $\theta$ on certain plane controlled by another parameter $\varphi$

After applying this decomposition, the bending, twisting torque and shear forces exerted on particle 2 are (expressed in $X_2'Y_2'Z_2'$ co-ordinates)

$$\begin{aligned}
\boldsymbol{\tau}_b' &= K_b\theta(-sin\varphi\,\boldsymbol{i}+cos\varphi\,\boldsymbol{j})\\
\boldsymbol{\tau}_t' &= K_t\psi\,\boldsymbol{k}\\
\boldsymbol{f}_s' &= -K_s\frac{|\boldsymbol{r}_b|\theta}{2}(cos\,\varphi\,\boldsymbol{i}+sin\varphi\,\boldsymbol{j})\\
\boldsymbol{\tau}_s' &= K_s\frac{|\boldsymbol{r}_b|^2\theta}{4}(sin\varphi\,\boldsymbol{i}-cos\varphi\,\boldsymbol{j})
\end{aligned} \tag{17}$$

$\boldsymbol{\tau}_s'$ is the torque generated by $\boldsymbol{f}_s'$. It should be pointed out that Eqs. (14) and (17) are based on assumption of two equal-sized spheres. In case of different sized spheres, torques should be changed accordingly. For example, the torque in Eq. (14) should be replaced by $\boldsymbol{\tau}_{s1}^t = |\boldsymbol{r}_b||\boldsymbol{f}_s^t|\,\hat{\boldsymbol{t}}\,R_1/(R_1+R_2)$ and $\boldsymbol{\tau}_{s2}^t = |\boldsymbol{r}_b||\boldsymbol{f}_s^t|\,\hat{\boldsymbol{t}}\,R_2/(R_1+R_2)$, where $R_1$ and $R_2$ are radii of two particles.

Torques and forces can easily be transformed back to $X_2Y_2Z_2$ frame.

$$\boldsymbol{\tau}_b^r = m\boldsymbol{\tau}_b' m^{-1}, \quad \boldsymbol{\tau}_t^r = m\boldsymbol{\tau}_t' m^{-1}$$

$$\boldsymbol{f}_s^r = m\boldsymbol{f}_s' m^{-1}, \quad \boldsymbol{\tau}_s^r = m\boldsymbol{\tau}_s' m^{-1}. \tag{18}$$

The total forces and torques of each particle are expressed as sums of contributions from all particles bonded to it

$$\boldsymbol{f} = \sum A^T(\boldsymbol{f}_r + \boldsymbol{f}_s^t + \boldsymbol{f}_s^r) \tag{19}$$

$$\boldsymbol{\tau}^b = \sum(\boldsymbol{\tau}_s^t + \boldsymbol{\tau}_s^r + \boldsymbol{\tau}_b^r + \boldsymbol{\tau}_t^r) \tag{20}$$

where

$$\boldsymbol{A} = \begin{pmatrix} q_0^2 + q_1^2 - q_2^2 - q_3^2 & 2(q_1q_2 + q_0q_3) & 2(q_1q_3 - q_0q_2) \\ 2(q_1q_2 - q_0q_3) & q_0^2 - q_1^2 + q_2^2 - q_3^2 & 2(q_2q_3 + q_0q_1) \\ 2(q_1q_3 + q_0q_2) & 2(q_2q_3 - q_0q_1) & q_0^2 - q_1^2 - q_2^2 + q_3^2 \end{pmatrix}$$

is the rotation matrix, representing the rotation specified by quaternion $q = q_0 + q_1 i + q_2 j + q_3 k$.

Since $\boldsymbol{f}$ in Eq. (1) is expressed in space-fixed frame, but $\boldsymbol{f}_r$, $\boldsymbol{f}_s^t$ and $\boldsymbol{f}_s^r$ in Eqs. (11), (13) and (18) are evaluated in the body-fixed frame, a second transfer is required in Eq. 19. Once forces and torques are gained, Eqs. (1, 2 and 3) can be integrated for each particle.

To test our algorithm described above, we performed simulations using a simple model which consists of seven bonded cubes. Slow constant velocities or angular velocities are exerted on the first and the last cube to make the whole chain deform (i.e. stretching, twisting, bending or combination of deformations). We then compare the relative errors incurred using the incremental method (which is used by the most DEMs) and using our algorithm for different time step sizes *dt* (Wang, 2008b). The relative error is defined as (kinetic energy + potential energy – external work)/(external work). We found that the errors in the incremental method increased much faster with increasing time step than those in our algorithm, indicating that our algorithm is more accurate and stable over a greater range of time step sizes (Fig. 4).

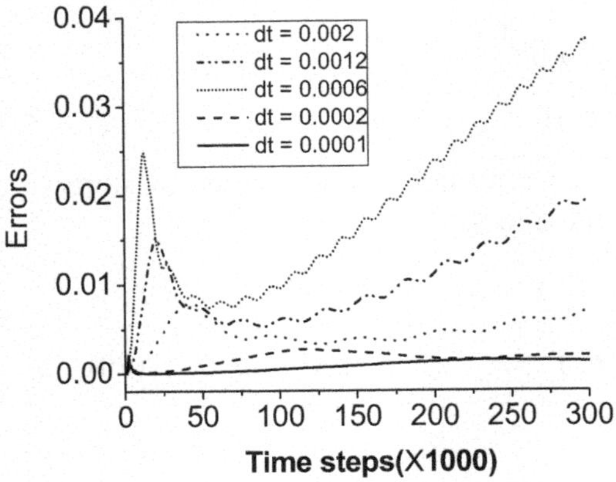


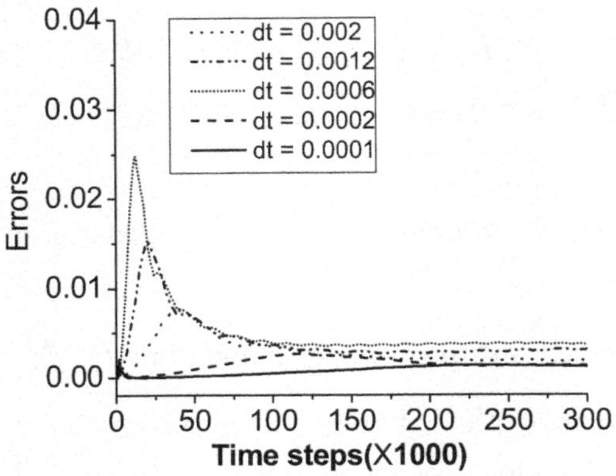


**Fig. 4** Evolution of Relative errors with time in the incremental method (*above*) and in our algorithm (*down*) for different integration time steps dt (from 0.0001 to 0.002). The errors increase much faster in the incremental method than that in our algorithm with increasing integration time step dt

### *3.1.3 Criterion for Bond Breakage*

In the ESyS_Particle model, a bond breaks under either of the following conditions:

If the pure extensional force exceeds the threshold $f_r \geq F_{r0}$ (but it does not break under pure compression)

If the pure shear force $|f_s| \geq F_{s0}$

If the pure twisting torque $|\tau_t| \geq \Gamma_{t0}$

If the pure bending torque $|\tau_b| \geq \Gamma_{b0}$.

When all the interactions exist at the same time, the following empirical criterion is used to judge whether or not a bond is going to break

$$\frac{f_r}{F_{r0}} + \frac{|f_s|}{F_{s0}} + \frac{|\tau_t|}{\Gamma_{t0}} + \frac{|\tau_b|}{\Gamma_{b0}} \geq 1 , \tag{21}$$

where $\boldsymbol{f}_s = \boldsymbol{f}_{s1} + \boldsymbol{f}_{s2}, \boldsymbol{\tau}_b = \boldsymbol{\tau}_{b1} + \boldsymbol{\tau}_{b2}$ (in 3-D). In the 2-D case the third term in Eq. (21) drops. We set $f_r$ positive under extension and negative under compression such that it is more difficult for a bond to break under compression than under extension, therefore the effects of normal force on breakage of the bond has been taken into account.

## 3.2 Solely Normal Repulsive Interaction

When two particles contact elastically, only the normal force $\boldsymbol{f}_r$ exists, and Eq. (11) holds only when $d < R_1 + R_2$, here $d$ is the distance between the two particles, $R_1$ and $R_2$ are radii of two particles.

## 3.3 Cohesionless Frictional Interaction

In the case of the cohesionless frictional interaction, forces are transmitted in both normal and tangential directions when $d < R_1 + R_2$. The normal force $\boldsymbol{f}_r$ is dealt exactly the same way it is dealt in case of the solely normal repulsive interaction. In the current ESyS_Particle model, frictional forces are updated in an incremental fashion, similar to the most DEMs. The reason we do not use a Finite Deformation Scheme (such as bonded interactions) is that particles may be frequently in and out of contact due to the change of normal forces. It is not convenient to store the positions and orientations of the particle pairs each time they contact.

Stick-slip style frictional forces are employed in tangential directions. At time $t$, if it is in a stick phase ($|\boldsymbol{f}_s| < |\boldsymbol{f}_r| \mu_s$, here $\mu_s$ is the static frictional coefficient), the frictional force at time $t+dt$ can be calculated as follows

$$\boldsymbol{f}_s(t+dt) = \boldsymbol{f}_s(t) + \Delta \boldsymbol{f}_s , \tag{22}$$

where the increment $\Delta \boldsymbol{f}_s = \Delta \boldsymbol{f}_{s1} + \Delta \boldsymbol{f}_{s2}$.

The first term, $\Delta \boldsymbol{f}_{s1} = K_s \Delta \boldsymbol{u}_s$, is the contribution from incremental shear displacements $\Delta \boldsymbol{u}_s = \boldsymbol{v}_s dt$, where

$$\boldsymbol{v}_s = \boldsymbol{\omega}_2 \times \boldsymbol{r}_{p2} - \boldsymbol{\omega}_1 \times \boldsymbol{r}_{p1} + \boldsymbol{v}_2 - \boldsymbol{v}_1 - ((\boldsymbol{v}_2 - \boldsymbol{v}_1) \cdot \boldsymbol{n})\boldsymbol{n}$$

is the relative transverse velocity due to relative motion at contact point (Fig. 5), where $\boldsymbol{v}_1$ and $\boldsymbol{v}_2$ are the velocity vectors of the two particles. The second term has two contributors

$$\Delta \boldsymbol{f}_{s2} = dt \boldsymbol{\omega}_{s1} \times \boldsymbol{f}_s + dt \boldsymbol{\omega}_{s2} \times \boldsymbol{f}_s, \tag{23}$$

where

$$\boldsymbol{\omega}_{s1} = (\boldsymbol{r}_2 - \boldsymbol{r}_1) \times (\boldsymbol{v}_2 - \boldsymbol{v}_1) / (\boldsymbol{r}_2 - \boldsymbol{r}_1)^2 \tag{24}$$

and

$$\boldsymbol{\omega}_{s2} = (((\boldsymbol{\omega}_1 + \boldsymbol{\omega}_2) \cdot \boldsymbol{n})\boldsymbol{n})/2 \tag{25}$$

are the angular velocity due to rotation of the line between two center and the angular velocity due to spin of the line between two centers respectively. Here $\boldsymbol{r}_1$ and $\boldsymbol{r}_2$ are the position vectors of the two particles, and $\boldsymbol{\omega}_1$ and $\boldsymbol{\omega}_2$ are the angular velocities.

Since the shear force $\boldsymbol{f}_s$ is stored in the global coordinate system, the rotation of the contact plane caused by movements of two particles as a rigid body change the direction of the stored shear forces. Therefore the two terms specified in Eqs. (24) and (25) have to be taken into account in the 3-D case. It should be pointed out that in the most literatures, the second term in Eq. (23) is missing (Hart et al., 1988; Monterio-Azevedo et al., 2005). In the 2-D case that term drops automatically, but in 3-D this term can not be ignored theoretically.

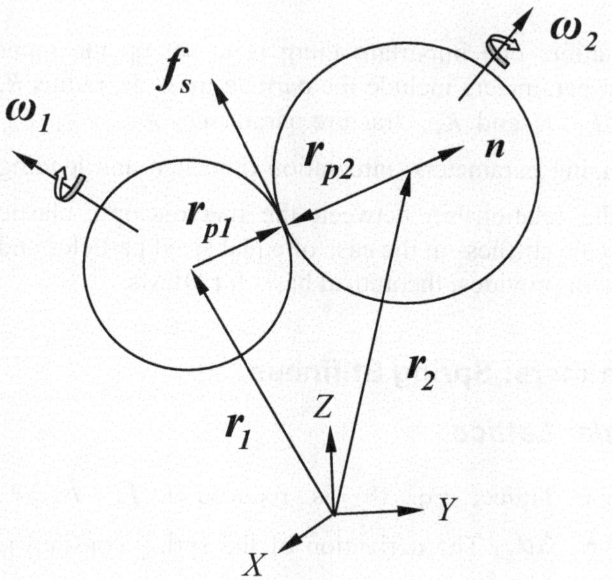


**Fig. 5** Incremental method to update shear forces

At time *t*, if it is in slip phase, $\boldsymbol{f}_s = -\mu_d |\boldsymbol{f}_r| \boldsymbol{v}_s / |\boldsymbol{v}_s|$, here $\mu_d$ is the sliding (or dynamic) frictional coefficient. Equation (22) still holds, but the increment only includes the contribution from the rotation of the contact plane caused by movements of two particles as a rigid body, or $\Delta \boldsymbol{f}_s = \Delta \boldsymbol{f}_{s2}$, here $\Delta \boldsymbol{f}_{s2}$ is still calculated using Eqs. (23, 24 and 25).

The transition between slip and stick is captured by the following criteria:

stick to slip: when $|\boldsymbol{f}_s| \geq \mu_s |\boldsymbol{f}_r|$,

slip to stick: when $|\boldsymbol{v}_s| = 0$.

Generally we choose $\mu_d < \mu_s$ such that unstable slip can be modeled. If $\mu_s = \mu_d$, it is similar to the Coulomb friction implementation.

## 4 Parameter Calibration

When starting a simulation, one important thing is to set up the input parameters. DEM input parameters include the particle mass *M*, radius *R*, rigidity parameters $K_r$, $K_s$, $K_b$ and $K_t$, fracture parameters $F_{r0}$, $F_{s0}$, $\Gamma_{t0}$ and $\Gamma_{b0}$, artificial damping parameters, integration time step and loading rate etc. We derive the relationship between the macro-scopic elastic constants and particle scale stiffness in the case of equal sized particles and regular lattices, which will provide a theoretical basis for DEMs.

### 4.1 Elastic Parameters: Spring Stiffness

#### 4.1.1 2-D Triangular Lattice

For the 2-D triangular lattice, Eq. (8) is reduced to $\boldsymbol{f}_r = K_r \Delta \boldsymbol{r}$, $\boldsymbol{f}_s = K_s \Delta \boldsymbol{s}$, and $\boldsymbol{\tau}_b = K_b \Delta \boldsymbol{\alpha}_b$. The derivation of the spring constant is based on a comparison between the discrete lattice model and the continuum model (Bathurst et al., 1988; Griffiths et al., 2001; Ostoja-Starzewski, 2002). First we let the displacement of every lattice point equal that of a corresponding point in the continuum model, then we choose a unit cell which is a periodically repeating part of the network in both models, lastly we let the energy stored in the unit cell of lattice equal the strain energy stored in the unit cell of the continuum model. If the model is arranged in Fig. 6 and the strain is uniform (no rotation components), we have (Wang et al., 2008c),

$$K_r = \frac{\sqrt{3}E}{3(1-\nu)},$$

$$K_s = \frac{1-3\nu}{1+\nu} K_r . \qquad (26)$$

The result (Eq. (26)) is invariant with respect to the orientional angle $\alpha$ (Fig. 6), in other words, the model arranged in Fig. 6 yields macroscopic isotropic elasticity.

To determine the bending stiffness, we consider the special lattice orientation $\alpha = 0$, in which both the continuum model and hexagonal arrangement are subject to the stress field: $\sigma_{11} = Hy$, $\sigma_{22} = 0$ and $\sigma_{12} = 0$, where H is a small constant. This stress field corresponds to macroscopic bending (in this case, rotation is involved). Similarly, by comparing the strain energy density in the continuum model with the equivalent energy density in discrete grids, we get (Wang et al., 2008c)

$$K_b = \frac{\sqrt{3}(1+\nu)(1-2\nu)ER^2}{36(1-\nu)}, \tag{27}$$

where $R$ is the radius of the particles.

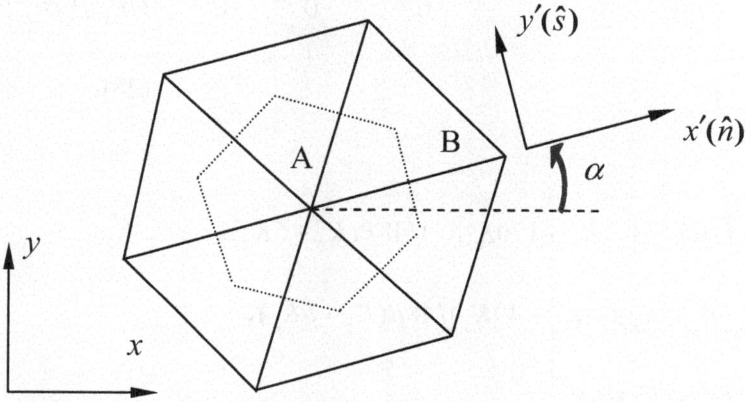


**Fig. 6** 2D close-packed hexagonal lattice and unit cell (the area enclosed by *dashed lines*)

## 4.1.2 3-D Lattices: HCP and FCC

In the 3-D case, the densest packing of equal-sized spheres are hexagonal closed packing (HCP) lattice (Fig. 7) and face-centered cubic (FCC) lattice (Fig. 8). However, both HCP and FCC structures are intrinsically anisotropic, or direction variant. When particles are arranged in HCP lattice (Fig. 7), we have

$$\begin{pmatrix} \sigma_{11} \\ \sigma_{22} \\ \sigma_{33} \\ \sigma_{23} \\ \sigma_{13} \\ \sigma_{12} \end{pmatrix} = \begin{pmatrix} C_{11} & C_{12} & C_{13} & 0 & 0 & 0 \\ C_{12} & C_{11} & C_{13} & 0 & 0 & 0 \\ C_{13} & C_{13} & C_{33} & 0 & 0 & 0 \\ 0 & 0 & 0 & C_{44} & 0 & 0 \\ 0 & 0 & 0 & 0 & C_{44} & 0 \\ 0 & 0 & 0 & 0 & 0 & C_{66} \end{pmatrix} \begin{pmatrix} \varepsilon_{11} \\ \varepsilon_{22} \\ \varepsilon_{33} \\ 2\varepsilon_{23} \\ 2\varepsilon_{13} \\ 2\varepsilon_{12} \end{pmatrix} \text{ and}$$

$$\begin{pmatrix} \varepsilon_{11} \\ \varepsilon_{22} \\ \varepsilon_{33} \\ 2\varepsilon_{23} \\ 2\varepsilon_{13} \\ 2\varepsilon_{12} \end{pmatrix} = \begin{pmatrix} 1/E_x & -\nu_{yx}/E_y & -\nu_{zx}/E_z & 0 & 0 & 0 \\ -\nu_{xy}/E_x & 1/E_y & -\nu_{zy}/E_z & 0 & 0 & 0 \\ -\nu_{xz}/E_x & -\nu_{yz}/E_y & 1/E_z & 0 & 0 & 0 \\ 0 & 0 & 0 & 1/\mu_{yz} & 0 & 0 \\ 0 & 0 & 0 & 0 & 1/\mu_{xz} & 0 \\ 0 & 0 & 0 & 0 & 0 & 1/\mu_{xy} \end{pmatrix} \begin{pmatrix} \sigma_{11} \\ \sigma_{22} \\ \sigma_{33} \\ \sigma_{23} \\ \sigma_{13} \\ \sigma_{12} \end{pmatrix} \tag{28}$$

where

$$C_{11} = \sqrt{2}\,(29K_r^2 + 89K_s^2 + 170K_rK_s)/48R(K_r + 5K_s),$$

$$C_{12} = \sqrt{2}\,(K_r - K_s)(11K_r + 49K_s)/48R(K_r + 5K_s),$$

$$C_{13} = \sqrt{2}\,(K_r - K_s)/6R,$$

$$C_{33} = \sqrt{2}\,(2K_r + K_s)/3R,$$

$$C_{44} = \sqrt{2}\,(K_r + 2K_s)/6R,$$

$$C_{66} = \sqrt{2}(3K_r^2 + 23K_s^2 + 22K_rK_s)/16R(K_r + 5K_s) = (C_{11} - C_{12})/2.$$

This is transversely isotropic, which is a special case of an orthotropic solid, with $x_1x_2$ plane of isotropy and generally 5 independent constants (Daniel et al., 1994). Here the Young's modulus, Poisson's ratio and shear modulus are respectively

$$E_x = E_y = \frac{3\sqrt{2}K_r(K_r + K_s)(3K_r^2 + 23K_s^2 + 22K_rK_s)}{R(18K_r^3 + 119K_r^2K_s + 128K_rK_s^2 + 23K_s^3)},$$

$$E_z = \frac{3\sqrt{2}K_r(K_r + K_s)}{R(5K_r + K_s)},$$

$$\nu_{xy} = \nu_{yx} = \frac{(K_r - K_s)(6K_r^2 + 23K_s^2 + 31K_r K_s)}{18K_r^3 + 119K_r^2 K_s + 128K_r K_s^2 + 23K_s^3},$$

$$\nu_{zx} = \nu_{zy} = \frac{K_r - K_s}{5K_r + K_s},$$

$$\nu_{xz} = \nu_{yz} = \frac{(K_r - K_s)(3K_r^2 + 23K_s^2 + 22K_r K_s)}{18K_r^3 + 119K_r^2 K_s + 128K_r K_s^2 + 23K_s^3},$$

$$\mu_{xy} = \frac{\sqrt{2}(3K_r^2 + 23K_s^2 + 22K_r K_s)}{16R(K_r + 5K_s)},$$

$$\mu_{xz} = \mu_{yz} = \frac{\sqrt{2}(K_r + 2K_s)}{6R}. \tag{29}$$

In the case of FCC (see Fig. 8), we have

$$\begin{pmatrix} \sigma_{11} \\ \sigma_{22} \\ \sigma_{33} \\ \sigma_{23} \\ \sigma_{13} \\ \sigma_{12} \end{pmatrix} = \begin{pmatrix} C_{11} & C_{12} & C_{12} & 0 & 0 & 0 \\ C_{12} & C_{11} & C_{12} & 0 & 0 & 0 \\ C_{12} & C_{12} & C_{11} & 0 & 0 & 0 \\ 0 & 0 & 0 & C_{44} & 0 & 0 \\ 0 & 0 & 0 & 0 & C_{44} & 0 \\ 0 & 0 & 0 & 0 & 0 & C_{44} \end{pmatrix} \begin{pmatrix} e_{11} \\ e_{22} \\ e_{33} \\ 2e_{23} \\ 2e_{13} \\ 2e_{12} \end{pmatrix} \text{ and}$$

$$\begin{pmatrix} \varepsilon_{11} \\ \varepsilon_{22} \\ \varepsilon_{33} \\ 2\varepsilon_{23} \\ 2\varepsilon_{13} \\ 2\varepsilon_{12} \end{pmatrix} = \begin{pmatrix} 1/E & -\nu/E & -\nu/E & 0 & 0 & 0 \\ -\nu/E & 1/E & -\nu/E & 0 & 0 & 0 \\ -\nu/E & -\nu/E & 1/E & 0 & 0 & 0 \\ 0 & 0 & 0 & 1/\mu & 0 & 0 \\ 0 & 0 & 0 & 0 & 1/\mu & 0 \\ 0 & 0 & 0 & 0 & 0 & 1/\mu \end{pmatrix} \begin{pmatrix} \sigma_{11} \\ \sigma_{22} \\ \sigma_{33} \\ \sigma_{23} \\ \sigma_{13} \\ \sigma_{12} \end{pmatrix} \tag{30}$$

where

$$C_{11} = \sqrt{2}(K_r + K_s)/2R,$$

$$C_{12} = \sqrt{2}(K_r - K_s)/4R\,,$$
$$C_{44} = \sqrt{2}(K_r + K_s)/4R = C_{11}/2\,.$$

From Eq. (30), we know that this is cubic material, the simplest case for an orthotropic solid. The 3 independent macroscopic elastic constants are

$$E = \frac{\sqrt{2}(K_r + 3K_s)K_r}{R(3K_r + K_s)}\,,$$
$$\nu = \frac{K_r - K_s}{3K_r + K_s}\,,$$
$$\mu = \frac{\sqrt{2}(K_r + K_s)}{4R}\,. \tag{31}$$

$K_r$ and $K_s$ can be expressed in terms of $E$ and $\nu$ as follows:

$$K_r = \frac{\sqrt{2}ER}{2(1-2\nu)}\,,$$
$$K_s = \frac{1-3\nu}{1+\nu}K_r\,. \tag{32}$$

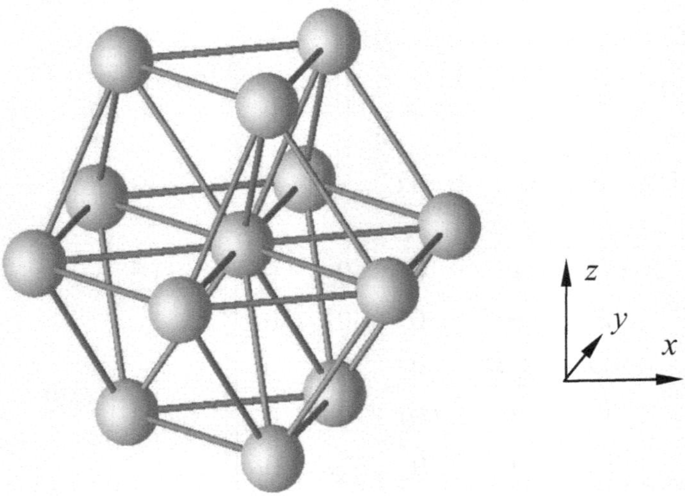


**Fig. 7** 3D HCP packing, the *central* particle is placed on the origin of coordinate

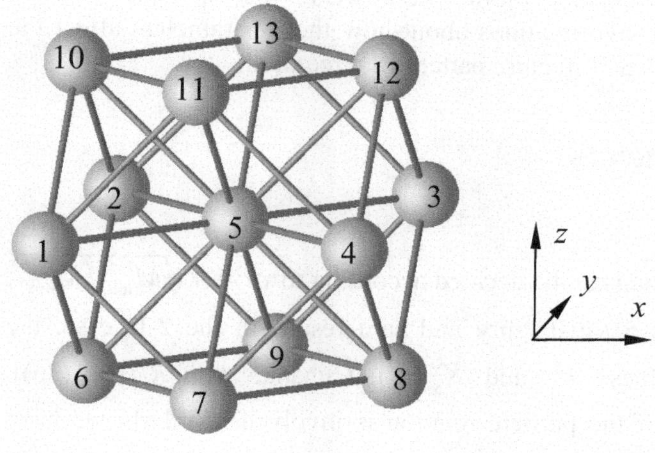


**Fig. 8** FCC lattice and coordinate system. The *center* of the FCC structure is placed on the origin of the coordinates

To have an estimation of $K_b$ and $K_t$, we consider an infinite FCC continuum subjected to the pure bending stress: $\sigma_{11} = Hz$, $\sigma_{22} = \sigma_{33} = 0$, $\sigma_{12} = \sigma_{23} = \sigma_{13} = 0$. Similarly, by comparing the strain energy distribution and the distribution of moments with respect to the $y$ axis between the continuum and discrete lattice, we have (Wang et al., 2008c),

$$K_b = \frac{\sqrt{2}ER^3}{48(1-\nu)} = \frac{(1-2\nu)R^2}{24(1-\nu)} K_r ,$$
$$K_t = \frac{1-3\nu}{1+\nu} K_b = \frac{(1-2\nu)R^2}{24(1-\nu)} K_s . \tag{33}$$

## 4.2 Fracture Parameters

It is very difficult to have a theoretical analysis on how to choose fracture parameters $F_{r0}$, $F_{s0}$, $\varGamma_{t0}$ and $\varGamma_{b0}$. However this problem can be investigated numerically. Some numerical tests have been done to reproduce the fracture of brittle rocks under uni-axial compression (Wang et al., 2006; 2008a). We find that $k = F_{s0}/F_{r0}$ is an important parameter to control the fracture behaviors and macro-scopic strength, and $k = 1$ to 4 gives reasonable brittle behaviors. Our preliminary studies also suggest that $\varGamma_{b0} = F_{r0}R/3$ in 2-D, or $\varGamma_{t0} = F_{s0}R/2$, $\varGamma_{b0} = F_{r0}R/4$ in 3-D will generate satisfactory results. Further comparisons between the different criteria and detailed investigations about how these parameters affect the macroscopic strength and fracture patterns are required.

## 4.3 Other Parameters

### 4.3.1 Time Step

Generally the time step can be decided according to $dt = \alpha\sqrt{M_{\min}/K_{\max}}$, here $\alpha \sim 0.1$ gives a satisfactory and safe result. In the 2-D case, the mass $M \propto R^2$, stiffness $K_r$ and $K_s$ do not change with $R$ (Eq. (26)), therefore $dt \propto R$. If the particle rotation is involved, similarly we have $dt = \alpha\sqrt{I_{\min}/K^r_{\max}}$. Since the moment of inertia $I \propto MR^2 \propto R^4$, $K^r$ ($K_b$ and $K_t$) $\propto R^2$ (Eq. (27)), therefore $dt \propto R$. In the 3-D case, $M \propto R^3$, $K_r$ and $K_s \propto R$ (Eq. (32)), we have $dt \propto R$ in the translational case. The rotational parameters $I \propto MR^2 \propto R^5$, $K_b$ and $K_t \propto R^3$ (Eq. (33)), we still have $dt \propto R$ in the rotational case. This means that if we keep the same macroscopic Young's modulus $E$, it means that a smaller time step is required for smaller particle sizes.

### 4.3.2 Artificial Damping

In the ESyS_Particle model, two types of damping are employed. The first one is global damping, $\boldsymbol{f}_1^{\nu} = -\nu_1 M\boldsymbol{v}$, here $\boldsymbol{v}$ is the absolute velocity of the particle, and $\nu_1$ is the damping coefficient. $\nu_1$ can not be significantly large such that in one time step a particle changes its direction of motion purely by the damping force, that is, $f_1^{\nu} dt < Mv$, then we have $\nu_1 < 1/dt$ as the upper limit.

The second kind of the damping force is $\boldsymbol{f}_2^{\nu} = -\nu_2 K \boldsymbol{v}_{ij}$, here $\boldsymbol{v}_{ij}$, $K$ and $\nu_2$ are the relative velocity between two particles, stiffness at the contact and damping coefficient. This kind of damping encourages the particles to move as rigid bodies, but discourages relative motion between the particles (such as wave propagation). Similarly we have $f_s^V dt < Mv_{ij}$, or $v_2 < dt/\alpha$, here $\alpha \sim 0.1$.

### 4.3.3 Loading Rate

Within one time step, particles should not be moved a distance larger than its size, or $V_{load} dt << R$. This principle should also hold true for highly dynamic processes such as the impact or blast simulations, in which case time steps are required be very small due to the large velocities.

# 5 Some Recent Simulation Results

Here we give some results from recent simulations using the ESyS_Particle model.

## 5.1 2-D Tests

### 5.1.1 Uni-Axial Tests

Figure 9a, 9b and 9c show simulations of fractures of brittle rock-like material under uni-axial compression. Random sized particles are arranged into rectangles of different aspect ratios (from 1 to 3). In the beginning, the neighbouring particles are bonded together to model the intact material. Two loading walls at the top and bottom of the model move slowly to compress the sample. On the right of each figure are the laboratory tests (Andreev, 1995). The figures suggest that the simulations are quite similar to the laboratory tests.

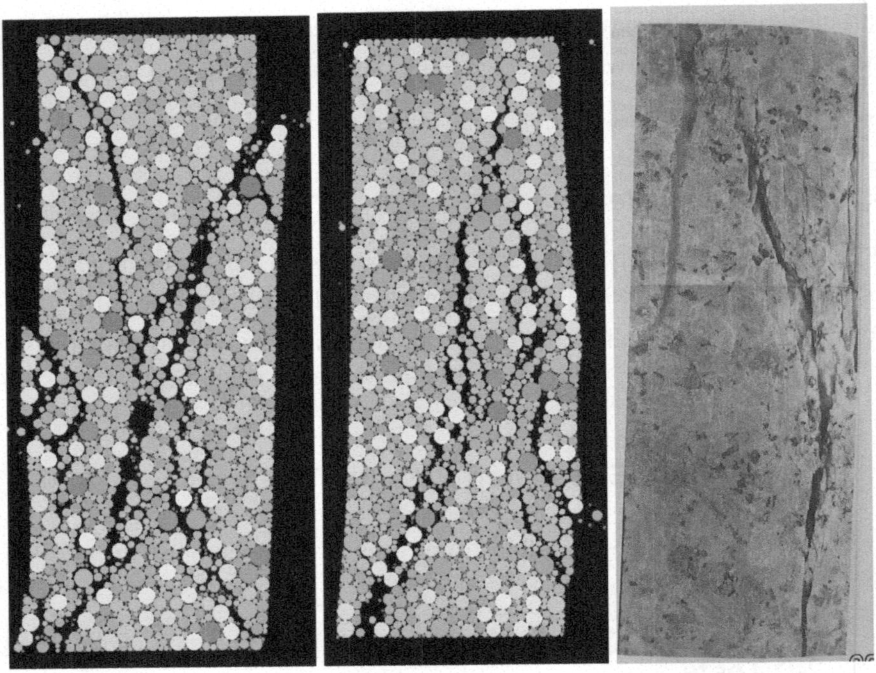

a

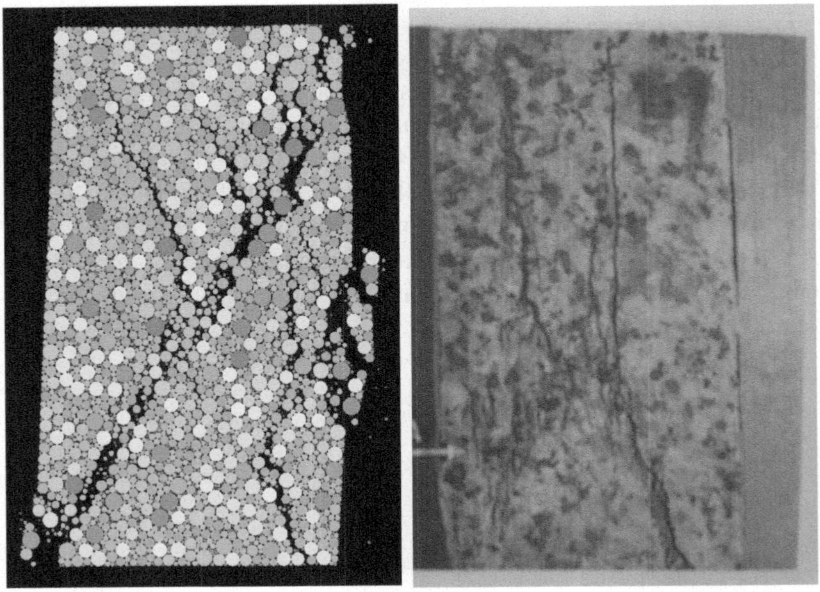

b

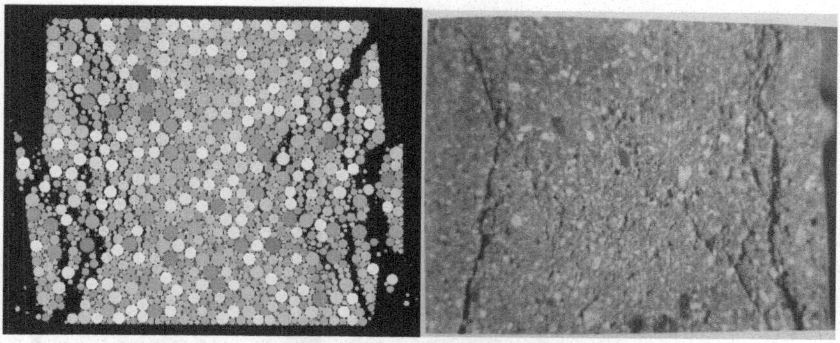

c

**Fig. 9** 2D simulation of brittle fracture under uni-axial compression for different aspect rations (3 in Fig.9a, 2 in Fig. 9b and 1 in Fig. 9c). The photos from the *right* of each picture are laboratory tests (Andreev, 1995)

### *5.1.2 Wing Crack Extension*

Figure 10 and Movie 1 show the 2-D wing crack extensions. In this figure there is a pre-existing oblique crack (the bonds are removed) in the beginning. Both regular and random particle arrangements are used. Stable tensile cracks are reproduced with an increase of loading. The left of Fig. 10 is a sketch of the laboratory test (Brace, 1960).

It is found that when only normal stiffness exists between two bonded particles, or when the single particle rotation is prohibited, or when single particle rotation exists but rolling stiffness is not permitted between two particles, the realistic pattern of crack propagation observed in laboratories can not be reproduced. Only when normal, shear and rolling stiffness exist and particle rotation is permitted, is it possible to reproduce laboratory tests (Wang et al., 2008a). We conclude that particle rotation and rolling resistance play a significant role and can not be neglected while modeling such phenomenon.

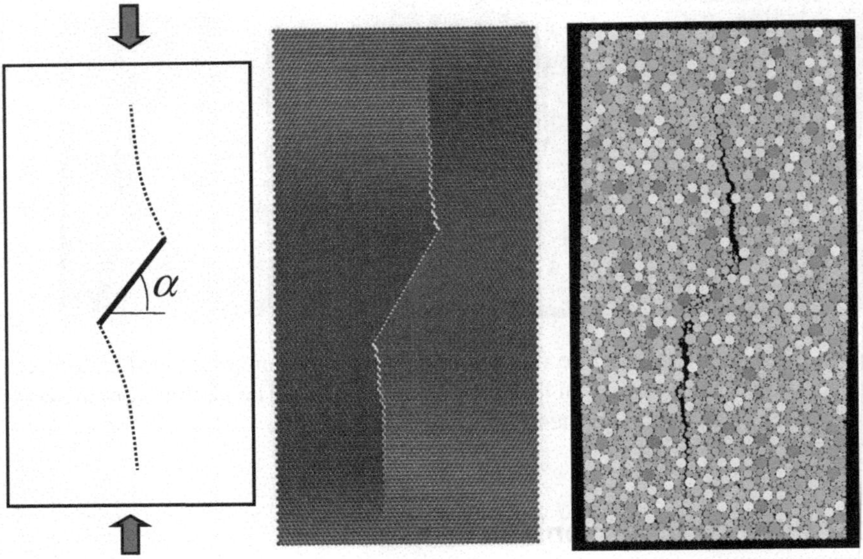


**Fig. 10** 2D wing crack extensions, *left*: a sketch of the laboratory test (Brace, 1960); middle: same sized particles; right: random sized particles

### *5.1.3 Shearing and Crushing of Aggregates*

Figure 11 and Movie 2 show the crushing of soil grains. The simulation models circular aggregates consisting of small particles bonded together, sheared by two elastic layers of bonded particles. The constant pressure is applied while shearing is modeled by moving the two elastic blocks. A periodical boundary is used in the horizontal direction. Rotations of the aggregates as rigid bodies can be observed, suggesting that frictional forces between the particles from the different aggregates take into effect. The fracture of some aggregates can be seen clearly.

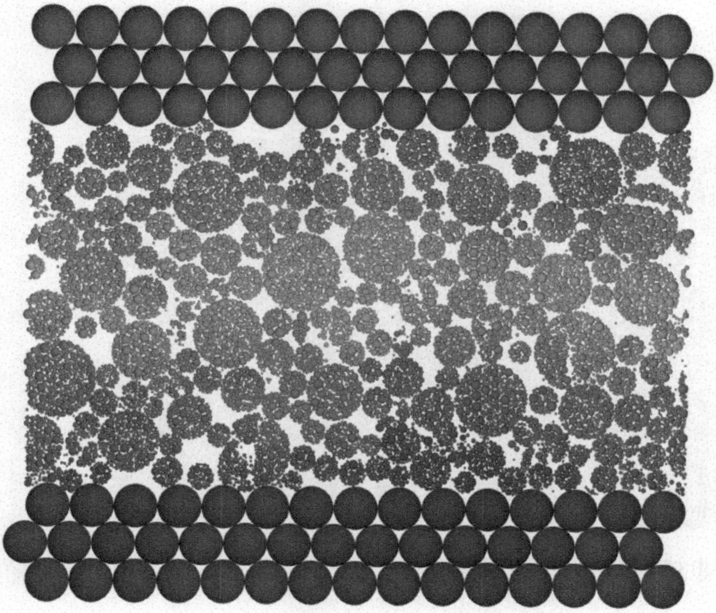

**Fig. 11** Simulation of crushing of grains under shearing and the constant pressure

### *5.1.4 Simulation of Brittle Fracture by Dynamic Impact*

Figure 12 and Movie 3 show the simulation of cracks generated in a brittle rock after an impact. The brittle rock is modeled by bonded particles of different sizes. The large ball falls under gravity and hits the rock and cracks are formed. The propagation of brittle cracks and elastic waves are reproduced.

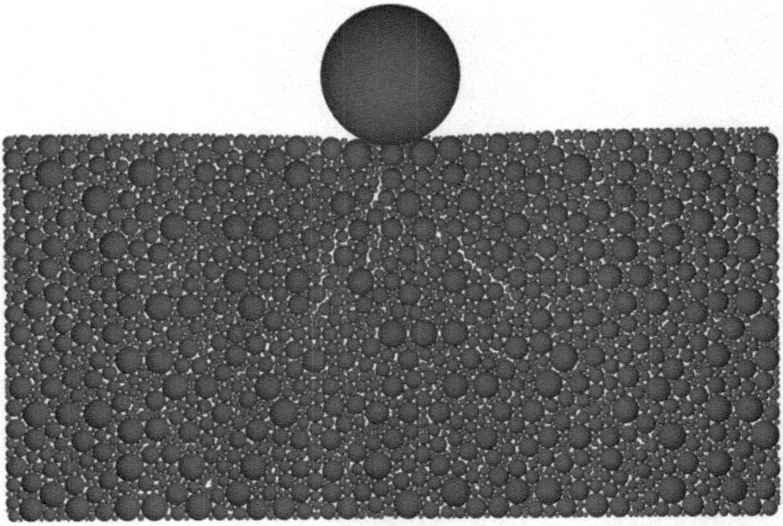

**Fig. 12** Simulation of brittle crack extension under impact

## 5.2 3-D Tests

### *5.2.1 Uniaxial Test*

Figure 13 and Movie 4 show the progressive fractures in a 3-D brittle rock. In this example, equal sized particles are arranged into FCC lattice which is subjected to uni-axial compression. The colors represent vertical displacement. Discontinuities in colors mean the formation of fractures which is difficult to be captured in laboratory tests since this process always occurs very fast. After the peak stress, the main faults are formed and two intact cores can be clearly observed with more fragile parts shattering away from four sides, which is always observed in rock fracture tests.

**Fig. 13** Progressive fractures in 3D brittle rock under uni-axial compression

### *5.2.2 Wing Crack*

Figure 14 Movie 5 present 3-D simulations of the extension of wing cracks in a brittle rock under slow uni-axial compression. The sample consists of equal-sized particles arranged into FCC cubic. Similarly, at the beginning of the simulation an oblique pre-existing circular crack is generated by removing the bonds in the cracked regions (small particles representing the centers of the removed bonds). The bigger particles in the plot are the centers of the fractured bonds, representing the modeled events. The laboratory tests are also given (Dyskin et al., 2003) in Fig. 15. Not only the basic shapes in laboratory tests, the curved wings (wrapping), are well reproduced, but also another interesting fact is also observed in our simulation: 3-D wing crack growth under uni-axial compression is limited, opposite to the 2-D case (Dyskin et al., 2003).

As far as we know, this is the first effort to model wing cracks using the Discrete Element Model.

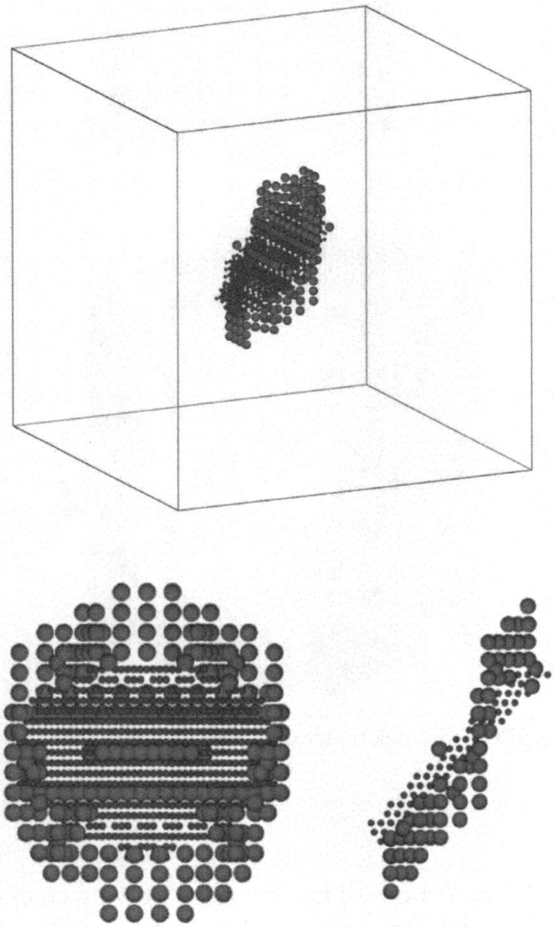

**Fig. 14** Simulated 3D wing crack extension. *Small particles* are the centers of the removed bonds (pre-existing fault). The *bigger particles* are the centers of the fractured bonds, representing the modeled events.

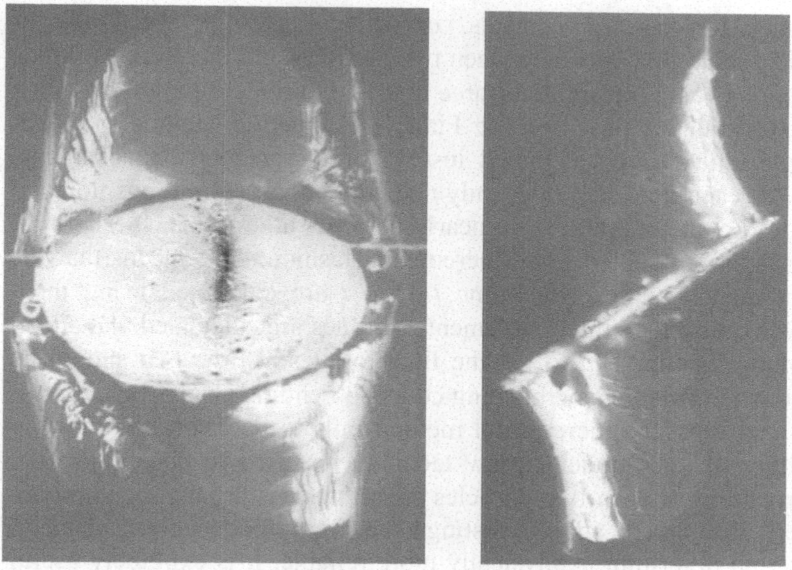

**Fig. 15** Laboratory tests of 3D wing crack extension under uni-axial compression (Dyskin et al., 2003).

## 6 Discussion: Major Differences of the ESyS_Particle Compared with the Other Existing DEMs

Compared with the other DEMs, the ESyS_Particle is different and advantageous in the following aspects:

Firstly we make use of unit quaternion to explicitly represent the orientation of a particle. In the most existing DEMs, particle orientations are **implicitly** represented using three angular velocities around three orthogonal axes. In 2-D cases this does not present a problem, since an angle can be easily integrated from angular velocities for each rigid body at each time step, thereby defining the orientation. However, in 3-D, this is not the case. Strictly speaking, one can not extract the exact orientation of a particle by simply integrating three angular velocities. The physical principle which limits this is: **finite rotations in 3-D are order dependent.** Consequently, for the same three finite angles integrated from angular velocities, different orders of rotations result in different final orientations (there are 12 possibilities). This salient feature of finite rotations in 3-D is one which is ignored by most modelers, and in this paper we attempt to address and highlight this discrepancy.

Secondly, for bonded contacts, our model permits six kinds of independent relative motions between two particles and therefore six kinds of interactions being transmitted (three in 2-D). Our algorithm to calculate forces and torques is based on the Finite Deformation Method in which total displacements are calculated instead of incremental displacements (Eq. (8)). At each time step *t*, we only need the initial, current position and orientation (i.e. at $t = 0$ and *t*). In nearly all other algorithms, shear forces and torques are computed in an incremental fashion. This means that the three incremental angles from time *t-dt* to *t* are computed using three angular velocities, then three incremental torques are calculated and added to the torques at time *t-dt* to get the final torques at time *t*. It should be noted that when twisting and bending co-exist, bending changes the axis of twisting, and thus, the incremental method fails to decouple the twisting and bending. In our model, a new technique is used to decompose the relative rotation between two particles such that twisting and bending are completely decoupled, strictly distinguished and sequence-independent. Therefore our algorithm is physically more reliable. It is extremely useful to apply our method when twisting and bending stiffnesses are different.

Thirdly, from a theoretical standpoint, only infinitesimal rotations in 3-D are order independent. Hence, in order to gain accurate results, the incremental method requires very small time steps. This computational inefficiency is highlighted by Wang (Wang, 2008b), where the errors in the incremental method are shown to increase much faster than those in our algorithm with increasing time step (Fig. 4).

Lastly, other differences include criterion to judge breakage of a bond, stick-slip friction and theoretical studies on how to choose model parameters.

## 7 Conclusions

What has been presented in this paper is a detailed description of the ESyS_Particle model. Particular focus is on the basic physical ingredients or mechanisms which a DEM should have.

We highlight the major differences between our model and most existing DEMs, which include the representation of particle rotations, fully decoupled and order-independent twisting and bending, a state-of-the-art algorithm to calculate forces and torques between bonded particles, unique criterion to judge breakage of a bond, stick-slip friction and detailed theoretical studies on how to choose model parameters. Specifically, single particle rotation and full interactions between particles make our model geometrically and theoretically complete, and studies suggest that the algorithm to decouple twisting and bending is physically reliable and numerically accurate because the basic physical principle, ***finite rotations in 3-D are order dependent,*** which is ignored by most modelers, is respected and obeyed in our model.

The problem of how to choose model parameters is discussed. These parameters include elastic stiffnesses, fracture parameters, integration time step, artificial damping parameters and loading rate.

The applicability of the ESyS_Particle is illustrated through several numerical simulations. Most qualitative features of rock fracture observed in laboratory tests are well reproduced, these include, fracture of brittle materials under compression, wing crack extension both in 2-D and 3-D, crushing of aggregates and fracture caused by dynamic impact. Although these examples have mostly focused on the simulations of fractures of intact materials, the ESyS_Particle is also very suitable to model "discrete" material (Latham et al., 2006).

It is found that single particle rotation and rotational stiffnesses play very important roles in reproducing realistic laboratory tests. The reason is that particle rotation effects local elasticity, especially stress fields near the crack tips, which is an important factor for crack extension.

The future direction of the ESyS_Particle will include the coupling with other physical processes. Currently, the model has been extended to include heat transfer, thermal expansion, friction generated heat, pore flow (or Darcy flow) and full coupling between mechanical and pore effects. In future Smooth Particle Hydrodynamics (SPH) will be implemented to model the interaction of liquids and solids. However, if multi-physics are involved, dimensionless analysis and scaling laws are required to determine the different physical parameters. More applications, either from engineering or scientific fields, are required to further validate and improve the model.

**Acknowledgements** Funding support is gratefully acknowledged by the Australian Computational Earth Systems Simulator Major National Research Facility, The University of Queensland and SGI. The ACcESS MNRF is funded by the Australian Commonwealth Government and participating institutions (Univ. of Queensland, Monash U, Melbourne U., VPAC, RMIT) and the Victorian State Government.

The first author would like to thank Mr. William Hancock for checking the manuscript.

## References

Abe S, Mora P, Place D (2000) Extension of the Lattice Solid Model to incorporate temperature related effects. Pure Appl Geophys 157:1867–1887

Abe S, Dieterich J, Mora P, Place D (2002) Simulation of the influence of rate- and state- dependent friction on the macroscopic behavior of complex fault zones with the lattice solid model. Pure Appl Geophys 159:1967–1983

Allen MP, Tildesley DJ (1987) Computer simulation of liquids. Oxford Science Press, Oxford

Anandarajah A (1994) Discrete-Element Method for simulating behavior of cohensive soil. J Geotech Engrg 120:1593–1613

Andreev GE (1995) Brittle failure of rock materials: test result and constitutive models. Rotterdam AA Balkema

Bardet JP, Proubet J (1991) A numerical investigation of the structure of persistent shear bands in granular media. Geotechnique 41:599–613

Bathurst RJ, Rothenburg L (1988) Micromechanical aspects of isotropic granular assemblies with linear contact interactions. J Appl Mech 55:17–23

Boutt DF, Mcpherson BJOL (2002) Simulation of sedimentary rock deformation: lab-scale model calibration and parameterization. GRL 29:10.1029/2001GL013987

Brace WF (1960) An extension of the Griffith theory of fracture to rocks. J G R 65:3477–3480

Buss SR (2000) Accurate and efficient simulation of rigid-body rotations. J Comp Phys 164:377–406

Campbell CS, Cleary PW Hopkins MA (1995) Large-scale landslide simulations: global deformation, velocities, and basal friction. J G R 100:8267–83

Chang CS (1992) Discrete Element Method for slope stability analysis. J Geotech Eng 118:189–1905

Chang SH, Yun KJ, Lee CI (2002) Modeling of fracture and damage in rock by the bonded-particle model. Geosys Eng 5:113–120

Cheng YP, Nakata Y, Bolton MD (2003) Discrete element simulation of crushable soil. Geotechnique 53:633–641

Cleary PW, Campbell CS (1993) Self-lubrication for long run-out landslides: examination by computer simulation. J G R 98:21911–21924

Cleary PW (2000) DEM simulation of industrial particle flows: case studies of dragline excavators, mixing in tumblers and centrifugal mills. Powder Technol 209:83–104

Cleary PW, Sawley ML (2002) DEM modeling of industrial granular flows: 3D case studies and the effect of particle shape on hopper discharge. Appl Math Model J 26:89–111

Cundall PA, Strack O (1979) A discrete element model for granular assemblies. Geotechnique 29:47–65

Cundall PA (1988) Formulation of a three-dimensional distinct element model – Part I, A scheme to detect and represent contacts in a system composed of many polyhedral blocks. Int J Rock Mech 25:107–116

D'Addetta GA, Kun F, Ramm E (2002) On the application of a discrete model to the fracture process of cohesive granular materials. Granul Matter 4:77–90

Daniel IM, Ishai O (1994) Engineering mechanics of composite material. Oxford University Press

Davie CT, Bicanic N (2003) Failure criteria for qusi-brittle materials in lattice-type models. Commun Numer Meth Eng 19:703–713

Delenne JY, Youssoufi MSE, Cherblanc F, Benet JC (2004) Mechanical behavior and failure of cohesive granular materials. Int J Numer Anal Mech Geomech 28:1577–1594

Donze FV, Bouchez J, Magnier SA (1997) Modeling fractures in rock blasting. Int J Rock Mech Min Sci 34:1153–1163

Dorby R, Ng T-T (1992) Discrete modeling of stress-strain behaviour of granular media at small and large strains. Eng Computation 9:129–143

Dullweber A, Leimkuhler B, McLachlan R (1997) Symplectic splitting methods for rigid-body molecular dynamic. J Chem Phys 107:5840–5851

Dyskin AV, Sahouryeh E, Jewell RJ, Joer H, Ustinov KB (2003) Influence of shape and locations of 3-D cracks on their growth in uniaxial compression. Engrg Fract Mech 70:2115–2136

Eberhardt E, Thuro K, Luginbuehl M (2005) Slope instability mechanisms in dipping interbedded conglomerates and weathered-the 1999 Rufi landlide, Switzerland. Eng Geol 77:35–56

Evans DJ (1977) On the representation of orientation space. Mol Phys 34:317–325

Evans DJ, Murad S (1977) Singularity free algorithm for molecular dynamic simul lation of rigid polyatomice. Mol Phys 34:327–331

Feng YT, Han K, Owen DRJ (2004a) Discrete element simulation of the dynamics of high energy planetary ball milling processes. Mater Sci Engrg A 375–377:815–819

Feng YT, Owen DRJ (2004b) A 2D polygon/polygon contact model: algorithmic aspects. Engrg Comput 21:265–277

Fincham D (1992) Leapfrog rotational algorithm. Molec Simul 8:1165

Fleissner F, Gaugele T, Eberhand P (2007) Applications of the discrete element method in mechanical engineering. Multibody Syst Dyn 18:81–94

Ghaboussi J, Barbosa R (1990) Three-dimensional discrete element method for granular materials. Int J Numer Anal Meth Geomech 14:451–472

Goldstein H (1980) Classical mechanics. 2nd edn., Addison-Wesley

Griffiths DV, Mustoe GGW (2001) Modeling of elastic continuum using a grillage of structural elements based on discrete element concepts. Int J Numer Meth Eng 50:1795–1775

Guo YG, Morgan JK (2004) Influence of normal stress and grain shape on granular friction: results of discrete element simulations. J G R 109:B12305

Hart R, Cundall PA, Lemos J (1988) Formulation of a three-dimensional distinct element model – Part II. Mechanical calculations for motion and interaction of a system composed of many polyhedral blocks J Rock Mech Min Sci Geomech Abstr 25:117–125

Hazzard JF, Collins DS, Pettitt WS, Young RP (2000a) Simulation of unstable fault slip in granite using a bonded-particle model. Pure Appl Geophys 159:221–245

Hazzard JF, Young RP (2000b) Simulation acoustic emissions in bonded-particle models of rock. Int J Rock Mech Min Sci 37:867–872

Hazzard JF, Young RP, Maxwell SC (2000c) Micromechanical modelling of cracking and failure in brittle rock. J G R 105:16683–16697

Hazzard JF, Young RP (2002) Moment tensors and micromechanical models. Tectonophysics 356:181–197

Hazzard JF, Mair K (2003) The importance of the third dimension in granular shear. G R L 30:doi:10.1029/2003GL017534

Hazzard JF, Young RP (2004) Dynamic modeling of induced seismicity. Int J Rock Mech 41:1365–1376

Hentz S, Daudeville L, Donze FV (2004a) Identification and validation of a discrete element model for concrete. J Eng Mech 130:709–719

Hentz S, Donze FV, Daudeville L (2004b) Discrete Element modeling of concrete submitted to dynamics loading at high strain rates. Comput Struct 82: 2509–2524

Hogue C (1998) Shape representation and contact detection for discrete element simulations of arbitrary geometries. Eng. Comput 15:374–390

Holst JMFG, Rotter JM, Ooi JM, Rong GH (1999) Numerical modeling of silo filling. II: Discrete element analysis. J Engrg Mech 125:104–110

Hu JC, Angelier J (2004) Stress permutations: three-dimensional distinct element analysis accounts for a common phenomenon in brittle tectonics J G R 109:doi:10.1029/2003JB002616

Huang HY, Detournay E, Bellier B (1999) Discrete element modeling of rock cutting. Rock mechanics for industry. Amadei, Kranz, Scott and Smeallie (eds) Balkema Rotterdam

Hunt SP, Meyers AG, Louchnikov V (2003) Modelling the Kaiser effect and deformation rate analysis in sandstone using the discrete element method. Comput Geotech 30:611–621

Issa JA, Nelson RB (1992) Numerical analysis of micromechanical behaviour of granular materials. Engrg Comput 9:211–223

Iwashita K, Oda M (1998) Rolling resistance at contacts in simulation of shear band development by DEM. J Eng Mech 124:285–292

Iwashita K, Oda M (2000) Micro-deformation mechanism of shear banding process based on modified distinct element method. Powder Technol 109:192–205

Jensen RP, Bosscher PJ, Plesha ME, Edil TB (1999) DEM simulation of granular media – structure interface: effects of surface roughness and particle shape. Int J Numer Anal Meth Geomech 23:531–547

Jiang MJ; Yu HS, Harris D (2006) Discrete element modelling of deep penetration in granular soils. Int J Numer Anal Meth Geomech 30:335–361

Johnson KL (1987) Contact mechanics. Cambridge University Press, Cambridge

Johnson SM, Williams JR, Cook BK (2007) Quaternion-based rigid body rotation integration algorithms for use in particle methods. Int J Numer Meth Engrg 74:1303–1313

Kol A, Laird BB, Leimkuhler BJ (1997) A symplectic method for rigid-body molecular simulation. J Chem Phys 107:2580–2588

Krysl P, Endres L (2005) Explicit Newmark/Verlet algorithm for time integration of the rotational dynamics of rigid bodies. Int J Numer Meth Engrg 62: 2154–2177

Kuhn MR, Bagi K (2004) Contact rolling and deformation in granular media. Int J Solid Stru 41:5793–5820

Kuipers JB (1998) Quaternion and rotation sequences. Princeton University Press, Princeton, New Jersey

Langston PA, Al-Awamleh MA, Fraige FY, Asmar BN (2004) Distinct element modeling of non-spherical frictionless particle flow. Chem Eng Sci 59: 425–435

Latham S, Abe S, Mora P (2006) Parallel 3D simulation of a fault gouge using the lattice solid model. Pure Appl Geophys 163:1949–1964

Li Y, Xu Y, Thornton C (2005) A comparison of discrete element simulations and experiments for "sandpile" composed of spherical particles. Powder Tech 160:219–228

Lin X, Ng TT (1997) A three-dimensional discrete element model using arrays of ellipsoids. Geotechnique 47:319–329

Lu M, McDowell GR (2007) The importance of modelling ballast particle shape in the discrete element method. Granul Matter 9:69–80

Magnier SA, Donze FV (1998) Numerical simulations of impacts using a discrete element method. Mech Cohes-Frict Mater 3:257–276

Matsuda Y, Iwase Y (2002) Numerical simulation of rock fracture using three-dimensional extended discrete element method. Earth Planets Space 54: 367–378

Matuttis HG, Luding S, Herrmann H (2000) Discrete element simulations of dense packings and heaps made of spherical and non-spherical particles. Powder Tech 109:278–293

McDowell GR, Harireche O (2002) Discrete element modelling of soil particle fracture. Geotechnique 52:131–135

Miller III TF, Eleftheriou M, Pattnaik P, Vdirango, Newns AD (2002) Symplectic quaternion scheme for biophysical molecular dynamic. J Chem Phys 116:8649–8659

Mindlin RD, Deresiewicz H (1953) Elastic spheres in contact under varying oblique forces. J Appl Mech 20:327–344

Monterio-Azevedo N, Lemos JV (2005) A generalized rigid particle contact model for fracture analysis. Int J Numer Anal Meth Geomech 29:269–285

Mora P, Place D (1993) A lattice solid model for the nonlinear dynamics of earthquakes. Int J Mod Phys C4:1059–1074

Mora P, Place D (1994) Simulation of the frictional stick-slip instability. Pure Appl Geophys 143:61–87

Mora P, Place D (1998) Numerical simulation of earthquake faults with gouge: towards a comprehensive explanation for the heat flow paradox. J G R 103:21067–21089

Mora P, Place D (1999) The weakness of earthquake faults. G R L 26:123–126

Mora P, Place D, Abe S, Jaume S (2000) Lattice solid simulation of the physics of earthquakes: the model, results and directions, in GeoComplexity and the Physics of Earthquakes (Geophysical Monograph series; no. 120),Rundle JB, Turcotte DL and Klein W (eds) American Geophys Union, Washington DC, pp 105–125

Mora P, Place D (2002a) Stress correlation function evolution in lattice solid elasto-dynamic model of shear and fracture zones and earthquake prediction. Pure Appl Geophys 159:2413–2427

Mora P, Wang YC, Yin C, Place D (2002b) Simulation of the Load-Unload Response Ratio and critical sensitivity in the Lattice Solid Model. Pure Appl Geophys 159:2525–2536

Morgan JK (1999) Numerical simulations of granular shear zones using the distinct element method: II. The effect of particle size distribution and interparticle friction on mechanical behavior. J G R B 104:2721–2732

Morgan JK, Boettcher MS (1999) Numerical simulations of granular shear zones using the distinct element method: I. Shear zone kinematics and micromechanics of localization. J G R B 104:2703–2719

Morgan JK (2004) Particle dynamics simulations of rate and state dependent frictional sliding of granular fault gouge. Pure Appl Geophys 161:1877–1891

Morrison RD, Cleary PW (2004) Using DEM to model ore breakage within a pilot scale SAG mill. Min Engrg 17:1117–1124

Morrison RD, Shi F, Whyte R (2007) Modelling of incremental rock breakage by impact-for use in DEM models. Min Engrg 20:303–309

Munjiza A, Owen DRJ, Bicanic N (1995) A combined finite/discrete element method in transient dynamics of fracturing solids. Engrg Comput 12:145–174

Munjiza A, Latham JP, John NWM (2003) 3D dynamics of discrete element systems comprising irregular discrete element-integration solution for finite rotations in 3D. Int J Numer Meth Eng 56:35–55

Mustoe G (1992) A generalized formation of the discrete element method. Engrg Comput 9:181–190

Ng TT, Dobry R (1994) A nonlinear numerical model for soil mechanics. J Geotech Engrg 120:388–403

Ning Z, Boerefijn R, Ghadiri M, Thornton C (1997) Distinct element simulation of impact breakage of lactose agglomerates. Adv Powder Tech 8:15–37

Oda M, Konishi J, Nemat-Nasser S (1982) Experimental micromechanical evaluation of the strength of granular materials: effects of particle rolling. Mech Mater 1:269–283

Oda M, Kazama H (1998) Microstructure of shear bands and its relation to the mechanisms of dilatancy and failure of dense granular soils. Geotechnique 48:465–481

Oda M, Iwashita K (2000) Study on couple stress and shear band development in granular media based on numerical simulation analyses. Int J Eng Sci 38:1713–1740

Omelyan IP (1998a) Algorithm for numerical integration of the rigid-body equations of motion. Phys Rev E 58:1169–1172

Omelyan IP (1998b) On the numerical integration of motion for rigid polyatomics: the modified quaternion approach. Comp Phys 12:97–103

Ostoja-Starzewski M (2002) Lattice models in micromechanics. Appl Mech Rev 55:35–60

Owen DRJ, Feng YT (2001) Parallelised finite/discrete element simulation of multi-fracturing solids and discrete systems. Engrg Comput 18:557–576

Place D, Mora P (1999) A lattice solid model to simulate the physics of rocks and earthquakes: incorporation of friction. J Comp Phys 150:332–372

Place D, Mora P (2000) Numerical simulation of localization in a fault zone. Pure Appl Geophys 157:1821–1845

Place D, Mora P (2001) A random lattice solid model for simulation of fault zone dynamics and fracture process, in Bifurcation and Localization Theory for Soil and Rock'99, Muhlhaus HB, Dyskin AV, Pasternak E, and Balkema AA (eds) Rotterdam/Brookfield

Place D, Lombard F, Mora P, Abe S (2002) Simulation of the micro-physics of rocks using LSM earth. Pure Appl Geophys 159:1933–1950

Potyondy D, Cundall P, Lee CA (1996) Modelling rock using bonded assemblies of circular particles. Rock Mechanics, Aubertin M, Hassani F, Mitri H (eds) Balkema, Rotterdam

Potyondy D, Cundall P (2004) A bonded-particle model for rock. Int J Rock Mech Min Sci 41:1329–1364

Prochazka PP (2004) Application of discrete element methods to fracture mechanics of rock bursts. Eng Fract Mech 71:601–618

Rapaport DC (1995) The art of molecular dynamic simulation. Cambridge University Press

Sakaguchi H, Muhlhaus H (2000) Hybrid modeling of coupled pore fluid-solid deformation problems. Pure Appl Geophys 157:1889–1904

Schlangen E, Garboczi EJ (1997) Fracture simulations of concrete using lattice models: computational aspects. Engrg Fract Mech 57:319–332

Schwarz OJ, Horie Y, Shearer M (1998) Discrete element investigation of stress fluctuation in granular flow at high strain rates. Phys Rev E 57:2053–2061

Scott D (1996) Seismicity and stress rotation in a granular model of the brittle crust. Nature 381:592–595

Sheng Y, Lawrence CJ, Briscoe BJ, Thornton C (2004) Numerical studies of uniaxial powder compaction process by 3D DEM. Engrg Comp 21:304–317

Sitharam TG (2000) Numerical simulation of particulate materials using discrete element modeling. Curr Sci 78:876–886

Thornton C (2000) Numerical simulations of deviatoric shear deformation of granular media. Geotechnique 50:43–53

Thornton C, Zhang L (2003) Numerical simulations of the direct shear test. Chem Eng Technol 26:153–156

Thornton C, Zhang L (2006) A numerical examination of shear banding and simple shear non-coaxial flow rules. Phil Mag 86:3425–3452

Ting JM, Khwaja M, Meachum LR, Rowell JD (1993) An ellipse-based discrete element model for granular materials. Int J Num Anal Meth Geomech 17:603–623

Toomey A, Bean CJ (2000) Numerical simulation of seismic waves using a discrete particle scheme. G J I 141:595– 604

Tordesillas A, Walsh DCS (2002) Incorporating rolling resistance and contact anisotropy in micromechanical models of granular media. Powder Tech 124:106–111

Tordesillas A, Peters J (2004) Role of particle rotations and rolling resistance in a semi-infinite particulate solid indented by a rigid flat punch. The 12th Biennial Computational Techniques and Applications Conference, 27 Sep–1 Oct 2004. The University of Melbourne, Victoria, Australia

Wang YC, Yin XC, Ke FJ, Xia MF, Peng KY (2000) Numerical simulation of rock failure and earthquake process on mesoscopic scale. Pure Appl Geophys 157:1905–1928

Wang YC, Mora P, Yin C, Place D (2004) Statistical tests of load-unload response ratio signals by Lattice Solid Model: implication to tidal triggering and earthquake prediction. Pure Appl Geophys 161:1829–1839

Wang YC, Abe S, Latham S, Mora P (2006) Implementation of particle-scale rotation in the 3-D Lattice Solid Model. Pure Appl Geophys 163:1769–1785

Wang YC, Mora P (2008a) Modeling wing crack extension: implications to the ingredients of Discrete Element Model. Pure Appl Geophys, 165: 609–620

Wang YC (2008b) A new algorithm to model the dynamics of 3-D bonded rigid bodies with rotations. Acta Geotechnica, in press

Wang YC, Mora P (2008c) Elastic properties of regular lattices and calibration of 3-D Discrete Element Model. J Mech Phys Solids, in press

Yang RY, Zou RP, Yu AB (2000) Computer simulation of the packing of fine particles. Phys Rev E 62:3900–3908

Young RP, Hazzard JF, Pettitt WS (2000) Seismic and Micromechanical Studies of Rock Fracture. G R L 27:1667–1670

Zhang D, Whiten WJ (1999) A new calculation method for particle motion in tangential direction in discrete element simulations. Powder Tech 102:235–243

Zhou YC, Wright BD, Yang, RY Xu BH, Yu AB (1999) Rolling friction in the dynamic simulation of sandpile formation. Physica A 269:536–553

# VII. The TeraShake Computational Platform for Large-Scale Earthquake Simulations

Yifeng Cui,[1] Kim Olsen,[2] Amit Chourasia,[1] Reagan Moore,[1] Philip Maechling[3] and Thomas Jordan[3]

[1] San Diego Supercomputer Center, 9500 Gilman Drive, MC0505, La Jolla, CA 92093, USA
[2] San Diego State University, 5500 Campanile Drive, San Diego, CA 92182, USA
[3] University of Southern California, 3651 Trousdale Parkway, Los Angeles, CA 90089, USA

**Abstract** Geoscientific and computer science researchers with the Southern California Earthquake Center (SCEC) are conducting a large-scale, physics-based, computationally demanding earthquake system science research program with the goal of developing predictive models of earthquake processes. The computational demands of this program continue to increase rapidly as these researchers seek to perform physics-based numerical simulations of earthquake processes for larger geographical regions, at high resolution, and for higher frequencies. To help meet the needs of this research program, a multiple-institution team coordinated by SCEC has integrated several scientific codes into a numerical modeling-based research tool we call the TeraShake computational platform (TSCP). A central component in the TSCP is a highly scalable earthquake wave propagation simulation program called the TeraShake anelastic wave propagation (TS-AWP) code. In this chapter, we describe how we extended an existing, stand-alone, well-validated, finite-difference, anelastic wave propagation modeling code into the highly scalable and widely used TS-AWP and then integrated this code into the TeraShake computational platform that provides end-to-end (initialization to analysis) research capabilities. We also describe the techniques used to enhance the TS-AWP parallel performance on TeraGrid supercomputers, as well as the TeraShake simulations phases including input preparation, run time, data archive management, and visualization. As a result of our efforts to improve its parallel efficiency, the TS-AWP has now shown highly efficient strong scaling on over 40K processors on IBM's BlueGene/L Watson computer. In addition, the TSCP has

H. Xing, *Advances in Geocomputing*, Lecture Notes in Earth Sciences 119,
DOI 10.1007/978-3-540-85879-9_7

developed into a computational system that is useful to many members of the SCEC community for performing large-scale earthquake simulations.

# 1 Introduction

Earthquakes are among the most complex terrestrial phenomena, and modeling of earthquake dynamics is one of the most challenging computational problems in science. Recent advances in earthquake science, combined with the increased availability of terascale computing resources, have made it practical to create three-dimensional (3D) simulations of seismic wave propagation. As analytical methods do not provide solutions for realistic models, numerical methods are necessary. The numerical methods in modeling earthquake motion include finite-difference (FD), finite-element, spectral element and the pseudo-spectral method (Chaljub et al. 2006). Each of these computational methods differs one from another by their ranges of applicability, their accuracy and efficiency. None of these techniques is universally applicable to all medium-wavefield configurations with sufficient accuracy and efficiency (Moczo et al. 2007). However, among the numerical methods used for large scale ground motion simulations, the FD method is still the most widely used method and is becoming increasingly important in the seismic industry and structural modeling fields, as it can be relatively accurate and computationally efficient.

Large-scale 3D ground motion simulations are used to estimate seismic risk, to plan public earthquake emergency response activities, and to design the next generation of earthquake-resistant structures. By helping geoscientists, emergency response groups, and civil engineers better understand seismic hazards, these modeling efforts can potentially save lives and properties. However, the computational challenges in current and future 3D ground motion modeling are daunting. Accurate simulations must span an enormous range of scales, from meters near the earthquake source to hundreds of kilometers across entire regions, and time scales from hundredths of a second – to capture the higher frequencies that affect the most common types of buildings – to hundreds of seconds for the ground motions to propagate across the entire affected region. Adding to these challenges, high-resolution runs may produce data sets that are many tens of terabytes in size. Recent simulations of this kind have been run on some of the world's largest and fastest supercomputers.

A collaborative and inter-disciplinary research team that includes researchers from the Southern California Earthquake Center (SCEC), San Diego Supercomputer Center (SDSC), and San Diego State University (SDSU) is pursuing a research program in earthquake system science that utilizes physics-based numerical modeling to conduct hazard assessments of earthquakes that may occur in Southern California. Specifically, analyses are made of the magnitude, distribution, and duration of the ground shaking that occurs when seismic waves from the earthquake ruptures propagate across the region. This research group has integrated several scientific codes into a community-accessible tool for earthquake wave propagation research that we call the TeraShake computational platform (hereafter referred to as TSCP). The TSCP includes an anelastic wave propagation modeling code that numerically solves the partial differential equations of the elastic wave equation with attenuation included by an anelastic coarse-grained method (Day 1998; Day and Bradley 2001), as well as other codes and tools that support earthquake simulation research including initialization codes, data analysis codes, data management codes, and visualization tools. The TSCP began as standalone "personal" research code, but over the last few years, the SCEC research group has transformed it into to a collection of community codes that are useful to a wide range of researchers working on different research projects. In order to make the TSCP more broadly useful, we developed special software tools to support the full research lifecycle, including data management, analysis, and visualization.

Our experience with the TSCP indicates that high levels of expertise from many disciplines are required to solve the computational and data management challenges that emerge when running earthquake wave propagation simulations at very large scales. In this chapter, we describe some of the details of the optimization techniques that we used to implement the TSCP and the data management tools used to support the TeraShake simulations. We look at the challenges we faced porting the codes to National Science Foundation (NSF) TeraGrid resources, optimizing the application performance, optimizing the run initialization, and optimizing the I/O. Scalability of the optimized application is presented on multiple HPC resources including IBM TJ Watson BG/L. We then discuss the data management challenges and publication of the results in a digital library, as well as the visualization expertise used for analyzing the terabytes of results. We then summarize the major scientific findings of the TeraShake simulations. At the end, we discuss the lessons learned while performing large-scale earthquake simulations, and draw some conclusions about the support required for massively parallel computations.

## 2 The TeraShake Computational Platform

The TeraShake computational platform is the SCEC Community Model Environment's (SCEC/CME) first capability-computing platform. The TeraShake team, led by Kim Bak Olsen at San Diego State University, has tackled several large-scale simulations of earthquakes on the southern San Andreas fault based on multiple research codes (Cui et al. 2007b, 2007c; Olsen et al. 2006a, 2006b, 2007; http://www.scec.org/cme).

One geographical area of high interest is the southern portion of the San Andreas Fault, between Cajon Creek and Bombay Beach in the state of California in the United States, which has not seen a major event since 1690, and has accumulated a slip deficit of 5–6 m (Weldon et al. 2004). The potential for this portion of the fault to rupture in an earthquake with a magnitude as large as Mw7.7 is a major component of seismic hazard in southern California and northern Mexico.

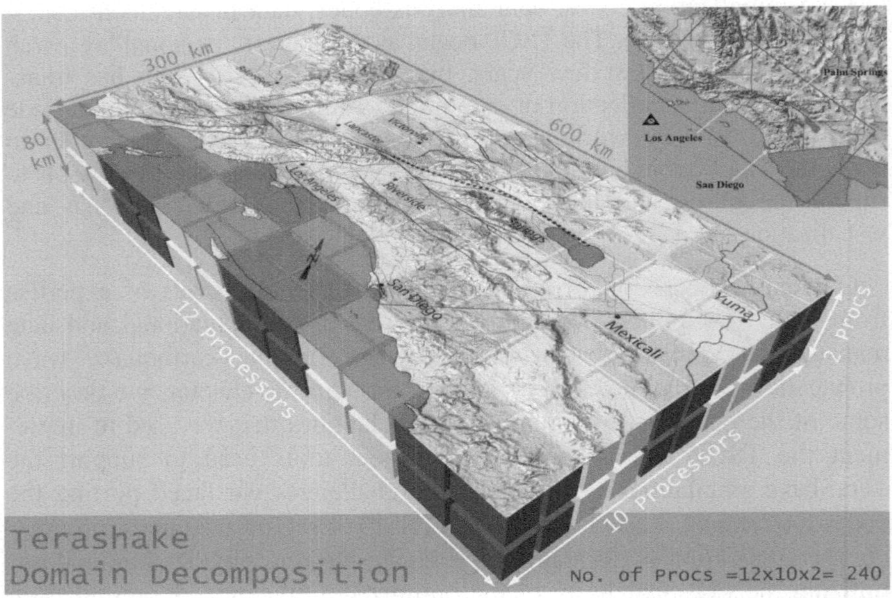


**Fig. 1** The *top right* inset shows the simulation region 600 km long and 300 km wide, indicated by the *red rectangle*. In the *center* of the topography, *fault lines*, and city locations are visible. This also shows the domain decomposition of the region into 240 processors (see *Movie* 1, available on accompanying DVD)

The TeraShake-1 simulations (hereafter referred to as TS1) used a 3,000 by 1,500 by 400 grid-point mesh that divided the simulation volume into

1.8 billion cubes with a spatial resolution of 200 m (Fig. 1). A mesh of this resolution supports finite difference simulations up to a maximum frequency of 0.5 Hz. A series of simulations were run using similar configurations and we refer to any of the simulations in the series as a TeraShake simulation. The large simulation volume and the relatively high frequency content of TS1 resulted in some of the largest and most detailed earthquake simulations of this region at the time they were executed (Fig. 1). TS1 used up to 2048 processors on the NSF funded TeraGrid computer resources (http://teragrid.org/about/), and produced up to 43 TB of time-varying volumetric data output in a single run. The simulation results were registered in a digital library, managed by San Diego Supercomputer Center's Storage Resource Broker (SRB) (Moore et al. 2005), which archived a second copy into SDSC's IBM High Performance Storage System (HPSS).

The TS1 used relative simple kinematic descriptions of the earthquake rupture (a rupture description is also known as a "source description") derived from the 2002 M7.9 Denali, Alaska, earthquake, with a constant rupture velocity and slip contained in 6 consecutive pulses. Using these simple source descriptions, the TS1 simulations produced new insights into how rupture directivity – the tendency for energy to be focused in the direction of rupture propagation – can couple seismic waves into sedimentary basins. The TS1 simulations showed how the chain of sedimentary basins between San Bernadino and downtown Los Angeles form an effective waveguide that channels surface waves along the southern edge of the Transverse Ranges. The details of the TS1 can be found in (Olsen et al. 2006a).

The TeraShake-2 simulations (hereafter referred to as TS2) added a physics-based dynamic rupture to the simulation. The TS2 research involved two separate simulation runs. First, the researchers run a dynamic rupture simulation that produces an output file containing a source description. Then the modelers use this source description file as the rupture definition for the second simulation, which is an earthquake wave propagation simulation. The dynamic rupture simulation was run at a very high 100 m resolution, to create an earthquake source description of 200-m for frequencies up to 0.5 Hz for the San Andreas Fault. The dynamic rupture simulations include additional physical constraints including friction laws that are not followed by the kinematic source descriptions used in the TS1 simulations. With the exception of the source descriptions, the configuration parameters used by the TS1 and TS2 wave propagation simulations were identical, including the 3D velocity model, fault geometry, and grid intervals in time and space. We adjusted the initial stress state for dynamic simulations to obtain a predominantly subshear average rupture velocity, resulting in average rupture velocities that are nearly the same in the two

(TS1 and TS2) cases (Olsen et al. 2007). Details of the TS2 methodologies can be found in (Olsen et al. 2007). The TS2 peak-ground motions were smaller than the TS1 peak-ground motions by a factor of 2–3 in certain areas of the Los Angeles region. We believe this significant reduction in peak ground motions is due to a less coherent wavefield radiated from the complex dynamic source.

In work that continues and extends the TeraShake simulations, SCEC is currently collaborating with United States Geological Survey (USGS) and the City of Los Angeles Emergency Preparedness Department on an earthquake preparedness exercise to improve public awareness and readiness for the "Big One", the next great earthquake along the southern San Andreas Fault. This public earthquake preparedness exercise is called the Great Southern California ShakeOut (www.shakeout.org). The ShakeOut exercise defines a scientifically plausible earthquake on the southern San Andreas and then asks emergency management groups, and the public, to practice their response to the event as if it has actually occurred. The scientifically plausible scenario earthquake used in the ShakeOut exercise is an Mw7.8 rupture that initiates near Bombay Beach by the Salton Sea and propagates unilaterally 300 km toward the northwest up to near Lake Hughes. Simulating this scenario earthquake poses a considerable computational challenge due the very large outer scale length of the problem and frequency content up to 1 Hz.

## 3 TeraShake Application: Anelastic Wave Model

To compute the propagation of the seismic waves that travel along complex paths from a fault rupture across an entire domain, the Finite Difference (FD) method is still the dominant method, due to accuracy, flexibility and computational efficiency.

Anelastic wave propagation (AWP) models are used here to denote numerical codes that can simulate earthquake waves propagating through heterogeneous materials. AWP codes are used in several types of geophysical research. They are used to predict ground motions for scenario earthquakes, in data inversion studies of geological models, and for probabilistic seismic hazard studies. As SCEC researchers simulate larger geographical regions and also higher frequencies, the computational requirements of the AWP codes increase rapidly in order to support the computational requirements of current and future SCEC research.

The anelastic wave propagation modeling code used in the research described in this paper was developed by Kim Bak Olsen (Cui et al. 2006a;

Day and Bradley 2001; Marcinkovich and Olsen 2003; Olsen 1994; Olsen et al. 2003, 2006a) currently at San Diego State University. As the original Olsen's AWP code was modified for use on the TeraShake project, it became known as the TeraShake AWP (hereafter referred to as TS-AWP). The TS-AWP is the core high performance modeling code that forms the foundation of the TeraShake computational platform. The TS-AWP modeling code solves the 3D velocity-stress wave equation explicitly by a staggered-grid FD method, fourth-order accurate in space and 2nd-order accurate in time. One of the significant advantages of the code is the option of using the efficient Perfectly Matched Layers to implement absorbing boundary conditions on the sides and bottom of the grid, and a zero-stress free surface boundary condition at the top (Marcinkovich and Olsen 2003).

The 3D isotropic, elastic velocity-stress system of equations is given as

$$\begin{aligned}
\frac{\partial v_x}{\partial t} &= \frac{1}{\rho}\left(\frac{\partial \sigma_{xx}}{\partial x} + \frac{\partial \sigma_{xy}}{\partial y} + \frac{\partial \sigma_{xz}}{\partial z}\right) \\
\frac{\partial v_y}{\partial t} &= \frac{1}{\rho}\left(\frac{\partial \sigma_{xy}}{\partial x} + \frac{\partial \sigma_{yy}}{\partial y} + \frac{\partial \sigma_{yz}}{\partial z}\right) \\
\frac{\partial v_z}{\partial t} &= \frac{1}{\rho}\left(\frac{\partial \sigma_{xz}}{\partial x} + \frac{\partial \sigma_{yz}}{\partial y} + \frac{\partial \sigma_{zz}}{\partial z}\right) \\
\frac{\partial \sigma_{xx}}{\partial t} &= (\lambda + 2\mu)\frac{\partial v_x}{\partial x} + \lambda\left(\frac{\partial v_y}{\partial y} + \frac{\partial v_z}{\partial z}\right) \\
\frac{\partial \sigma_{yy}}{\partial t} &= \lambda\frac{\partial v_x}{\partial x} + (\lambda + 2\mu)\frac{\partial v_y}{\partial y} + \lambda\frac{\partial v_z}{\partial z} \\
\frac{\partial \sigma_{zz}}{\partial t} &= \lambda\left(\frac{\partial v_x}{\partial x} + \frac{\partial v_y}{\partial y}\right) + (\lambda + 2\mu)\frac{\partial v_z}{\partial z} \\
\frac{\partial \sigma_{xy}}{\partial t} &= \mu\left(\frac{\partial v_x}{\partial y} + \frac{\partial v_y}{\partial x}\right) \\
\frac{\partial \sigma_{xz}}{\partial t} &= \mu\left(\frac{\partial v_x}{\partial z} + \frac{\partial v_z}{\partial x}\right) \\
\frac{\partial \sigma_{yz}}{\partial t} &= \mu\left(\frac{\partial v_y}{\partial z} + \frac{\partial v_z}{\partial y}\right)
\end{aligned} \qquad (1)$$

where $\mathbf{v}(\mathbf{x},t)$ is the velocity vector field, a function of both position $\mathbf{x}$ and time $t$, $\sigma(\mathbf{x},t)$ is the stress tensor, $\lambda(\mathbf{x})$ and $\mu(\mathbf{x})$ are Lame's elastic constants, and $\rho$ is density.

TS-AWP is written in Fortran 90 and message passing is done in this code with the Message Passing Interface (MPI) using domain decomposition. Each processor is responsible for performing stress and velocity calculations for its portion of the simulation volume grid, as well as dealing with boundary conditions at the external edges of each volume (Fig. 2). Ghost cells, comprising a two-cell-thick padding layer, manage the most recently updated wavefield parameters exchanged from the edge of the neighboring subgrid. I/O is done using MPI-I/O, and the velocity output data are written to a single file. The code was extensively validated for a wide range of problems, from simple point sources in a half-space to dipping propagating faults in 3D crustal models (Olsen et al. 2003).

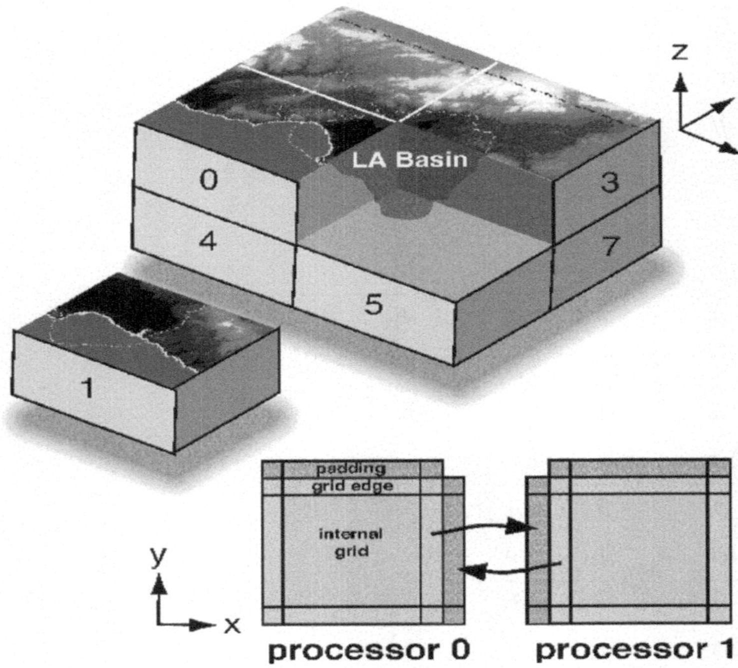


**Fig. 2** Illustration of TS-AWP communication between neighboring subgrids over 8 processors

# 4 Enhancement and Optimization of the TeraShake Application

The TS-AWP code has been through many enhancements for effective use of memory and I/O, effective communications between parallel tasks, and providing portable programming. These optimizations are essential to reduce production time required for TeraShake scale simulations.

## 4.1 Porting and Optimizations

Porting and optimizing the parallel code to new architectures is a challenging task from software perspective. Besides the efficient development of the program, an optimal execution is of paramount importance. We therefore use a few examples to discuss the portability issues and application specific optimizations we carried out for the TS-AWP code.

To maintain portability of the code between computer architectures, compilers, and MPI libraries, some structural and functional changes to the code were necessary. We identified and fixed bugs related to use of MPI message passing between nodes and use of MPI-IO for writing to disk that caused the code to hang on some target platforms. We identified a wrongly assigned process rank in the communicator that specifies the Cartesian location to which topology information was being attached. We also found that the original design of the MPI-IO data type in the code that represents count blocks was defined at each time step, which caused a memory leak problem. Our improved TS-AWP defines an indexed data type once only at the initialization phase, and effectively sets new views for each task of a file group to obtain efficient MPI-IO performance.

These bugs are illustrative of the types of problems that must be addressed for massively parallel simulations. The memory leak did not affect the runs at smaller scale, but was a controlling factor in the ability at large scale to effectively use memory.

Compiler changes are often invisible, but can cause significant bottlenecks at large scales. For example, significant differences in the semantics of unformatted FORTRAN writes were found due to the Intel compiler upgrade to 8.0, which break most FORTRAN programs that write data in bytes. This is because the record length units *recl* in the OPEN statement have been changed from bytes to 4 byte words. Our solution is to use the *inquire* function to return the length of the output required for writing in processor-dependent units, which is portable in every architecture.

To enable easy compilation across different platforms, we created a portable makefile using a C-preprocessor. This adds a parameter to the compiling command which denotes the machine architecture, and the parameter is parsed in the C-preprocessor. One example of portability considerations is the need to define an indexed datatype. The MPI-2 function MPI_TYPE_CREATE_ INDEXED_BLOCK used in the application is not supported by every implementation of MPI. We added a more general MPI_TYPE_INDEXED call that uses a block size of 1, to enable use of indexed datatypes on the Cray XT3 and EM64T computers.

Enhancement and integration of new features is often necessary for codes used in large-scale simulations. Previous simulations of the Southern California region with the TS-AWP code were only tested up to a 592 by 592 by 592 mesh. As we began the TeraShake simulation work, new problems emerged in managing memory requirements for problem initialization. The TS-AWP code was enhanced from 32-bit to 64-bit addressing, and 64-bit integer variables are added to manage 1.8 billion mesh points. 32-bit MPI addresses have an inconvenient memory limit default of 256 MBs per task and 2GB maximum memory size. The 64-bit MPI version removes these barriers. Having portability in mind, we have maintained the option to use the 32-bit mode for 32-bit machines like Blue Gene/L.

The original TS-AWP code was used for forward modeling of ground motion only. As part of the TS2 effort, we incorporated dynamic rupture simulation mode into the code. This new feature models slip dynamically on the fault surface to generate a more realistic source than the kinematic source description used in TS1. When adding this feature, we implemented the changes so that users can freely select between dynamic rupture and wave propagation simulations providing users significant flexibility in how they use this code.

To benchmark code performance, we inserted performance measurements in the application to analyze the wall clock time spent for each stage of the execution, and identified some significant performance bottlenecks. For single-CPU optimization, we examined loop fusion, unrolling, and the avoidance of branching in a few critical inner loops. Some of the very time-consuming functions were in-lined, which immediately saved more than 50% of the initialization time. The reduction of the required memory size and tuning of data operations were necessary steps to enable execution of the TeraShake simulation.

A MPI-IO optimization was accomplished by modifying the collective writes. Instead of using an individual file pointer, we used an explicit offset to perform data access at the specific location given directly as an argument passed into the routine. These calls do not update any file pointers, thereby saving I/O interactions. The change to explicit offset positioning not only made large output generation possible, but also greatly improved the I/O performance.

## 4.2 Optimization of Initialization

The TS-AWP code uses several configuration files as inputs. These input configuration files include a velocity mesh, a source description, and a definition of general computing parameters such as number of time steps in the simulation. Some of these initialization files, including both the velocity mesh and the source description, can be quite large resulting in long initialization times as these files are read in.

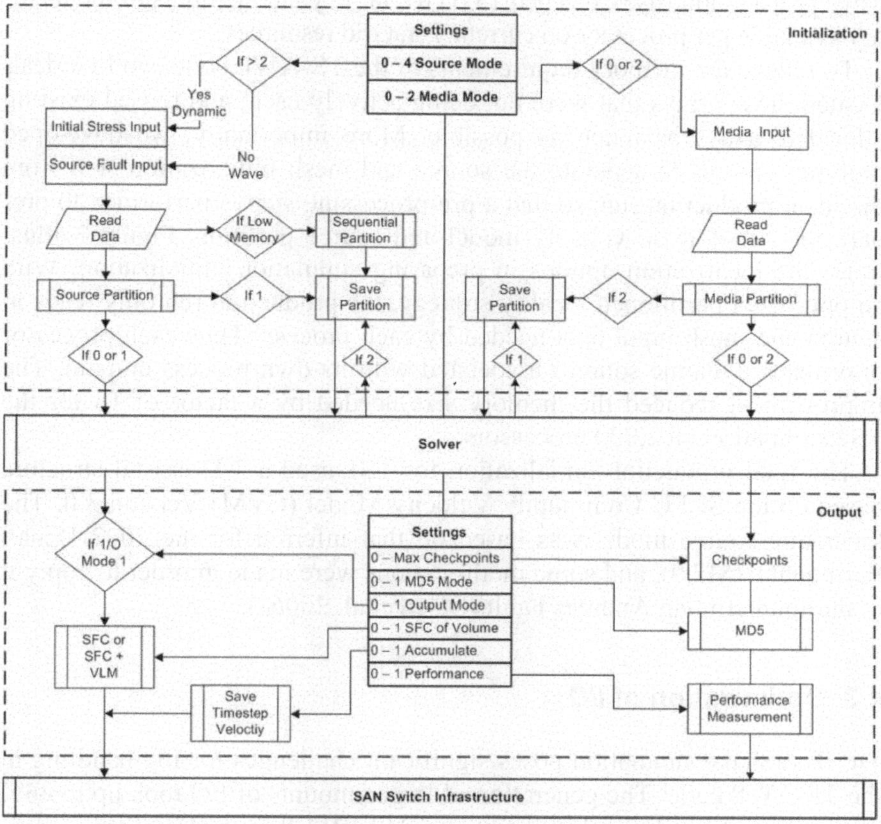


**Fig. 3** Mapping the TS-AWP code to different architectures using flexible settings. *Top* of the chart illustrates the options in preparing the initialization, *top left* the source partition process, *top right* the media partitions. *Bottom part* of the chart illustrates the simulation output options, varied from single step to accumulated steps. Options also include the possibility of turning I/O, benchmark measurements or checksum MD5 off completely, which help analysis and adjustment from architecture to architecture, based on memory and I/O bandwidth of target machines. Settings are modified at execution time

The TS-AWP initialization presented a significant challenge as we scaled up to the TeraShake problem size. The original TS-AWP code did not separate the source and mesh generation processing from the finite difference solver. This made it difficult to scale the code up to a large problem size due to memory constraints, as each process reads the entire domain source. An even worse memory problem was introduced by the increase in memory requirements from TS1 to TS2. TS1 used an extended kinematic source, which required 3GB. However, the local fault rupture memory requirement for TS2 wave propagation case exceeded 13.6 GB, far beyond the limits of the memory available per processor on current TeraGrid resources.

To reduce the memory requirements of the TS-AWP code, we first deallocated those arrays that were not being actively used, and reused existing allocated arrays as much as possible. More importantly, we developed multiple options to separate the source and mesh initialization step from the main production run, so that a pre-processing step is performed to prepare the sub-domain velocity model and source partition. Figure 3 illustrates the localization options in preparing simulation initialization. With an optimized partitioned locality strategy, the production run only reads in source and mesh input data needed by each process. Thus each processor only reads dynamic sources associated with its own process domain. The improvement reduced the memory size needed by a factor of 18 for the TS2 simulation on 2,000 processors.

The final production initialization for TS1 used a 3-D crustal structure based on the SCEC Community Velocity Model (CVM) Version 3.0. The kinematic source model was based on that inferred for the 2002 Denali Earthquake (M7.9), and some modifications were made in order to apply it to the southern San Andreas Fault (Olsen et al. 2006a).

### 4.3 Optimization of I/O

The TeraShake simulation poses significant challenges for I/O handling in the TS-AWP code. The generation of large amounts of I/O took up to 46% of the total elapsed simulation time on 240 IBM Power4 DataStar processors. The I/O performance saturated quickly as the number of processors increased to more than a few hundred.

Storage devices perform best when transferring large, contiguous blocks of data. The TS-AWP originally performed I/O consisting of many small serial reads or parallel writes rather than a few large reads or writes. To improve the disk write performance, we carefully calculated the runtime memory utilization needed for writes, and accumulated output data in a memory buffer until it reached an optimized size before writing the data to disk. This optimization gave the best tradeoff between I/O performance and memory overhead. This single optimization reduced the I/O time by a factor of ten or more on target problem size, resulting in a very small surface-velocity write time compared to the total elapsed time (Fig. 4).

To improve disk read performance, we optimized the TS-AWP code by reading data in bulk. We aggressively read data from disks beyond those attached to the local processor. We calculated the actual location of the data and then assigned the read to the corresponding processors, which improved the disk read performance by a factor of 10 or more.

In the original TS-AWP code, the media velocity model – SCEC CVM3.0 in this case – is read by a master processor using an implied do loop, essentially causing each set of 8 data elements or 32-bytes to be read once. The data structure is in a large 3-dimensional array, and the data being accessed on disk were not stored contiguously. We changed the access pattern to read the media data in a large contiguous block before assigning the 32-byte data sets to corresponding processors. With this change in place, we observe a significant speedup on some platforms such as the IA-64, reducing the media data reading time from 20 h to 2 h as a result.

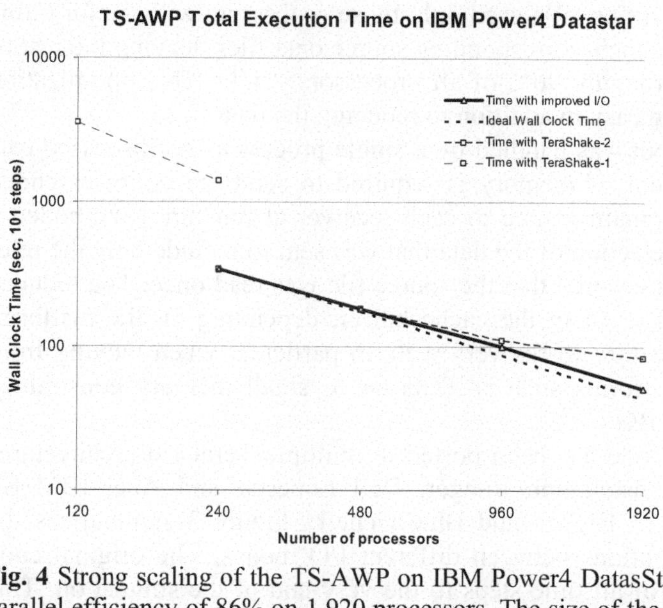


**Fig. 4** Strong scaling of the TS-AWP on IBM Power4 DatasStar at SDSC, with a parallel efficiency of 86% on 1,920 processors. The size of the TeraShake domain is 600 km by 300 km by 800 km with a resolution of 200 m. The *dashed line* with *filled squares* is the scaling of TeraShake-1; the *dashed line* with *filled diamonds* is the scaling of TeraShake-2 after optimizations; the *bold solid line* with *triangles points* is scaling of the most-recently updated code with improved I/O; the *dotted line* is ideal scaling case. All measurements are for surface velocity outputs only

## 4.4 Mapping TS-AWP to Different TeraGrid Architectures

We carried out the TeraShake simulations on a range of different TeraGrid systems. We used different TeraGrid computers at the different stages of the simulations. One size does not fit all. There is an adaptation procedure to meet specific requirements of each architecture. Our experience indicates that it is important to determine how fundamental system attributes affect application performance.

We present an adaptive approach in the TS-AWP that enables the simultaneous optimization of both computation and communication at run-time using flexible settings (Fig. 3). These techniques optimize initialization, source/media partition and MPI-IO output in different ways to achieve optimal performance on the target machines, which have proven to be highly effective for use on multiple architectures.

We added code to write each set of media data in parallel to a processor-specific file once the initialization was complete (Fig. 3). Using this scheme, any restart of the simulation could read in each processor's media file directly, and save the entire media initialization time.

Source data partition initialization was implemented in a similar way to the media data partition. We used only those processors with useful information to generate their corresponding source data files, leaving the rest of the processors – roughly 90% of all processors – idle. This optimization made bookkeeping easy in addition to reducing the data size.

Source data input was handled by a single processor. As discussed earlier, a large amount of memory is required to send the entire extended time-dependent domain source to each receiver at run time. We added a setting to adjust selection of the data that was sent to include only the relevant domain. This ensured that the source file was read once. The reduced size was adjusted to fit in the cache buffer, depending on the platform. This portability feature works very well, in particular when moving from large memory platforms such as Datastar to small memory constrained platforms such as BG/L.

The TS-AWP code has been ported to multiple TeraGrid architectures including Sun Constellation Ranger, Dell Lonestar and Abe, IBM SP Datastar, Cray XT3 Bigben and Blue Gene/L. Figure 3 summarizes the run-time switch options between different I/O modes. The original code held the velocity of all time steps to the very end of the simulation. This works, however, only at small scale. We added an option to select certain time steps for output, as well as a specification of the number of steps to be accumulated. These settings made the application flexible enough for job scheduling on different resources. We also added a feature to split surface and volume output, with each generating selected output at an independent time step. Typically, surface output is generated at each time step for detailed visualization analysis, while volume data is generated at each 10th to 100th time step.

To enable performance evaluation and benchmarking on different platforms, we added an additional option to turn the I/O and performance measurements off completely. Performance measurements collect information on how much time is spent in each section of the code including initialization, computing and I/O time. This information is very helpful for performance analysis at both hardware and software levels.

## 4.5 Scaling the Code up to 40k Processors

As a result of many enhancements and optimizations discussed above, the TS-AWP code scales up to a large number of processors, as illustrated in Fig. 4. The figure also shows the improvement in results obtained through single CPU tuning as well as use of machine-specific optimization flags. Scaling results for TS-AWP on TeraGrid computing resources indicate that the scalability will saturate due to higher interconnect latency and network topology limitations.

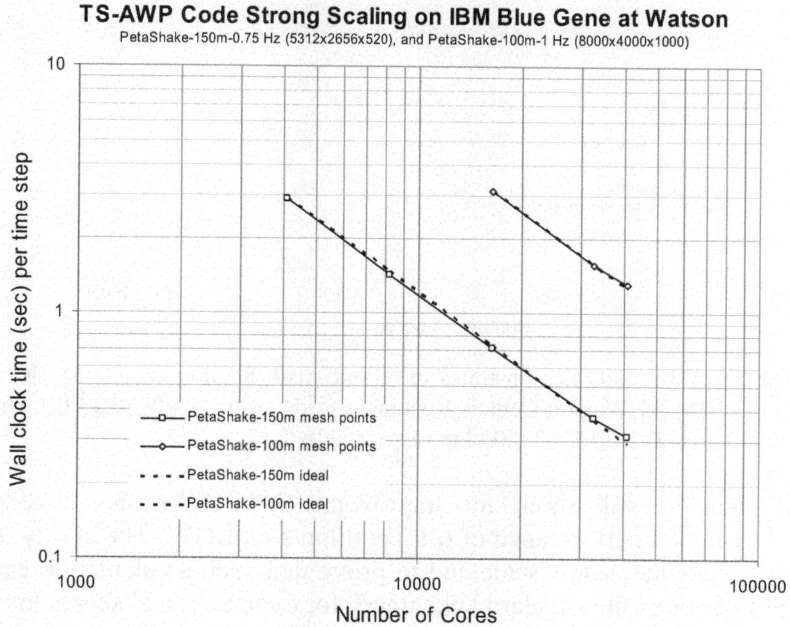


**Fig. 5** TeraShake TS-AWP code achieves 96% parallel efficiency of strong scaling between 4,096 and 40,960 BG/L processors at IBM TJ Watson Center

Excellent strong scaling of TS-AWP has been demonstrated on the Blue Gene/L machine at IBM TJ Watson Research Center (BGW), with an outstanding 96% parallel efficiency of strong scaling between 4,096 and 40,960 cores (Fig. 5). We benchmarked two PetaShake-sized simulations to demonstrate strong scaling (fixed total problem size) at Watson: the PetaShake-100 m with 8,000×4,000×1,000 mesh nodes, and the PetaShake-150 m with 5,312×2,656×520 mesh nodes. The weak scaling (the problem size grows proportionally with the number of processors) is also nearly linear up to 32,768 processors (Fig. 6). The weak scaling is demonstrated on $150^3$ km and 800 km × 400 km × 100 km with four different grid resolutions (Fig. 6).

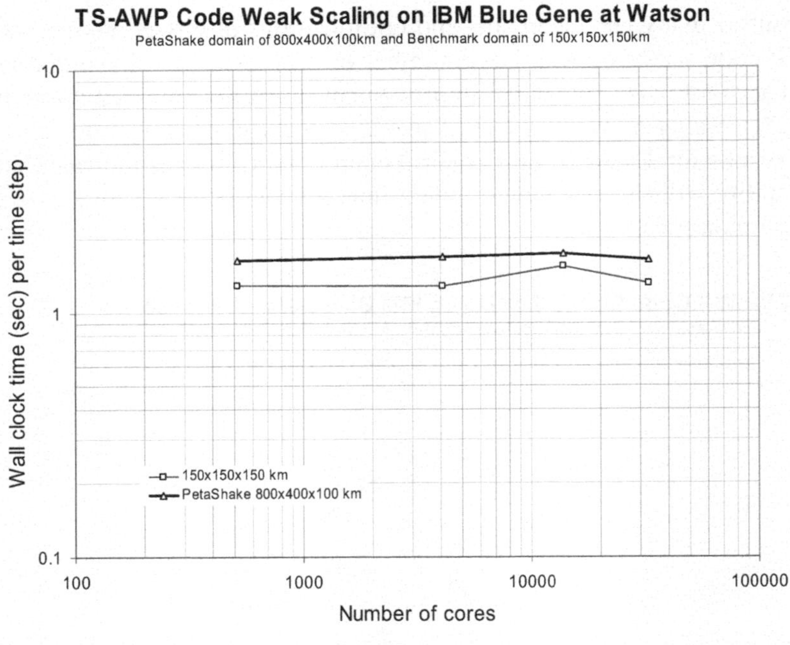


**Fig. 6** The TS-AWP code demonstrates excellent weak Scaling on 32,768 BG/L processors at IBM TJ Watson Center. Rectangle grids are at 0.4/0.2/0.13/0.1 km, triangle grids are at 0.2/0.1/0.07/0.05 km respectively

While there is still room for improvement, TS-AWP has already achieved sustained performance of 6.1 TeraFlop/s on BGW. The ability to benchmark the code at this scale and to prove that such a calculation can be done has defined the standard benchmark for computational seismology and is of enormous benefit to the seismological community because it establishes a high level of performance and performance expectations for these types of code from this time forward.

The reasons for the impressive scaling on the BlueGene/L are mainly hardware architecture related. The TS-AWP requires nearest-neighbor point-to-point communication and frequently uses the MPI_Barrier operation. The Blue Gene/L's 3-D torus communication network and extremely fast MPI_Barrier operation, assisted by hardware, was essential in achieving the perfect scaling. In addition to this, BGW dedication of compute nodes to running the user application also made a significant difference. The BGW compute nodes run a very lightweight kernel, and implement only the strictly necessary functionality, which minimizes any perturbations of the running process.

## 4.6 Preparing for TeraShake Executions

Checkpoint and restart capabilities were not available in the original TS-AWP code. This was not a critical issue as long as the code was only used to run small simulations that completed in a few hours. However, a single TeraShake simulation takes multiple days to complete. As we began to run larger and longer simulations, we realized that checkpointing and restart capabilities were essential requirements for the TS-AWP code. We integrated and validated these capabilities, partly prepared by Bernard Minster's group at Scripps Institution of Oceanography. Subsequently, we added more checkpoint/restart features for the initialization partition, as well as for the dynamic rupture algorithms. To prepare for post-processing and visualization, we separated the writes of volume velocity data output from writes of velocity surface data output. The latter was output at each time step. To track and verify the integrity of the simulation data collections, we generated MD5 checksums in parallel at each processor, for each mesh sub-array in core memory. The parallelized MD5 approach substantially decreased the time needed to checksum several terabytes of data.

The TS1 runs required a powerful computational infrastructure as well as an efficient and large-scale data handling system. We used multiple TeraGrid compute resources for the production runs at different stages of the project. Multiple post-processing steps were required to generate derived data products including surface seismograms and visualizations. The optimal processor configuration was a trade-off between computational and I/O demands. Volume data was generated at each 10th–100th time step for the runs. Surface data were archived for every time step. Checkpoint files were created at each 1000th step in case restarts were required due to reconfigurations or eventual run failures. The model computed

22,728 time steps of 0.011 sec duration for the first 250 sec of the earthquake scenario.

As we mentioned earlier, the TS2 simulations actually involved two simulations each, (1) a dynamic rupture simulation and then (2) a wave propagation simulation. The TS2 dynamic rupture simulations used a mesh size of 2,992 × 800 × 400 cells at grid spacing of 100 m, after the appropriate dynamic parameters were determined from several coarse-grid simulations with 200 m cells. Figure 7 shows the execution and data flows between compute resources at SDSC and the National Center for Supercomputing Applications (NCSA) and the local file systems at SDSC, a Global Parallel File System running between SDSC and NCSA, and an archival storage system (High Performance Storage System) at SDSC.

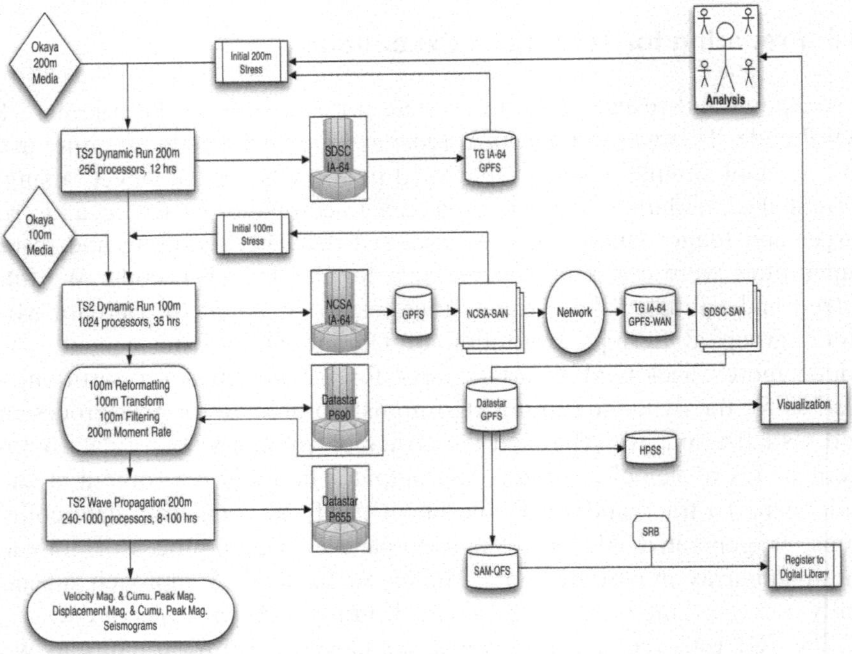


**Fig. 7** TeraShake-2 simulation data flow. Different TeraGrid resources were used at different stages of the simulation. The pre- and post-processing used fat nodes on SDSC 32-way IBM SP p690, the computational-intensive dynamic rupture runs used National Center for Supercomputing Applications (NCSA) IA-64, the data-intensive wave propagation simulations used well-balanced San Diego Supercomputer Center (SDSC) IBM SP 8-way DataStar p655 nodes. The files were accessed and transferred using fast gridFTP, and registered to SCEC digital library using SDSC Storage Resources Broker (SRB), backed up nearline on SAM-QFS file system and offline on IBM High Performance Storage System (HPSS)

The TS2 initial conditions were generated in multiple stages. The 200 m resolution runs were conducted on 256 TeraGrid IA-64 processors at SDSC. The 100 m resolution runs were conducted on 1,024 TeraGrid IA-64 processors at NCSA. The TS2 wave propagation runs were executed on up to 2,000 processors of Datastar, determined to be the most efficient available processor. The simulation output data were written to the DataStar GPFS parallel disk cache, archived on the Sun SAM-QFS file system, and registered into the SCEC Community Digital Library supported by the SDSC SRB.

The most recent ShakeOut simulation increased the upper frequency limit of the deterministic ground-motion prediction to 1 Hz from the 0.5 Hz supported in the TeraShake simulations. One hundred meter grid spacing on a regular grid results in 14.4 billion grid points, eight times more grid-points than the TeraShake scenarios. The simulation used up to 12k cores on the Texas Advanced Computing Center (TACC) Ranger computer and took 25–65 h to compute 240 sec of wave propagation. Ranger is a 504 TeraFlop/s Sun Constellation Linux Cluster configured with 3,936 16-way SMP compute-nodes (blades), 123 TB of total memory and 1.73 PB of global disk space, currently the largest computing systems for open science research. The pre-processing input partition and post-processing analysis were performed on the SDSC DataStar nodes.

## 4.7 Maintenance and Additional Techniques for the TeraShake Platform

While the large-scale wave propagation simulations were done using the TS-AWP code, additional codes were developed to automate and support the job of setting up, running, and analyzing the TS-AWP simulations. We consider these additional codes separate from the highly optimized TS-AWP but part of the TeraShake computational platform. In this section, we discuss some additional techniques we developed to support the TSCP.

**Scripts to enable workflow management.** We developed a set of post-processing tools to derive many terabytes simulation output to support scientific analysis and visualization. We improved the performance of the post-processing scripts for generating velocity magnitude and other quantities mainly through reductions in memory requirement. This resulted in a 6x speedup of the TeraShake post-processing calculations.

Terascale to petascale computation runs involve a large number of similar simulations with different parameter settings. To achieve better productivity and efficiency, a workflow mechanism is often necessary to automate the whole process and enable efficient management of output data files. We have built a large number of scripts with which a user can easily prepare the pipeline steps including pre-processing, batch jobs, post-analysis of the output and comparison of multiple simulations. These scripts form the foundation and components for automatic workflow mechanisms.

**Implementation of a parallel staggered-grid split-node method.** Dalguer and Day at SDSU integrated a parallel staggered-grid split-node (SGSN) method (Dalguer and Day 2007) into TS-AWP code. This SGSN approach was formulated to adapt the traction-at-split-node method in the velocity-stress staggered-grid finite difference scheme. This method provides an efficient and accurate means to simulate spontaneous dynamic rupture simulation. We have added MPI-IO and checkpointing/restart capabilities to support the new SGSN features.

**Automatic validation of the TeraShake application.** The research codes including the TS-AWP are constantly under development. It is necessary to validate every change to ensure that the new changes do not affect the validity of the scientific results. We have developed and implemented an automatic validation package into the application. This automatic validation package contains a small, well-defined set of TS-AWP input configuration files, and correct, or "reference", output seismogram files. We run the TS-AWP code using these known input files, and when the TS-AWP code produces output seismograms, we run a program that performs a least-squares difference between the "reference" seismograms and the new output files. This capability to quickly re-verify the TS-AWP after a software change has proven to be very helpful.

**Maintenance of CVS access for the TeraShake code.** We have made multiple codes that are part of the TSCP, including the TS-AWP code, accessible through the CVS repository to the SCEC community. We update the CVS repository across all code modifications, after validation tests are conducted. We update, document, integrate and validate all code changes such as SGSN, Earthworks-command, I/O changes, initialization changes, 64/32-bit switch and more. Included in the CVS package are: source code, documents, scripts, settings, testing cases, executables, and automatic validation scripts for benchmark with homogeneous media and point sources.

## 5 Data Archival and Management

The TeraShake and ShakeOut data management was highly constrained by the massive scale of the simulations. On the TS1 simulations, we saved both surface and volume velocity data. The output from the TS1 seismic wave propagation was migrated onto both a Sun SAM-QFS file system and the HPSS archive as the run progressed in order to sustain a data transfer rate over 120 MB/sec. This enabled the storage of 10 TB/day of simulation output to tape drives that sustained up to 120 MB/sec I/O rates. With current tape drive technology, data rates that are 4 times larger are sustainable.

The TS1 and TS2 simulations comprise hundreds of terabytes of binary output and more than one million files, with 90,000 – 120,000 files per simulation. We organized each simulation as a separate sub-collection in the SRB data grid and published the sub-collections through the SCEC community digital library. We labeled the files with metadata attributes which define the time steps in the simulation, the velocity component, the size of the file, the creation date, the grid spacing, and the number of cells, etc. (Moore et al. 2005). Researchers can access all files registered into the data grid by their logical file name, independently of whether the data were on the parallel file system GPFS, the SAM-QFS archive, or the HPSS archive. General properties of the simulation such as the source characterization are associated as metadata with each simulation collection. Integrity information is associated with each file (MD5 checksum) as well as the location of replicas. Since even tape archives are subject to data corruption, selected files are replicated onto either multiple storage media or multiple storage systems. The SCEC digital library includes the digital entities (simulation output, observational data, and visualizations), metadata about each digital entity, and services that can be used to access and display selected data sets. The services have been integrated through the SCEC portal into seismic-oriented interaction environments (Moore et al. 2005; Olsen et al. 2006b). A researcher can select a location on the surface of an earthquake simulation scenario, and view the associated seismogram, by pointing and clicking over the interactive cumulative peak velocity map, or interact with the full resolution data amounting to 1 TB.

## 5.1 SCEC Data Grid

The SCEC community digital library integrates the GridSphere portal technology (www.gridsphere.org/) with the SRB data grid (http://www.sdsc.edu/srb/). The portal technology provides a standard web-based interface for applying seismogram extraction services on the simulation output, for presenting visualizations of the seismic wave propagation, and for browsing the SCEC collection. GridSphere is written in Java, which simplifies the porting of the portal onto different operating systems. Each service is implemented as a Java portlet that can be invoked through the GridSphere portal.

The SRB data grid provides the distributed data management needed for a shared collection that resides on multiple storage systems. The basic capabilities provided by a data grid can be characterized as data virtualization, trust virtualization, latency management, collection management, and federation management (Moore et al. 2005). Data virtualization is the management of data collection properties independently of the storage repositories where the data are stored. It is achieved by providing two levels of indirection between the GridSphere portal and the underlying storage system. The SRB maps from the functions specified by GridSphere through a Java class library interface to a standard set of operations that can be performed on data at remote storage systems. The SRB then maps from the standard operations to the specific protocol required by the remote storage system. This ensures that modern access methods for data manipulation can be used on all types of storage systems, whether file systems, tape archives, databases, or object ring buffers. The standard operations include Posix I/O calls such as open, close, read, write, seek, and stat.

The SRB uses a logical file name space to provide global persistent identifiers for each file. Since the files may reside in multiple storage systems, the SRB must map from the global persistent identifier used in the portal to the actual file location (IP address and file system pathname). The location of the file is stored as a metadata attribute attached to the logical file name. The SRB manages replicas of files, making it possible to put a copy on disk for interactive access while managing a copy in a tape archive for long term storage. The locations of the replicas are also attached to the logical file name as metadata attributes. In practice, all system state information required to manage each file is preserved as metadata attributes associated with the logical file name. The properties of the files (owner, size, creation date, location, checksum, replicas, aggregation in containers, etc.) are managed by the SRB independently of the storage repository. Users can also assign descriptive metadata to each file, including information about the simulation that created the file. A collection hierarchy is imposed on the logical file name space, with each sub-collection able to manage a different set of descriptive metadata attributes.

The SRB is middleware, consisting of federated software servers that are installed at each location where data may reside, and clients that support a wide variety of access mechanisms. The SRB servers are organized as a peer-to-peer environment. When a GridSphere portal accesses a SRB server, the request is forwarded to the SRB server that manages the metadata catalog holding the persistent state information and descriptive metadata attributes. The actual location of the requested file is resolved, and the specified operation is sent to the remote SRB server where the file is located. The operation is performed and the result sent back to GridSphere.

The structure of the SRB data grid is shown in Fig. 8. The SRB is generic infrastructure that supports a wide variety of access mechanisms, of which a GridSphere portal is one type of client. The three basic access mechanisms are a C library, a set of Unix shell commands, and a Java class library. All other interfaces are ported on top of one of these clients. Examples of other interfaces include digital library systems such as DSpace (http://www.dspace.org/) and Fedora (http://www.fedora.info/), workflow systems, load libraries, web browsers, and community specific services. The SRB supports access to a wide variety of storage systems. Storage resources are logically named, enabling collective operations such as load leveling across storage repositories. Separate drivers are written for each type of storage repository.

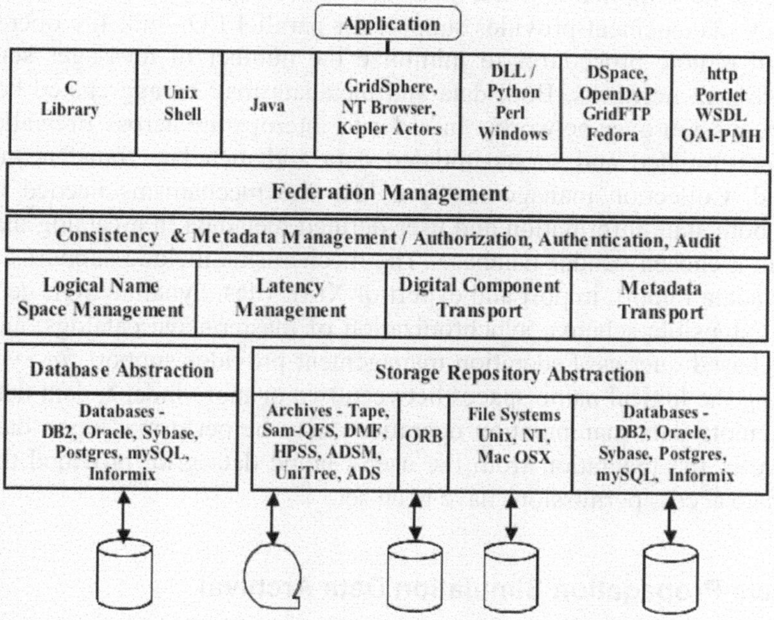


**Fig. 8** Storage Resource Broker (SRB) data grid components

The architecture is modular, enabling the addition of new drivers without having to modify any of the access mechanisms. Similarly, new access mechanisms can be added without having to modify any of the storage resource drivers. The peer-to-peer server architecture also allows the dynamic addition of new storage systems. Since the SRB manages all properties of the shared collection, data can be replicated onto the new storage device dynamically. This enables the migration of data onto new storage devices and the removal of obsolete storage systems without affecting the ability of users to access the shared collection.

Two other virtualization mechanisms are also important. Trust virtualization is the management of authentication, authorization, and audit trails independently of the storage repositories. The SRB accomplishes this by assigning ownership of the files to the SRB data grid. The SRB manages distinguished names for the users of the data grid independently of the storage system. Users may be aggregated in groups. Access controls by either user name or group membership are then managed as constraints imposed between the logical resource name, the logical file name, and the SRB distinguished user names. These controls remain invariant as the files are moved between storage systems under SRB control. Access controls can be separately applied on files, metadata, and storage systems.

Latency management provides support for parallel I/O, bulk file operations, and remote procedures to minimize the number of messages sent over wide area networks. Both data and metadata may be aggregated before transmission over networks. In order to interoperate across firewalls, both client-initiated and server-initiated data and metadata transfers are supported. Collection management provides the mechanisms needed to manage both state information and user-defined metadata in a catalog that resides in a chosen vendor database. The mechanisms include support for bulk metadata import, import and export of XML files, dynamic SQL generation, extensible schema, synchronization of master/slave catalogs, and attribute based queries. Federation management provides support for synchronizing the logical name spaces between two or more independent data grids. Remote data manipulation operations may be performed upon data anywhere in the federation from the user's home data grid, provided the appropriate access permissions have been set.

## 5.2 Wave Propagation Simulation Data Archival

Each time step of a TeraShake simulation produces a 20.1 GByte mesh snapshot of the entire ground motion velocity vectors. Surface data were archived for every time step totaling one (1) terabyte. Up to 133,000 files were generated per run. The data management challenges were further increased when, for one of the TeraShake runs, we decided to save a 4D wavefield containing 2,000 time steps, amounting to 40 TB of data. The available disk space during the run was only 36 TB. Thus as the data were generated, they had to be moved to an archive for storage. Two archival storage systems were used to ensure the ability to move 10 TB/day, for a sustained data transfer rate over 120 MB/sec from the disk to the archive.

### 5.2.1 SCEC Data Management Challenges

A data collection was created that annotated each file with metadata that defined the time step in the simulation, the velocity component, the size of the file, the creation date, the grid spacing, and the number of cells. General properties of the simulation such as the source characterization were associated as metadata for the simulation collection. The data are organized in a collection hierarchy, with a separate sub-collection for each simulation run.

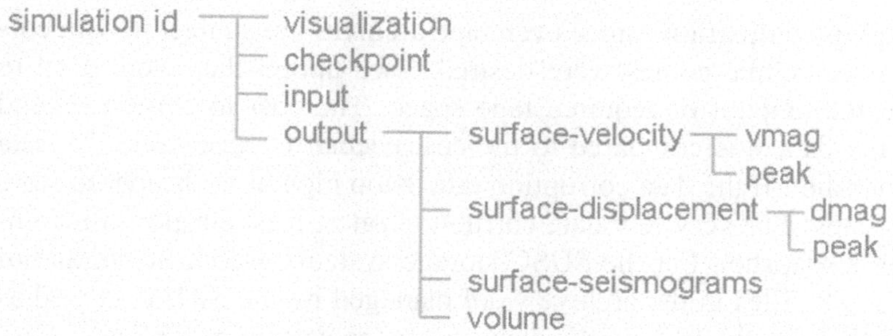


**Fig. 9** SCEC digital library collection hierarchy

Within the sub-collection, files are organized in a collection hierarchy shown in Fig. 9. The binary output are described using Hierarchical Data Format headers and each file is fingerprinted with MD5 checksums for future validation of data integrity. The HDF version 5 technology (http://hdf.ncsa.uiuc.edu/HDF5/) supports the creation of a separate HDF header file for each output file. The location of the HDF5 header file is then stored as an attribute on the logical file name in the SRB data grid. The simulation output and the derived data products were registered into the SCEC digital library. Users can access the SCEC public collections from a web browser, using the URL: http://www.scec.org/diglib.

The data management requirements after the generation of the simulation output were equally challenging. SDSC provides storage space to SCEC for over 250 TB of tape archive, and on average 4 TB of on-line persistent disk space. Cache disk space to support applications on average is about 5 TB. The challenges included:

**Providing a common logical name space across all storage systems.** The SRB data grid provides this automatically through the SRB logical name space. All files registered into the data grid can be accessed by their logical file name, independently of whether the data were on GPFS, SAM-QFS, or the HPSS archive.

**Managing data distribution.** Data were migrated from the archive to the disk cache to support visualizations and to compare results between runs. Derived data products such as surface seismograms were kept on the persistent disk. All data were archived in either HPSS or SAM-QFS. By registering each file into the SRB data grid, the migration between systems could be done using a common logical file name space. This required installing a SRB server on each storage system.

**Managing replication.** Since even tape archives are subject to data corruption, two archive copies were desired. Two approaches were used to minimize the amount of required tape space. The cost to store a second copy of the data was compared to the cost needed to regenerate the data from a snapshot. If the data corruption rate is too high, it is cheaper to store a second copy. For very low data corruption rates, it is cheaper to recompute from a snapshot. For the SDSC storage systems, we chose storage of snapshots. The files in the archive were managed by the SRB data grid as the originals, with the files on the disk systems registered as replicas.

**The second approach used syntactic replication.** We observed that the surface velocity from the simulation was transformed to create the surface seismogram files (space order versus time order). Thus the derived seismogram data provided the equivalent of a backup copy of the surface velocity data.

**Automating ingestion of file metadata.** The efficient management of MD5 checksums, HDF5 descriptive headers, and descriptive metadata for each file required bulk metadata loading into the SRB data grid. Administrative tools were written to simplify the management of the metadata generation and registration into a SRB collection.

**Managing access controls across the multiple storage systems.** The SRB supports multiple access roles, enabling the identification of a curator who manages the publication of data and metadata in the SCEC Digital Library. Each user of the system is given a "home" collection under which they can store simulation results. A curator account is used to control publication of selected data to the SCEC collection published at /home/sceclib.scec. An additional access role is defined for management operations to backup high priority datasets. Public access to the SCEC digital library is managed through a "public" account that is given read access permission.

**Validating collection integrity.** Since the risk of data loss is present for each storage system, periodic verification of checksums is necessary for each file. If a problem is detected, synchronization of valid replicas is used to automate error correction. The SRB data grid provides explicit administrative commands for such synchronization.

### *5.2.2 Comparison to Grid Technology*

The differences between the SRB data grid and other approaches for distributed data management are the degree of integration of the SRB software, the consistency assertions performed by the SRB, the generality of the SRB solution, and the SRB's ability to interoperate with all major data storage environments. Comparisons are usually made with Grid middleware (http://www.ogf.org/), in which a differentiated service with a separate state information catalog is created for each desired function (authentication, authorization, storage resource discovery, replication, metadata management, and data movement). Interactions between the Grid middleware services are managed by the calling application, including the decision of the order in which to make the service calls, and the coordination of consistency across the state information. From the perspective of the Grid middleware approach, the SRB is an application that provides consistent management of distributed data, as opposed to an access service to data residing in a file system.

With the SRB data grid, the coordination is built into the system. The user specifies a request through their preferred access mechanism (load library, browser, workflow actor, digital library, C library call, Unix shell command) and the SRB issues the required operations on the remote systems. The original SRB environment supported the traditional 16 Posix I/O operations. In order to support digital libraries and preservation environments, the number of operations has been extended to over 80 functions that can be executed at the remote storage system. The extensions include support for database queries, metadata registration, execution of remote procedures, integrity checking, synchronization, aggregation into containers, third party transfer, and server-initiated parallel I/O. These additional functions, the ability to simplify administration through use of an integrated system, and the support for digital libraries as well as preservation environments have driven the widespread use of the SRB technology.

## 5.3 SCEC Digital Library

The SCEC/CME collaboration and SDSC have developed digital library interfaces that allow users to interactively access data and metadata of ground motion collections, within a seismic oriented context. The Earthquake Scenario-Oriented Interfaces (available at the URLs: http://sceclib.sdsc.edu/TeraShake and http://sceclib.sdsc.edu/LAWeb) are built upon the WebSim seismogram plotting package developed by Olsen.

Surface seismograms are accessed through the SCEC seismogram service within the SCEC portal. A researcher can then select an earthquake simulation scenario and select a location on the surface, by pointing and clicking over an interactive cumulative peak velocity map. The portal accesses the correct file within the SCEC community digital library, and displays all three velocity components for the chosen site and scenario. Users can use this web application, shown in Fig. 10, to interact with the full surface resolution (3,000 × 1,500) data of a TeraShake scenario, amounting to 1 TB per simulation.

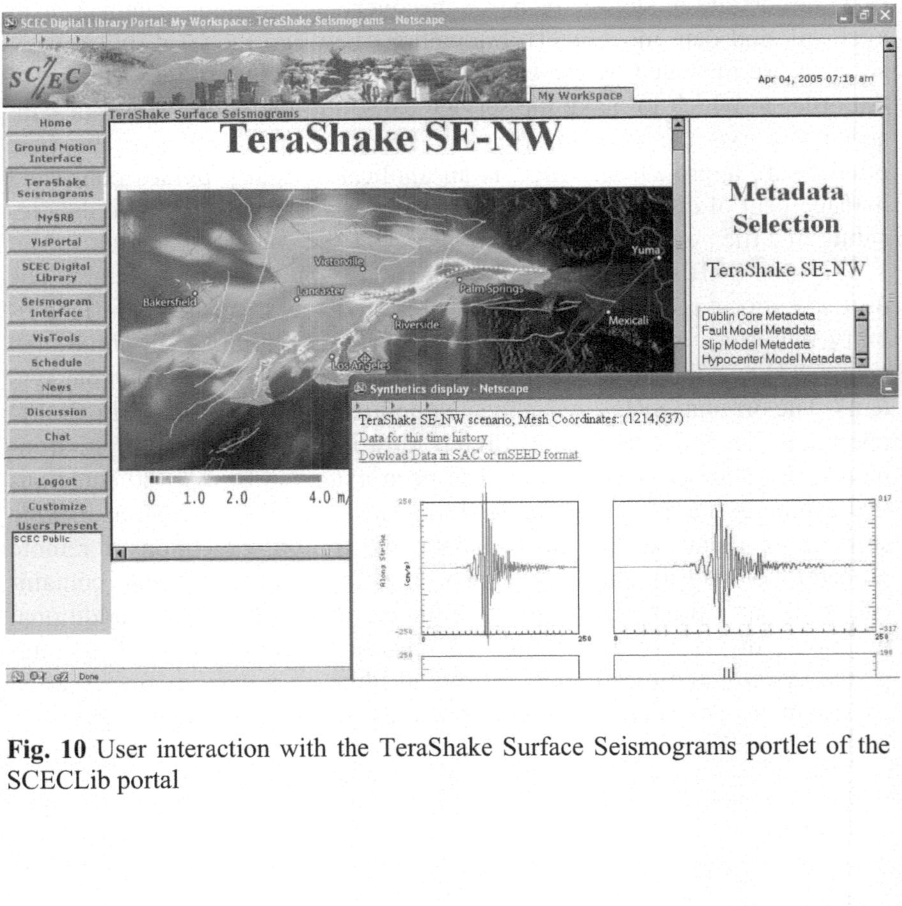


**Fig. 10** User interaction with the TeraShake Surface Seismograms portlet of the SCECLib portal

As more sophisticated simulations are executed in the future, the SCEC digital library will need to be updated to provide state-of-the-art results. This is the driving factor behind the creation of digital libraries of simulation output. The research of the future depends upon the ability to compare new approaches with the best approaches from the past, as represented by the digital holdings that the community has assembled.

## 6 TeraShake Visualization

Visualization has been a key in investigation of the TeraShake simulations. The large amount of data produced by the simulation requires new methods and techniques for analysis. Since the beginning of the study, the visualization process has been collaborative. At the very first stage, the surface data was color mapped, and it went through several iterations to capture and highlight the desired data range and features. Since most of the data is bidirectional (velocity components in positive and negative direction), the color ramp was chosen to have hot and cold colors on both ends, and the uninteresting range in the middle was chosen to be black (see Fig. 11a).

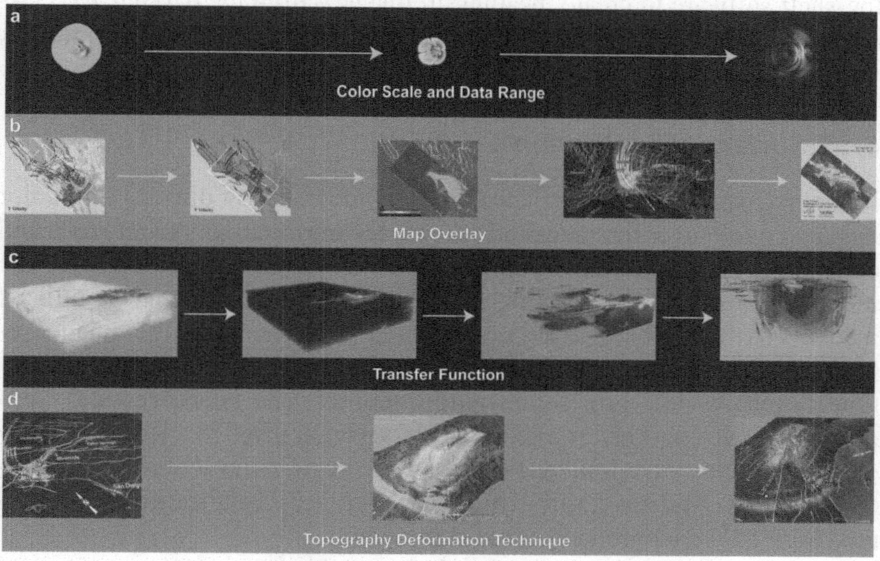


**Fig. 11** Iterative refinements of the visualization incorporating feedback from scientists

The next step was to provide contextual information; this process also underwent several revisions as better overlay maps become available with time (see Fig. 11b). Concurrently, different transfer functions were designed to capture features of interest through direct volumetric rendering (see Fig. 11c). Lastly, alternate methods like topography deformation (see Fig. 11d) and self contouring (see Fig. 12) were developed to aid analysis. There were several challenges to accomplish visualization; foremost difficulty was handling the sheer size of data ranging from 1 TB to 50 TB for each simulation. Keeping this data available for a prolonged period on a shared filesystem or moving it around was impractical. Interactive visualization tools for this capacity of temporal data are either virtually nonexistent or require specialized dedicated hardware. Furthermore, the disparate geographic location of scientists also made these tasks difficult. Thus, we settled on providing packaged animations through the web which were refined over time from feedback.

## 6.1 Visualization Techniques

We utilized existing visualization techniques and combined them with off-the-shelf software to create meaningful imagery from the dataset. We classify our visualization process into four categories: surface, topographic, volumetric and static maps.

### 6.1.1 Surface Visualization

In this technique, the 2D surface data are processed via direct 24-bit color map and overlaid with contextual geographic information. Annotations and captions provide additional explanation. The temporal sequence of these images is encoded into an animation for general dissemination. We utilized Adobe's After Effects™ for compositing and encoding the image sequences. Layering the geographic information with high resolution simulation results provides precise, insightful, intuitive and rapid access to complex information. This is required for seismologists to clearly identify ground motion wave-field patterns and the regions most likely to be affected in San Andreas Fault earthquakes. Surface visualizations were created for all 2D data products using this method.

Comparison of multiple earthquake scenarios was vital to understand rupture behavior. In Fig. 12 the left image shows the areas affected by a rupture traveling northwest to southeast; the right image shows the corresponding results for a rupture moving southeast to northwest. Both images have geographical and contextual information (fault lines, freeways) overlain.

Seismic information of time, peak velocity and instantaneous location are shown visually via a graphical cursor and in text. The goal is to gain an understanding of the region most heavily impacted by such an earthquake and the degree of damage. The side-by-side character of this visualization also provides the scientists an important visual intuition regarding similarities and differences between fault rupture scenarios. In particular, such comparison revealed significant differences in the ground motion pattern for different rupture directions, and in one case, wave-guide effects leading to strong, localized amplification. With use of animations the scientists were able in some instances to detect instabilities in the absorbing boundary conditions, and to identify alternate conditions to remedy the problem. In other instances, specific physical behaviors were observed, such as that the rupture velocity was exceeding the S wave speed at some locations. Such "supershear" rupturing is of great interest to seismologists, because it generates different types of ground motions than subshear ruptures (Dunham and Archuleta 2005).

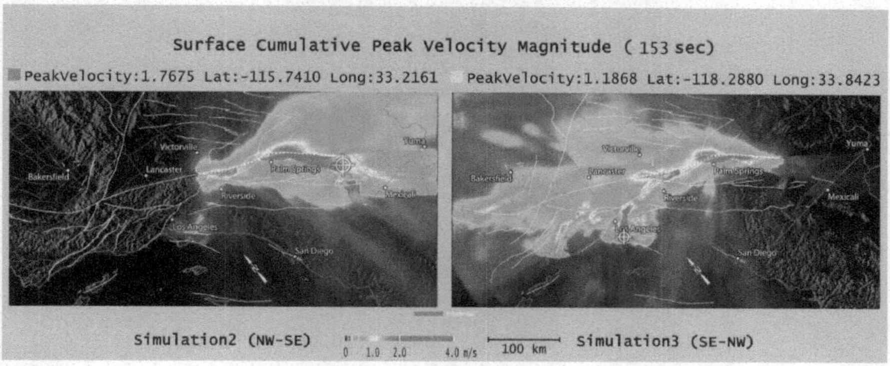


**Fig. 12** Image comparing maps of maximum ground velocity in two of the TeraShake simulated earthquake scenarios on the San Andreas Fault. TS1.2 (*left*) is for rupture toward the southeast, while TS1.3 (*right*) is for rupture toward the northwest (see *Movie* 2, available on accompanying DVD)

### *6.1.2 Topographic Visualization*

This process utilizes the dual encoding of the surface velocity data as both color mapping and as displacement mapping. The surface velocity data were used to create a color mapped image; the displacement magnitude calculated from the surface velocity data were used to generate a grey scale image. The grey scale image is used as a displacement map to create terrain deformation along the vertical axis (see Fig. 13a) using Autodesk's Maya™.[1]

The animation corresponding to Fig. 13a, allows the scientist to gain a better understanding of the kind of waves propagating in the model. Another example is use of across-sectional view (see Fig. 13b) that shows the surface on the southwest side of the fault is lowered to the bottom of the fault to allow the rupture on the fault to be viewed. This kind of visualization allows the seismologists to connect surprising features in the ground motion with features in the rupture propagation. Currently we are working to develop a method to display true three-dimensional deformations based on three components of surface velocity data to provide more realistic insight.

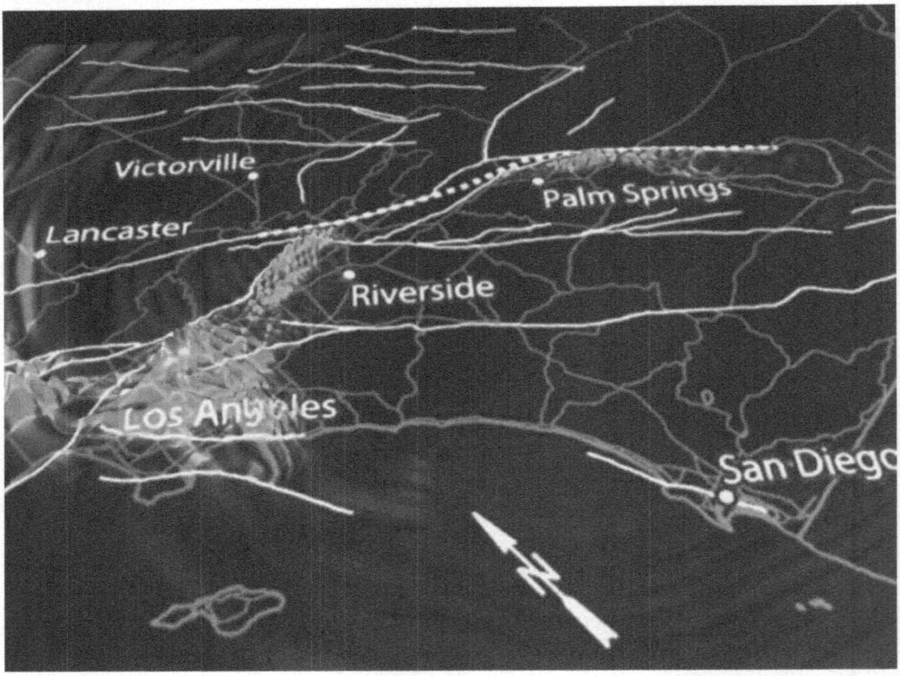


**Fig. 13a** Deformation along vertical axis of the terrain using displacement mapping to show velocity magnitudes along with color (see *Movie* 3, available on accompanying DVD)

[1] Maya is widely used animation software in films, which offers state of the art 3D modeling, lighting, rendering and animation capabilities. URL: http://www.autodesk.com/maya

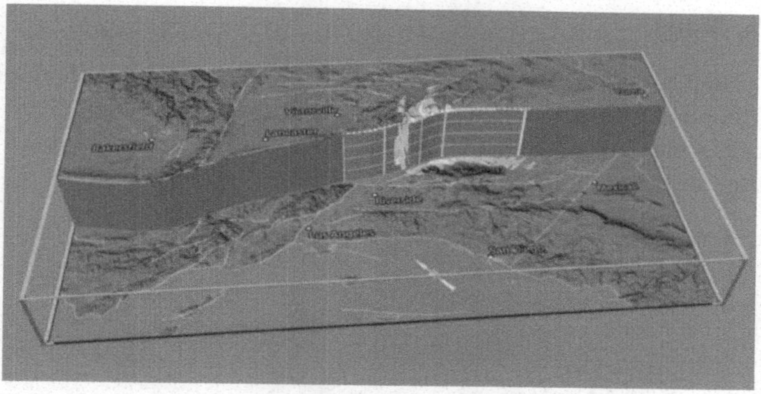

**Fig. 13b** A cross sectional view showing sliprate on the vertical fault and velocity magnitude of wave propagation on the horizontal surface (see *Movie* 4, available on accompanying DVD)

### 6.1.3 Volumetric Visualization

The largest data sets produced by the TSCP simulations are volumetric data sets. We performed direct volume rendering (Kajiya and Herzen 1984) of the volumetric dataset and composited it with contextual information to provide a holistic view of the earthquake rupture and radiated waves to the scientists (see Fig. 14). Our initial work has helped the seismologists to see the general depth extent of the waves. For example, dependent on the component of the wavefield, waves propagating predominantly in the shallow layers may be identified as surface waves. Such waves typically contain large amplitude and long duration and can be particularly dangerous to some types of structures. However, more research in addition to existing work (Chopra et al. 2002; Yu et al. 2004; Uemura and Watanabe 2004) needs to be done to represent multivariate data in a unified visual format. Additional challenge in volumetric visualization is how to visually present to the user a global understanding of the behavior of the seismic wave while at the same time allowing them to examine and focus on localized seismic activity. This challenge is important, as often only reviewing the global behavior of the seismic wave hides important localized behaviors while at the same time simply focusing on localized activity often hides how this localized motion impacts the overall behavior of the wave.

### 6.1.4 Static Maps

Some data products like spectral acceleration (depict the vibration characteristic of ground at different frequencies see Fig. 15), peak ground velocities and peak ground displacements are non-temporal and require visual representation for better understanding.

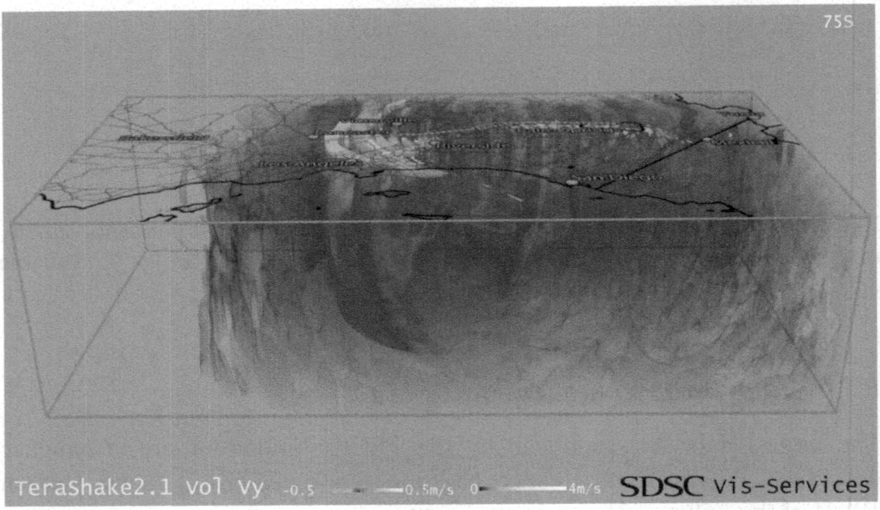


**Fig. 14** Snapshot showing volume rendered velocity in y direction (see *Movie* 5, available on accompanying DVD)

### 6.1.5 Self Contoured Maps

We developed a technique to highlight features in 2D by using bump mapping. The detailed treatment of this approach is described in detail by Wijk (Wijk and Telea 2001). Encoding the spectral acceleration levels using both color maps and bump maps reveals subtle transitions between ground motion levels within localized regions with similar spectral acceleration properties. The color and bump encoding technique brought out variations in the data that were not previously visible (see Fig. 15).

### 6.1.6 Map Service Portal for Surface Data

Scientists want to conduct hands on analysis in an attempt to gain a better understanding of the output data. The large amounts of TeraShake data pose a significant problem for accessibility and analysis. We developed a web front end (Fig. 16) where scientists can download the data and are able to create custom visualizations over the web directly from the surface data. The portal uses LAMP (Linux, Apache, MySQL, PHP) and Java technology for web middle-ware and on the back-end compute side relies on specialized programs to fetch data from the archive, visualize, composite, annotate and make it available to client browser. We have also added support to create geo-referenced images which could be viewed in Google Earth.[2]

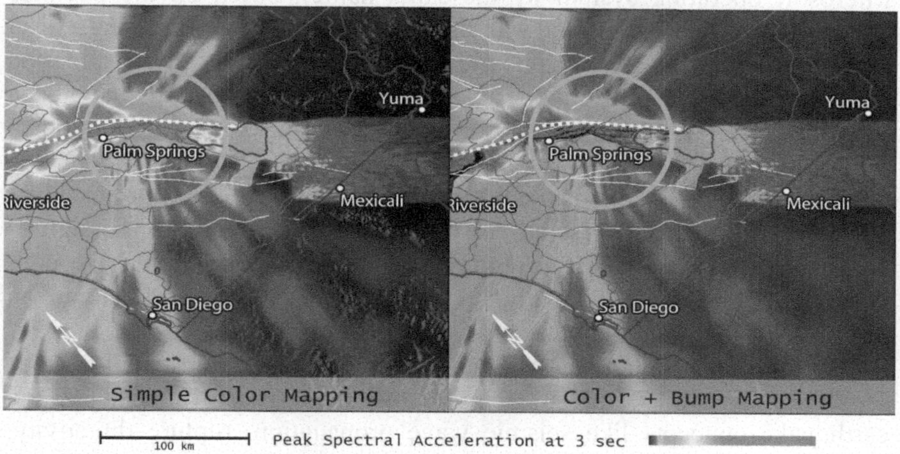


**Fig. 15** Maps showing the advantage of the self contouring technique in contrast to simple color mapping. Our innovative color encoding technique brought out variations in the data that were not previously visible. In the region highlighted by the *orange circle*, scientists identified a star burst pattern, indicating an unusual radiation of energy worthy of further investigation, which went unnoticed with simple color mapping

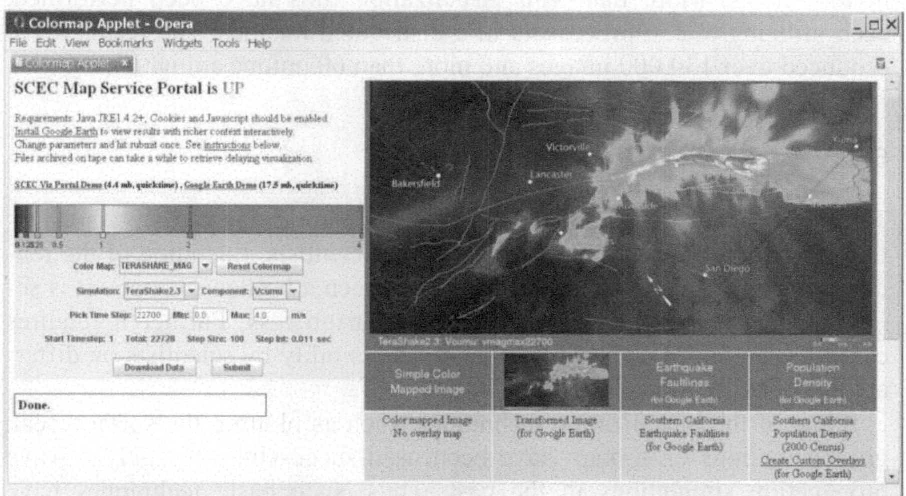


**Fig. 16** Screenshot of the map service portal

[2] Google Earth is an interactive 3d geo browser URL: http://earth.google.com/

## 6.2 Visualization Tools and Results

SDSC's volume rendering tool called Vista (not to be confused with the Microsoft operating system by the same name), based on the Scalable Visualization Toolkit (SVT), was used for visualization rendering. Vista employs ray casting (Kajiya and Herzen 1984) with early ray termination for performing volumetric renderings. Surface and volume data have been visualized with different variables (velocities and displacements) and data ranges in multiple modes. The resulting animations have proven valuable not only to domain scientists but also to a broader audience by providing an intuitive way to understand the results. Visualization required significant computational resources. TeraShake Visualizations alone have consumed more than 10,000 CPU hours and over 30,000 CPU hours on SDSC's DataStar and TeraGrid IA-64 respectively.

The use of multiple data sets with different visual representations (see Figs. 12, 13, 14, 15 and 16) helps the seismologists to understand key earthquake concepts like seismic wave propagation, rupture directivity, peak ground motions, and the duration of shaking. The strong visual impact leads the viewer from the global context of earthquake hazards to the hazards in a specific region and then into the details about a specific earthquake simulation. Viewing the earthquake simulation evolve over time leads viewers to gain insights into both wave propagation and fault rupture processes, and illustrates the earthquake phenomena in an effective way to non-scientists. More than 100 visualization runs have been performed, each utilizing 8–256 processors in a distributed manner. The results have produced over 130,000 images and more than 60 unique animations.[3]

## 6.3 Visualization Discussion

The role of contextual information has been pivotal; the surface visualizations have proven to be very helpful to the scientists. Encoding of the rendered image sequence into animations has been a bottleneck since it is serial, time consuming and lossy compression process. Further it requires careful selection of codecs for broader accessibility by scientists on different platforms.

Some of the visualization techniques, such as plotting the surface peak ground motions on a map; have been used successfully to analyze wave propagation simulations in the past. Thus, such basic techniques have

[3] TeraShake visualization webpage: http://visservices.sdsc.edu/projects/scec/TeraShake

proven to provide useful scientific information and are useful for initial assessment of simulations. However, without the exploratory visualizations applied to the TeraShake simulations, such as the color and bump encoding technique, the origin of some features in the results would have been unclear. These new methods should be considered in future large-scale simulations of earthquake phenomena. While these methods are under development, future efforts should concentrate on documentation of the procedures to promote widespread use.

In addition to the scientific insight gained, these methods can provide important instructional learning material to the public. For instance, a TeraShake animation is being used in teacher training in Project 3D-VIEW, a NASA funded program, potentially becoming a part of the curriculum in thousands of classrooms. One of the animations was featured in National Geographic channels documentary on "LA's Future Quake". Other example is, the composite of the fault plane and surrounding crust (Fig. 11c) illustrating the connection between the earthquake rupture and the shaking felt in the vicinity of the fault.

As large temporal datasets pose significant challenges for analysis, automation of visualization techniques and methods applied in a planned way is desirable. Interactive visualization of large temporal datasets seems useful but is non-trivial and often impractical. Domain specific feature capturing algorithms coupled with visualization can play an important role for analysis. Animations though non-interaction can often serve the purpose for gaining insight when created in a thoughtful manner. Off-the-shelf software's such as Maya™ can augment scientific visualization tools. Multiple representations of the same data products with different techniques can be valuable assets. Use of high dynamic range (HDR) imagery for amplifying fidelity and precision in visualization has seemed promising but lack of HDR display hardware and plethora of tone mapping methods make this task difficult.

## 7 Scientific Results of TeraShake-1 and TeraShake-2

Three TS1 simulations were carried out using a hypocentral depth of 10 km. One scenario starts at the northwestern end rupturing toward the southeast, and two start at the southeastern end and rupture toward the northwest. Fig. 12 shows the maximum root-sum-of-square peak ground velocity (PGV) for all components for the NW-SE and SE-NW(1) scenarios of TS1. The PGV distributions reveal a striking contrast in ground motion pattern between NW-SE versus SE-NW rupture scenarios. In addition to the expected rupture directivity in the rupture direction, the chain of sedimentary basins running westward from the northern terminus of the rupture to downtown Los Angeles forms a low-velocity structure that acts as a waveguide. The TeraShake simulations show that this waveguide may trap seismic energy along the southern edge of the San Bernardino and San Gabriel Mountains and channel it into the Los Angeles region. This guided wave is efficiently excited by both SE-NW rupture scenarios, but not appreciably by the NW-SE rupture scenario (Fig. 12). The waves amplified by the waveguide for the TS1 SE–NW scenarios generated PGVs (>3 m/sec) in localized areas, much larger than expected for the Los Angeles area.

To test whether the unexpectedly large ground motions obtained for the SE-NW TS1 simulations were related to the somewhat simplified kinematic source derived for the Denali earthquake we have carried out a series of simulations (TS2) with sources derived from spontaneous rupture models. Due to numerical limitations of the finite-difference method on a Cartesian grid, the dynamic rupture modeling with this method is limited to planar fault surfaces. For this reason, we used a two-step, approximate procedure to compute the ground motions from the segmented San Andreas fault rupture. Step one was a spontaneous dynamic rupture simulation for simplified, planar fault geometry. Step two was a separate, kinematic simulation, using as a source the space-time history of fault slip from step one, mapping the latter onto the segmented San Andreas fault geometry.

The dynamic rupture sources were modeled using a simple slip weakening friction law, implemented using the stress-glut method (Andrews 1999). The rupture was nucleated artificially in a small patch near the end of the fault. The initial shear-stress distributions were generated from a sequence of approximations of the dynamic inversion results for the M7.3 1992 Landers earthquake (Peyrat et al. 2001). Three different high-resolution (dx = 100 m) dynamic rupture simulations were generated and used as sources for three different TS2 wave propagation runs (dx = 200 m). The stress and friction parameters were tapered in the upper 2 km of the fault to avoid numerical artifacts.

Figure 17 compares the kinematic source in TS1 and dynamic source in TS2 by a snapshot of the sliprate. The dynamic source is characterized by strong variations in rupture speed as well as frequent separation into several slipping areas of highly varying shape and sliprate. In contrast, the kinematic rupture has constant rupture velocity and slip contained in six consecutive elementary pulses.

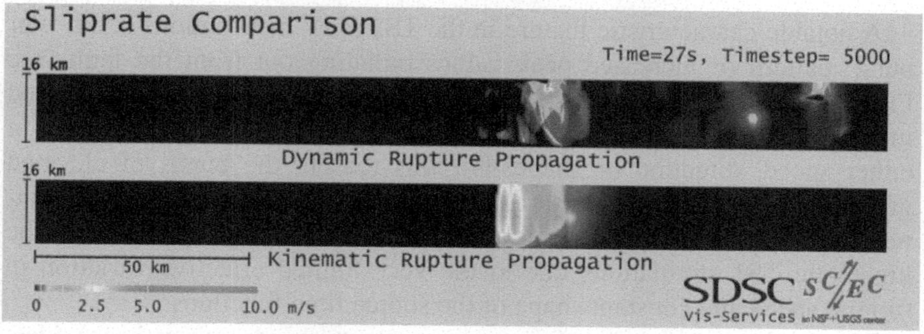


**Fig. 17** Comparison of snapshots of sliprate for dynamic (*top*) and kinematic (*bottom*) source (see *Movie* 6, available on accompanying DVD)

Figure 18 compares the PGV for TS1.3 (kinematic source) and TS2.1 (dynamic source). The PGV patterns from the dynamic source descriptions contain the same overall features as for the kinematic TS1 results, such as rupture directivity and localized amplification. However, the PGVs from the dynamic rupture are generally smaller than those from the kinematic source by a factor of 2–3. We believe that the smaller PGVs for the dynamic source models are mainly caused by less coherent wavefronts generated by the complex dynamic source, as compared to those from the much simpler kinematic rupture propagation. In particular, the abrupt changes in direction and speed observed in the spontaneous rupture propagation tend to decrease the amplitudes of the radiated wavefield in the forward-directivity direction.

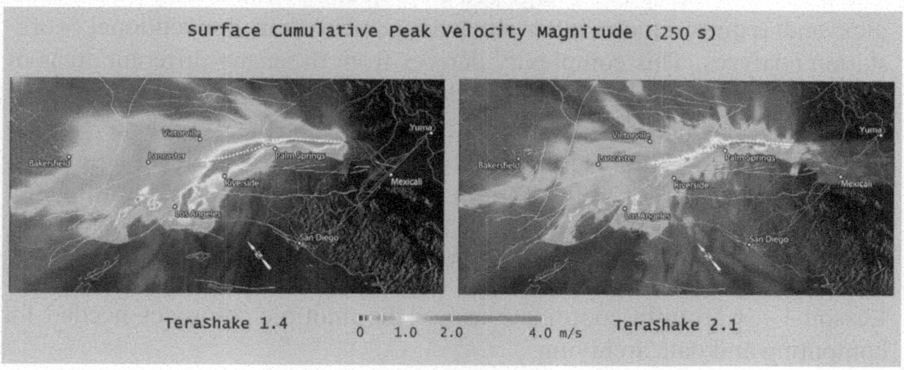


**Fig. 18** Comparison of PGVs for SE-NW scenarios using (*left*) kinematic and (*right*) dynamic sources (see *movie* 7, available on accompanying DVD)

A notable characteristic feature in the TS2 PGV distributions is the 'star burst' pattern of increased peak values radiating out from the fault (see Fig. 18, right). These rays of elevated peak ground motions are generated in areas of the fault where the dynamic rupture pulse changes abruptly in either speed, direction or shape. For this reason, the bursts of elevated ground motion are also correlated with pockets of large, near-surface slip-rates on the fault. Such pattern is absent from the PGV distributions for the kinematic TS1 simulations due to the very limited effective variation in rupture speed and constant shape of the source time functions.

Despite the significant differences in peak amplitudes, there is considerable similarity between the spatial patterns of ground motion excitation for the TS1 and TS2 scenarios with a common rupture direction. An important example is the wave-guide induced band of amplification extending into the Los Angeles basin for the SE-NW ruptures. Thus, the TS1 and TS2 results indicate that such sedimentary waveguide effects, where they exist, may have a large, systematic impact on long-period shaking levels.

A comprehensive set of images and animations based on the TeraShake simulations is presented on the SDSC visualization group website: http://visservices.sdsc.edu/projects/scec/. The science results can be found in (Olsen et al. 2006a, 2007).

## 8 Lessons Learned from Enabling Very-Large Scale Earthquake Simulations

The process of executing a large-scale application simulation is more complex and requires more intensive management than conventional workstation analyses. This complexity derives from the many different areas of expertise from multiple disciplines that jointly participate in the problem-solving. This is particularly true while performing the TeraShake simulations as we worked to resolve the large number of system-level issues that emerged as we increased the TeraShake computation scale. The following are lessons we learned running the TeraShake and ShakeOut simulations that might be useful to other applications for large-scale simulations:

Lesson 1 – Carefully coordinate use of the multiple resources needed for computing and data archiving.

The TeraShake simulation required scheduling of more than one resource pool so as to be able to efficiently use resources from multiple sites. At SDSC, the 32-processor IBM p690 fat memory nodes were used for testing, code validation, pre-processing and post-analysis. The pre-processing results are stored on TeraGrid GPFS-WAN, a global parallel file system-wide area network, so that a production run can restart from Blue Gene/L where relatively small amount of memory is available. The TeraGrid IA-64 resources were used for compute-intensive dynamic rupture simulations. The data-intensive wave propagation runs required resources with a balanced CPU and memory configuration like the 8-processor IBM p655 Datastar nodes. The latest ShakeOut simulations for frequency up to 1-Hz were executed on the largest NSF TeraGrid resource at TACC. The resource scheduling and collaboration across administrative boundaries has been critical to the success of the project.

The increasing size and complexity of massive data collections are unique challenges given limited data transfer rates. This requires careful consideration of not only the compute location, but whether associated storage systems are capable of handling the output. Efficient data transfer and access to large files is of the first priority. To ensure the datasets are safely archived, multiple copies of the dataset at multiple locations are necessary. These requirements are most easily met through use of data grid technology which provides the storage location transparency and the management of replicas.

Lesson 2 – High throughput systems are needed to support multiple-day execution runs.

Large-scale capability simulations often take multiple days to complete. For example, the ShakeOut-D simulation took 65 h on 2000 cores (Cui et al. 2007c). Currently, most supercomputer sites configure resources for a typical run that lasts less than 24 h. Longer execution time is a challenge considering the stability of large compute systems.

In addition to the basic simulation runtime, large-scale simulations also require significant time to prepare, and analyze. Specialists in different areas (e.g. data management, visualization) are often needed to support a simulation. Once the group has been coordinated and is ready to begin a simulation it is essential that the work progress as scheduled. Computing systems that can support rapid iterations, simulation starts, stops, and retries, are very valuable.

Then we combine these elements, the need for high throughput computing, in addition to high performance computing, becomes clear. Diverse numerical modeling research programs need both capability and capacity computing. Large numbers of capability and capacity simulations need high throughout systems to complete the simulations at a rate acceptable that can keep a coordinated group of specialists engaged before they get pulled away onto another project.

Lesson 3 – MPI-IO support needs to be verified on each production system at scale as early as possible.
MPI-IO and parallel file systems provide a convenient model for access and high throughout to storage. However, MPI-IO access to disk has been a challenge for both hardware and software level support. As part of the MPI-2 standard, MPI-IO specifies the syntax and semantics of the MPI routines, but does not include any specification of how these routines should be implemented. This means different choices for MPI implementation may be very different in terms of performance on different architectures. Currently, very few large-scale data-intensive simulations use MPI-IO. The performance and portability of these components must be improved, in particular in the area of throughput and scalability.

Lesson 4 – Algorithm adaptation is needed to manage different computer architectures.
The increased variety of compute resources requires porting of an application to multiple types of architecture. This is an adaptation procedure to meet specific requirements for each architecture. It is important to determine how fundamental system attributes affect application performance. Some factors, in addition to clock speed, are critical such as high memory bandwidth, high I/O bandwidth, low latency interconnect, full bisection bandwidth. Well adapted algorithms in an application will not only adjust easily in the porting process, but also enable planning ahead for new architectures. Our flexible program settings, as shown in Fig. 3, as well as distributed data management through data grids provide practical examples of how to modify simulation software to successfully use multiple compute architectures effectively, and we believe our results will benefit other geoscience disciplines as well.

For earthquake wave propagation simulations, source and mesh partitioning is needed in advance to prepare for efficient computations. This will greatly reduce memory requirements. In addition, memory needs to be partitioned between I/O management and computation. It is efficient to decrease the amount of memory devoted to the computation and to allocate memory buffers for I/O aggregation. This allows programs to scale both computation and I/O handling to a large number of processors. In the future, fully integrated MPI-IO is needed for sufficient efficiency with the AWP-like seismic applications.

Lesson 5 – Large-scale simulations typically produce large-scale data management challenges. Consider devoting equal time to computational and data management planning.
The size of the simulation output imposes unique data management challenges. Consider the management of a million files. Each file only has

meaning if its context can be defined. The context includes the parameters controlling the simulation that generated the file, the mapping of the file contents onto a coordinate system, the mapping of the coordinate system onto a geometry, and the assignment of physical variable names. Managing the context requires creating a collection, with the assignment of metadata attributes to each file through the use of digital library technology.

Consider the management of one hundred terabytes of data. Current storage systems have observed bit-error-rates on the order of $10^{-14}$. This means a one hundred terabyte collection will inherently contain multiple errors. In practice, other sources of data corruption are more important. Files might be lost through media failure (damaged tape or disk head crash), through operator error (file overwrite), through vendor product malfunction (bad microcode in a tape drive or disk controller), and through natural disasters (fire, earthquakes). The larger the collection, the higher the risk is that one of these sources of data corruption will damage the collection. Managing data integrity requires the replication of data onto multiple types of storage systems, preferably located at geographically remote locations, through the use of data grid technology.

The need to share data also imposes unique data management challenges. The researchers need local copies of data to support their research, and read access to the entire collection. The researchers that create derived data products need the ability to write data onto storage systems at the remote locations. All collaborators need the ability to read data from any of the storage systems that are used to hold the collection. The ability to manage access controls across multiple administrative domains is provided by data grid technology.

The manipulation of a million files also requires both discovery mechanisms to identify relevant data and standard services for generating visualizations and derived data products. The ability to browse collections, query attributes, and execute standard services is supported by digital library technology.

When moving to petascale computing, file systems alone are not sufficient for managing massive scientific data sets. Instead, a collection-based approach is essential for coupling context to data. The actual technologies needed to manage simulation data are preferably an integration of digital library and data grid technology. These combined systems provide the data virtualization and trust virtualization (manage collection properties independently of storage system) needed to share and manage distributed data. The fact that periodic management functions should be executed to assure data integrity illustrates a major shortcoming with current data management systems, namely the need to automate the execution of management policies. As collections become larger, the labor required to maintain the collection integrity also increases. At SDSC, data management systems are under development that automate the execution of management policies. The goal is to minimize the amount of labor required to manage petabyte-sized data collections. The iRODS (integrated Rule Oriented Data System) is a first step towards this goal. An initial open source version of the software is available at http://irods.sdsc.edu.

Lesson 6 – Data analysis and visualization are essential to large scale simulations and animations are often the most useful.
Post-processing of simulation output from the current set of 50 TeraShake runs comprising more than 150 TBs of data is very time-consuming. Future larger scale simulations will need to automate data analysis and visualization procedures jointly with the simulation using workflow systems.

The key lessons learned in visualization of the simulation output can be summed up as:

**animation** – encoding images to video clips made it easy to manage views and distribute;

**annotation** – adding contextual information like maps, text, etc. significantly aided understanding;

**multiple observation points** – camera angles/paths help in providing insights; multi mode, multi view visualization – creating renderings with different styles and techniques (2D, 3D and hybrid) combined with different viewing positions provide better insights;

**automation** – streamlining the process to reduce human intervention is essential at the TeraShake data scale. Collocated renderings can offer a method to visualize the data as it is created.

Planned future work includes online analysis of surface data by remote web clients plotting synthetic seismograms. Data mining operations, spectral analysis and data sub setting are planned as future work. The TeraShake simulation project has provided some insights on the IT infrastructure needed to advance computational geoscience, which we will examine further. We would like to use 16-bit imagery instead of current 8-bit imagery to increase visualization fidelity. In addition, integration of geographic information systems database to overlay a variety of contextual information would aid further analysis. Integration of surface imagery with virtual globes like Google Earth, NASA's World Wind, etc. is being tested currently.

Verification of the simulation progress at runtime and subsequent seismological data assessment computation was a major concern of the TeraShake project. Visualization techniques helped solve this problem (Chopra et al. 2002) by asynchronous rendering the output data during the simulation run. Animations of these renderings were instantly made available for analysis. SDSC's volume rendering tool Vista, based on the Scalable Visualization Toolkit (SVT), was used for visualizations. Vista employs ray casting for performing volumetric rendering. Surface data have been visualized with different variables (velocities and displacements) and data ranges in multiple modes. The resulting animations have proven valuable not only to domain scientists but also to a broader audience by providing an intuitive way to understand the TeraShake simulation results.

## 9 Summary

Earthquakes pose a great natural threat to many urbanized areas. Numerical simulations of large earthquakes are playing an increasingly important role in understanding earthquake hazard. The TeraShake simulation was one of the earliest high performance computing activities at SCEC that targeted capability computing. Although the project started with a simulation code for which the code accuracy had been extensively verified for anelastic wave propagation and dynamic fault rupture (Day et al. 2003), significant modifications were required for large-scale computation. The major result of the enhanced code was the identification of the critical role a sedimentary waveguide along the southern border of the San Bernardino and San Gabriel Mountains has in channeling seismic energy into the heavily populated San Gabriel and Los Angeles basin areas. The simulations have considerable implications for seismic hazards in southern California and northern Mexico.

The TeraShake simulations demonstrated that optimization and enhancement of major application codes are essential for using large resources (number of processors, number of CPU-hours, terabytes of data produced). TeraShake also showed that multiple types of resources are needed for large problems: initialization, run-time execution, analysis resources, and long-term data collection management. The improvements made to the AWP have created a community TS-AWP code that can be used by the wider SCEC community to perform large-scale earthquake simulations. The TS-AWP code is already being integrated for use in other SCEC projects such as the SCEC Earthworks Science Gateway and a full three dimensional tomography study of southern California. The TS-AWP is useful because it is integrated into the larger suite of supporting codes in the TeraShake computational platform.

A SCEC computational platform is defined as a vertically integrated collection of hardware, software, and people that provide a broadly useful research capability. Based on our experience with the TeraShake computational platform, development of community codes for seismology requires many phases including optimization of high performance software, integration of support codes, installation into a workflow and data management environment, and convenient hosting for use by a wider community.

SCEC has established a goal of developing a PetaShake platform that is capable of performing petascale capability simulations of dynamic ruptures and anelastic wave propagation. A representative capability computation of this kind will model an M8.1 earthquake with an 800 by 400 by 100 km$^3$ domain at 25-m resolution using almost 2 trillion volume elements and 160,000 time steps. This problem is over 9,100 times larger than the TeraShake domain in terms of computational work, requires a sustained petaflops supercomputer and petabytes of storage for archiving surface seismogram output file. Such a high-resolution simulation will provide better estimates of strong ground motions for the most dangerous rupture scenario at frequencies of interest to the emergency management and civil engineering communities. The excellent scalability of the TS-AWP shown on 40k BG/L processors has been a breakthrough in the field of earthquake ground motion simulation, and constitutes an important step towards petascale earthquake rupture and wave propagation computing.

**Acknowledgements** We would like to acknowledge the contribution of the following to the TeraShake platform: Bernard Minster and Geoffrey Ely from SIO/UCSD; Steve Day, Luis Dalguer, Victor Cruz-Atienza from SDSU, David Okaya from SCEC/USC, Eva Deelman from ISI/USC, Marcio Faerman, Yuanfang Hu and Jing Zhu from SDSC. SDSC support staff members include: Bryan Banister, Leesa Brieger, Sheau-Yen Chen, Dong Ju Choi, Giridhar Chukkapalli, Steve Cutchin, Larry Diegel, Nancy Wilkins-Diehr, Robert Harkness, Christopher Jordan, Tim Kaiser, George Kremenek, Amitava Majumdar, Jon Meyer, Arcot Rajasekar, Richard Moore, Donald Thorp, Paul Tooby, and Michael Wan. Dan Lapine and Jim Glasgow supported the TeraShake job submissions and data storage at NCSA. Tommy Minyard and Karl Schulz supported the ShakeOut porting and executions on TACC machines. John Urbanic supported the porting to PSC XT3. Rajeev Thakur from ANL supported MPI-IO improvement. SDSC's computational collaboration effort was supported through the NSF-funded SDSC Strategic Applications Collaborations (SAC) and Strategic Community Collaborations (SCC) programs (award SCI 0438741). The TeraGrid executions and porting are supported through TeraGrid Advanced Support for TeraGrid Applications (ASTA). The TeraShake Digital Library effort is funded by the National Science Foundation Information Technology Research program through the award EAR 0122464. SDSC's visualization effort was supported through the NSF-funded SDSC core program. This project was also funded through NSF Grant EAR-0122464 and SCI 0438741: The SCEC Community Modeling Environment (SCEC/CME): An Information Infrastructure for System-Level Earthquake Research, through the NSF Grant EAR-0623704 – SCEC Petascale Cyberfacility for

Physics-based Seismic Hazard Analysis research Project (PetaSHA-1), and through NSF Grant OCI-749313 – SCEC Enabling Earthquake System Science Through Petascale Calculations (PetaShake). The simulations used NSF TeraGrid resources including SDSC DataStar, IA-64, BG/L, NCSA IA-64, TACC Lonestar, and PSC XT3. Special Thanks to DOE INCITE Program and IBM TJ Watson Research Center for providing access and technical supports to their BG/L resource during Blue Gene Consortium Days in 2006/2007. This research was supported by the Southern California Earthquake Center. SCEC is funded by NSF Cooperative Agreement EAR-0106924 and USGS Cooperative Agreement 02HQAG0008. The SCEC contribution number for this paper is 1195.

## References

Andrews, D.J. (1999). Test of two methods for faulting in finite-difference calculations, *Bull. Seism. Soc. Am.*, vol. 89, pp. 931–937

Chaljub, E., D. Komatitsch, J.P. Vilotte, Y. Capdeville, G. Festa (2006). Spectral Element Analysis in Seismology. Advances in Wave Propagation in Heterogeneous Earth, Ru-Shan Wu and Valerie Maupin, eds., in the series *Advances in Geophysics,*, R. Dmowska, ed., Elsevier Academic Press. Vol. 48, 365–419

Chopra, P., J. Meyer, A. Fernandez (2002). Immersive volume visualization of seismic simulations: A case study of techniques invented and lessons learned, *IEEE Visual.*, pp. 171–178

Cui, Y., R. Moore, K. Olsen, A. Chourasia, P. Maechling, B. Minster, S. Day, Y. Hu, J. Zhu, A. Majumdar, T. Jordan (2007b). Enabling Very-Large Scale Earthquake Simulations on Parallel Machines, *ICCS* 2007, Part I, *Lecture Notes in Computer Science Series*, Vol 4487, pp. 46–53, Springer

Cui, Y., R. Moore, K. Olsen, J. Zhu, L. Dalguer, S. Day, V. Cruz-Atienza, P. Maechling, T. Jordan (2007c). Mapping PetaShake Applications to TeraGrid architectures, *Eos. Trans.*, vol. AGU 88, no. 52, Fall Meet Suppl, Abstract: IN21B-0483

Cui, Y., K. Olsen, Y. Hu, S. Day, L. Dalguer, B. Minster, R. Moore, J. Zhu, P. Maechling, T. Jordan (2006). Optimization and scalability of a large-scale earthquake simulation application, *Eos. Trans.*, vol. AGU 87, no. 52, Fall Meet Suppl, Abstract: S41C-1351

Dalguer, L.A., S. Day (2007). Staggered-grid split-node method for spontaneous rupture simulation, *J. Geophys. Res.*, vol. 112, B02302 DOI 10.1029/2006JB004467

Day, S.M. (1998). Efficient simulation of constant Q using coarse-grained memory variables, *Bull. Seism. Soc. Am.*, vol. 88, pp. 1051–1062

Day, S.M., J. Bielak, D. Dreger, R. Graves, S. Larsen, K.B. Olsen, A. Pitarka (2003). Tests of 3D elastodynamic codes: Final report for Lifelines Project 1A02, Pacific Earthquake Engineering Research Center

Day, S.M., C. Bradley (2001). Memory-efficient simulation of an-elastic wave propagation. *Bull. Seis. Soc. Am.* vol. 91, pp. 520–531

DSpace digital library, http://www.dspace.org/

Dunham, E.M., R.J. Archuleta (2005). Near-source ground motion from steady state dynamic rupture pulses, *Geopys. Res. Lett.*, vol. 32, L03302, DOI 10.1029/2004GL021793

Fedora digital object repository middleware, http://www.fedora.info/

GridSphere portal technology: http://www.gridsphere.org/

HDF5 – Hierarchical Data Format version 5, http://hdf.ncsa.uiuc.edu/HDF5/

Kajiya, J.T., B.P.V. Herzen (1984). Ray tracing volume densities, in *Proc. SIGGRAPH,* vol. 18, no. 3, pp. 165–174

Marcinkovich, C., K.B. Olsen (2003). On the implementation of perfectly matched layers in a 3D fourth-order velocity-stress finite-difference scheme, *J. Geophys. Res.*, 2002JB002235

Moczo, P., J. Kristek, M. Galis, P. Pazak, M. Balazovjech (2007). The finite-difference and finite-element modeling of seismic wave propagation and earthquake motion, *Acta Phys. Alovaca.*, vol. 57, no. 2, 177-406

Moore, R., A. Rajasekar, M. Wan (2005). Data grids, digital libraries and persistent archives: an integrated approach to publishing, sharing and archiving data, Special Issue of the *Proc. IEEE Grid Comput.,* vol. 93, no. 3,, 578–588

Olsen, K.B. (1994). Simulation of three-dimensional wave propagation in the Salt Lake Basin. Ph.D. thesis, The University of Utah, 157p

Olsen, K.B., S.M. Day, C.R. Bradley (2003). Estimation of Q for long-period (>2s) waves in the Los Angeles Basin. *Bull. Seis. Soc. Am.* vol. 93, pp. 627–638

Olsen, K., S.M. Day, J.B. Minster, Y. Cui, A. Chourasia, M. Faerman, R. Moore, P. Maechling, T. Jordan (2006a). Strong shaking in Los Angeles expected from Southern San Andreas earthquake. *Geophys. Res. Lett.*, vol. 33, 1–4

Olsen, K.B., J. Zhu, J. Talley (2006b). Dynamic user interface for cross-plot, filtering and upload/download of time series data. *Eos. Trans.*, vol. AGU 87, no. 52, Fall Meet. Suppl., Abstract, IN51B-0814

Olsen, K.B., S.M. Day, J.B. Minster, Y. Cui, A. Chourasia, D. Okaya, P. Maechling, T. Jordan (2007). Tera Shake 2: Simulation of Mw7.7 earthquakes on the Southern San Andreas fault with spontaneous rupture description, accepted to *Bull. Seis. Soc. Am.* vol. 98, pp. 1162–1185, DOI:10.1785/012007148

Open Grid Forum, http://www.ogf.org/

Peyrat, S., K.B. Olsen, R. Madariaga (2001). Dynamic modeling of the 1992 Landers earthquake, *J. Geophys. Res*. vol. 106, no. 26, 467–26, 482

SCEC/CME Web Site, http://www.scec.org/cme

TeraGrid Website, http://teragrid.org/about/

The SDSC Storage Resource Broker, http://www.sdsc.edu/srb/

Uemura, A., C.K.J. Watanabe (2004). Visualization of seismic wave data by volume rendering and its application to an interactive query tool, in *Proc. PDPTA*, pp. 366–372, CSREA Press

Weldon, R., K. Scharer, T. Furnal, G. Biasi (2004). Wrightwood and the earthquake cycle: What a long recurrence record tells us about how faults work. *Geol. Seismol. Am. Today*, vol. 14, pp. 4–10

Wijk, J.J.V., A.C. Telea (2001). Enridged contour maps, in *Proc. Conf. Visual.*, IEEE Computer Society, pp. 69–74
Yu, H., K.L. Ma, J. Welling (2004). A parallel visualization pipeline for terascale earthquake simulations, *SC04,* vol. 6, no. 12, p. 49

# VIII. Probabilistic Forecast of Tsunami Hazards along Chinese Coast

Yingchun Liu,[1,2] Yaolin Shi,[1] Erik O.D. Sevre,[2] Huilin Xing,[3] and David A. Yuen[2]

[1] Graduate University of Chinese Academy of Sciences, Beijing, China
[2] Department of Geology & Geophysics and Minnesota Supercomputing Institute, University of Minnesota at Twin Cities, Minneapolis, MN 55455, USA
[3] The University of Queensland, Earth Systems Science Computational Centre, QLD4072, Australia

**Abstract** There is indeed a potential non-negligible threat for Chinese coast from tsunamogenic earthquakes originating at the neighboring subducting plate boundaries of Eurasian plate and Philippine sea plate: Manila trench and the Okinawa trough. This finding comes from our newly devised method for determining the probabilistic forecast of tsunami hazard (*PFTH*), which finds this probability distribution from direct numerical simulation of the waves excited by hypothetical earthquakes in these zones. There are significant differences in the bottom bathymetry between the South China Sea bordering the southern province of Guangdong and the East China Sea and Yellow Sea adjacent to the provinces of Zhejiang, Jiangsu, and Shandong. We have verified that the linear shallow-water equations can be employed to predict with sufficient accuracy the travel time of tsunami waves in the South China Sea, while the nonlinear shallow-water equations must be used for the shallower seas next to the northern Chinese provinces. Distribution for the possibility of tsunami waves with above 2.0 m hitting the coast has been shown in eastern China sea area, the delta region of the Yangzi River, the north-eastern coast of Zhejiang province, and northern Taiwan island. The distribution has also been displayed in South China Sea area, along the southeastern coast of mainland and Southwestern Taiwan. In this century the probability of a wave with a height of over 2.0 m to hit Hong Kong and Macau is about 10.0%, 0.5% for Shanghai, 3.2% for Wenzhou, and 7.2% for Keelung. Cities on eastern Chinese coast are less vulnerable than those on the southern Chinese coast. We also have discussed the prospects of tsunamis coming from large earthquakes along the Manila trench and the Ryukyu-Kyushu arc region to the north, as they can impact many countries in Southeast Asia, besides China.

H. Xing, *Advances in Geocomputing*, Lecture Notes in Earth Sciences 119,
DOI 10.1007/978-3-540-85879-9_8

## 1 Introduction

The Sumatra seismogenic tsunami in December 26, 2004 with the death of 0.33 million people shocked the whole world. This event has alerted the attention of many countries surrounding the Pacific and Indian Oceans about the danger of tsunami waves. The need of warning systems is strongly revived by this tumultuous incident. Tsunamis occur around the world from various causes, principally from shallow earthquakes in subduction zones. Around 90% of the global undersea earthquakes take place around the circum-Pacific belt, and only shallow major earthquake could induce big tsunami hazard (Polet and Kanamori, 2000). It is not difficult to detect large earthquakes and issue alarms, however, we judge roughly whether they generate tsunami or not in 10 min. Therefore, the forecasting of potential tsunami hazard in long term and the pinpointing the vulnerable locations represent a very important goal for tsunami warning.

Now most of methodologies for potential tsunami hazard are based on the statistical method. This is similar to how seismic risk is assessed. Probability seismic hazard analysis has been used widely worldwide, which can be traced back to the method proposed by Cornell (1968). In recent decades, as the study of seismic risk matures, there is a resurgence of interest in carrying out tsunami risk assessment by means of a probabilistic approach (e.g. Probability tsunami hazard analysis, PTHA) (Geist and Parsons, 2006) in order to quantitatively analyze the tsunami hazard. It is similar to the current seismic risk analysis possibilities methods (Probability seismic hazard analysis, PSHA) (Speidel and Mattson, 1997). Its merits are that uncertainties and possibilities of tsunami hazard have been considered. As the natures are different between tsunami and earthquake disaster, and due to imperfect historical tsunami information in different area, specific implementation steps are different. Wong used geographic information systems (GIS), combining flood analysis model, to analyze tsunami risk (Wong et al., 2005). McAdoo studied the probability of big wave risk of Oregan with tsunami sediment data of nearly 2000 yr (McAdoo and Watts, 2004). Annaka introduced logic tree method with considering the uncertainty of tsunami hazard (Annaka et al., 2004). Rich historical seismogenic tsunami data could provide the most reliable basis for study of tsunami hazard. But in China, or some other Pacific countries such as Indonesia and Thailand, there is few existing scientific literature which deals with the analysis of tsunamis. Reliable numerical tsunami simulation generated from potential earthquake can make up for inadequate historical information of tsunamis hazard. We developed new method PFTH (Liu et al., 2007) for forecasting the potential tsunami risk from latent seismic sources. This approach combines the assessment of potential seismic hazard, tsunami simulation, and the computation of the probabilities tsunami hazard.

Being protected by the Ryukyu arc and the Philippine islands, the far-field tsunami waves, that come from the other side of Pacific Ocean, such as South America, exert a weak influence on the Chinese coast. The main tsunami hazard arise from local seismic hazards within a couple of thousand kilometers. There are records of 26 documented tsunamis in 2000 years. About 8 or 9 of them brought devastating results (Wang and Zhang (2005); Li (1981); Xu (1981); Bao (1991); Li and Xu (1999); Chen et al. (2007)). Table 1 is an excerpt of tsunami hazard records in Chinese ancient books. The most devastating tsunami in this region occurred 140 years ago (in 1867) in Keelung at the northern tip of Taiwan. Both the northeast and southwest coasts of the Taiwan Island are more likely to be affected by impending tsunamis because of their closeness to plate boundaries.

Once again on December 26, 2006 we were alerted to the danger of tsunami hazards and potential economic consequences along the South China Sea coast by the Pingtung earthquakes off southern Taiwan. This area is prone to large subduction related tsunamogenic earthquakes because of the nature of the complex subducting plate boundary, ranging from Taiwan in the north to the Manila trench in the south. Until the end of 2004, there was little awareness about the potential tsunami danger from shallow large earthquakes in this region with great economic importance. We must now be prepared to set up a suitable system for broadcasting tsunami warnings along Chinese coast.

In this work we will employ our newly developed method PFTH (Liu et al., 2007) to analyze the probability for tsunami waves of various heights to hit the cities along the Chinese coast in this century. These waves are assumed to be caused by large earthquakes originating from the dangerous Manila trench and Okinawa trough. This analysis becomes ever more crucial because of the sharp increase in the coastal population density in China, and the intensive growth of harbors and the exploration of mineral resources in the coastal areas, ranging from Xiamen in the north to Hainan in the south. In this paper we report our results on tsunami

**Table 1** Historical records of the tsunamis along China coast (Wang and Zhang (2005); Li (1981); Xu (1981); Bao (1991); Li and Xu (1999))

| Wave Incidence | Time | Epicenter | Magnitude and wave height |
|---|---|---|---|
| Gulf of Penglai, Bohai | Apr. 4, 171 | | |
| Gulf of Laizhou, Bohai | July, 173 | | |
| Jiangsu and Zhejiang | July, 9, 1496 | South Sea of Japan | M≥ 8 |
| Quanzhou, Fujian | Dec. 29, 1604 | (24.7ºN, 119.0ºE) | M=7 |
| Anping, Taiwan | Juan. 8, 1661 | East-Southern Ocean of Taiwan | |
| Qiantangjiang, Zhejiang | Oct. 28, 1707 | (33.2ºN, 135.9ºE) | M=8.4 |
| Tainan | Juan. 5, 1721 | West-Southern Ocean of Taiwan | |
| Keelong, Taiwan | Feb. 24, 1741 | Okinawa, Japan | |
| Kaoishong, Taiwan | Apr. 5, 1781 | Southern Ocean, Taiwan | |
| Tainan, Taiwan | Aug. 9, 1792 | (23.6ºN, 120.6ºE) | M=7 |
| Kaoishong, Taiwan | June 11, 1866 | Southern Ocean of Taiwan | |
| Keelong, Taiwan | Dec. 18, 1867 | (25.25ºN, 122.2ºE) | 7 m |
| Keelong, Taiwan | July 4, 1917 | (25.0ºN, 123.0ºE) | M=7.3, 3.7 m |
| Huangzhou, Fujian | Feb. 13, 1918 | (23.6ºN, 117.3ºE) | M=7.3 |
| Yantan, Shandong | July 13, 1923 | (31.0ºN, 130.5ºE) | M=7.2 |
| Huanlien, Ilan, Taiwan | Nov. 15, 1986 | (24.1ºN, 121.7ºE) | M=7.6 |
| Keelong, Huanlien | May 24, 1960 | Chile | Keelung 0.66 m, Huanlien 0.3 m |
| Huanlien, Taiwan | Oct. 13, 1963 | Kurile Islands | 0.1 m |
| Huanlien | Mar. 28, 1964 | Alaska, USA | 0.15 m |
| Taichung | Mar. 12, 1978 | East-Southern Ocean of Taiwan | |
| Xiamen, Kanmen | Feb. 29, 1988 | North Pacific Ocean | Kanmen 0.37 m, Xiamen 0.34 m |
| Eastern of Taiwan Island | Aug. 8, 1993 | Mariana Islands | Huanlien 0.29 m, Cheng Kung 0.27 m |
| Fujian, Taiwan | Sept. 16, 1994 | Taiwan Strait | Dongshan 0.18 m, Penghu 0.38 m |

hazards prediction along the South China Sea and eastern sea area bordering regions from potential earthquakes coming from the Manila trench and Okinawa trough. We hope this paper will spur greater interest from countries around the Pacific Ocean in fundamental research in earthquakes, tectonics and geodetics in this area with a population exceeding several hundred million people.

## 2 Geological and Geophysical Analysis

The bathymetric map and the main structural units of the China epicontinentalic sea region are presented in Fig. 1. China sea region consists of two major sea areas: South China Sea, and eastern China sea area, with the latter further composed of the East Sea, Yellow Sea, and Bohai Sea. The Chinese epicontinental Sea is located at the interacting region between the Eurasian plate and the Philippine sea plate. As suggested by Kreemer, the greatest accumulated deformations have been accommodated by seismic faulting along the Manila trench according to the geodetic study by assuming that geodetic deformation represents the tectonic loading in the brittle part of crust; and the interaction is the most active among the global subduction zones (Kreemer et al. (2002); Kreemer and Holt (2001)). The Chinese sea is experiencing a tandem suturing of a volcanic arc and continental crust to continental margin. As a result of the collision of Eurasian and Philippine Sea plates, rapid rates of horizontal and vertical deformation and an abundance of seismic activity are demonstrated in two belts (Fig. 2). Taiwan is located at the junction of two belts. To the south, the young Eurasian continental lithosphere of the South China Sea is subducting eastward beneath oceanic lithosphere of the Philippine Sea plate at a rate of about 80 mm/yr at the Manila trench (Yu et al., 1999). On the other hand, to the north, the polarity of subduction is the opposite, and extension is occurring at about 30–40 mm/yr in a back-arc region above the Ryukyu subduction zone (Nakamura (2004); Becker et al. (2000)).

The South China Sea (Fig. 1), which lies on the western part of the Pacific Ocean, is one of the largest marginal sea along the continental margin of East Asia, covering an area around 3,500,000 km$^2$, almost as large as three times that of the Bohai Sea, Yellow Sea and East China Sea combined together. The South China Sea, along with Taiwan and the Philippines island arc-trench to its east, constitute a very complex channel-basin structural system. Bordered by the Eurasia continent, Pacific and Indian Ocean, the South China Sea belongs to the transitional crust between the oceanic and continental crust tectonic zone. Complex geological structures are reflected in the large-scale mass movement along the horizontal directions in this region, which is often accompanied with extensive vertical movement. The South China Sea spreads from the center and subducts along the Manila trench (Liu et al., 1988).

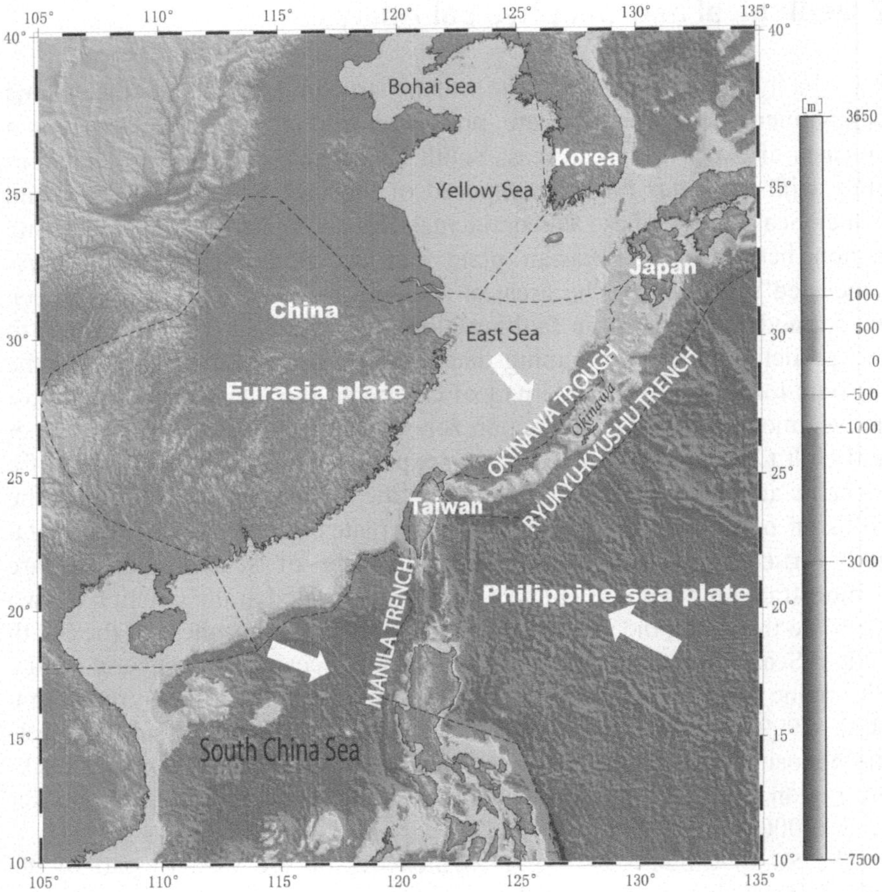


**Fig. 1** The China Sea and adjacent area

The crust of this region is under tremendous tectonic stresses from many directions due to the complex interactions among three plates mentioned earlier.

As illustrated in Fig. 1, in the strike belt between the south of Taiwan island and Philippines Islands, which is a complex deformation zone, interaction between two plates is complicated (Zang et al., 1990). The eastern boundary of deformation zone is from Taiwan Valley fault, passing the southeast Luzon trough, to the Philippine Trench; the west side of interaction zone starts at large South China Sea subduction zone on the southwestern of Taiwan Island, passing Manila trench, to Negros and Gedaba

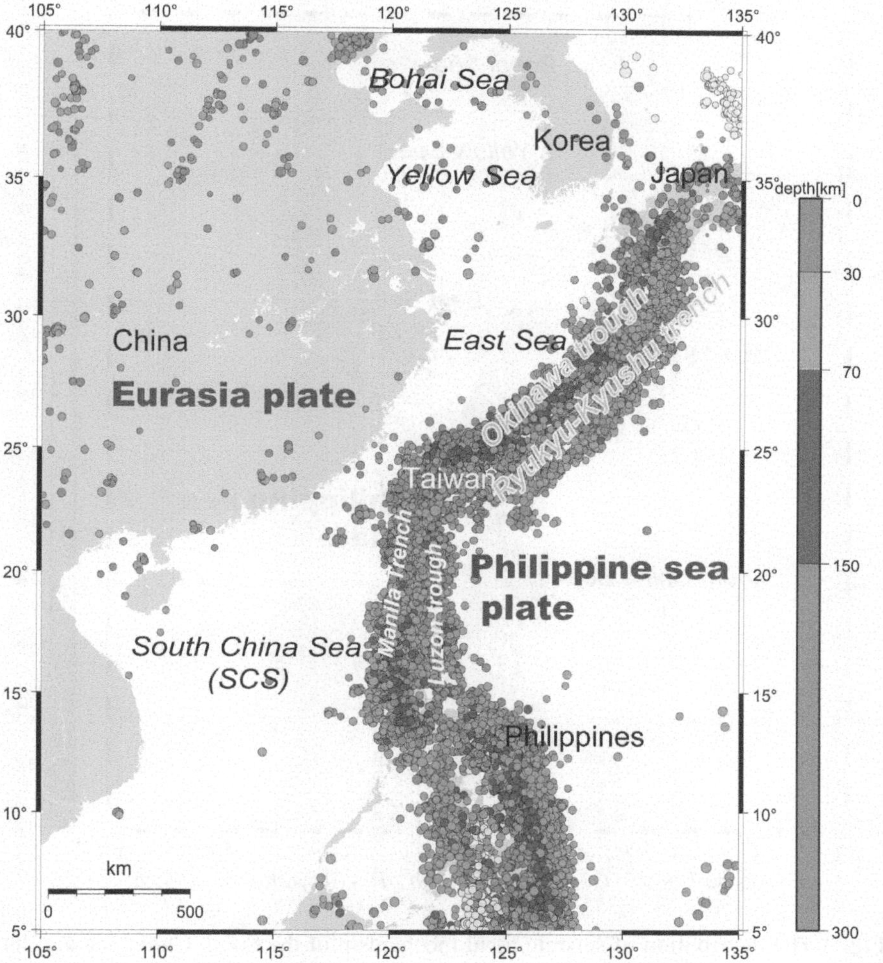


**Fig. 2** The historical earthquakes distribution in China Sea and adjacent area

subduction zones. The analysis of the focal mechanism solutions (Fig. 3) (Harvard CMT solutions) reveals that the stress states of the Manila trench and its adjacent region are different. In the northern part of the Manila trench, and the adjacent Philippine faults, the focal mechanism solutions show the evidence of the compressive-thrusting. On the other hand, in the southern region of the Manila trench, the stress distribution becomes very complex. Along the two sides of western Luzon trough close to the trench,

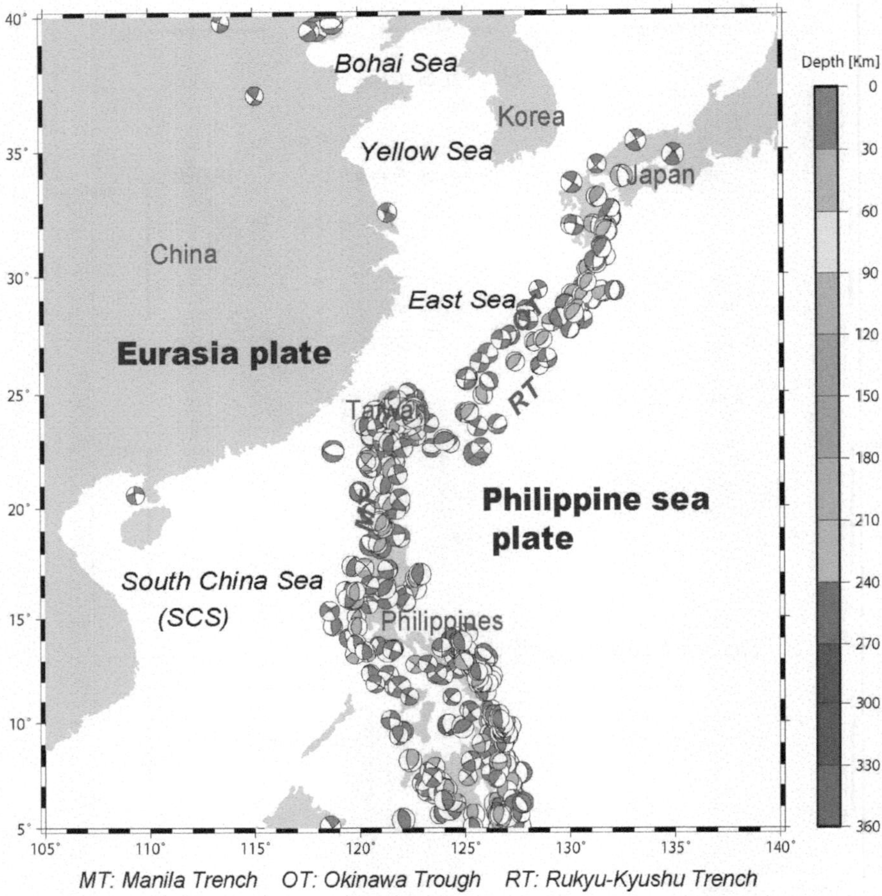


**Fig. 3** The distribution of seismic focal mechanism of the South China Sea and its adjacent regions. The magnitude is ≥6.0. Data derived from Harvard CMT solutions (1976–2006)

the focal mechanism solutions show an oblique-strike with normal faulting, and in front of the diving zone away from the trench they show a thrusting character. Local seismic focal mechanism solutions also indicate high cumulative moment rates from data taken over the last 30 years in the Manila subduction segments.

However, the present-day earthquakes are not distributed evenly along the plate boundaries and historical seismic records indicate that the Manila trench has been highly affected by major earthquakes in the north-western part of Philippines (shown in Fig. 2). Furthermore, major earthquakes along the Manila trench with a higher frequency have also influenced the stress state of southern Taiwan. Thus the geological evolution, the GPS velocity field, CMT background, and seismic distribution provide sufficient evidence that potential seismic energy is focusing along the Manila subduction in South China Sea region.

The eastern China Sea region (Fig. 1), including Bohai Sea, Yellow Sea, and East China Sea, lies next to the Asian continent and the circum-Pacific island arc and subsidence of tectonic belts. The long-term, complex interaction between the Eurasian plate and the Pacific plate induces a series of NE–SW uplift and subsidence tectonic belts. Their geological ages increase from west to east.

From the geophysical focal mechanism solutions, we can infer that the maximum principal stress and pressure in the Yellow Sea and East China Sea area lie along the same direction as that in eastern Chinese mainland (Yun et al., 1997). It could represent the extension of stress field in the mainland. By analyzing the character of borehole caves from the drilling of oil exploration in southern Yellow Sea and East China Sea, and combining shallow earthquake focal mechanism of the Ryukyu island arc and the Okinawa Trough region, Xu confirmed the Yellow Sea and the North China region have similar modern tectonic stress field characteristics (Xu and Zhong, 1997). This is further evidence that the greatest stress direction is NEE–SWW in East China Sea, while their smallest compressive stress direction is NNW–SSE (Fig. 3). The largest and smallest compressive stress in Okinawa trough is respectively same as that in East China Sea area. Therefore, the upper crust level of stress gap in East China Sea area is small, which causes weak seismic activity.

Recent studies on the spatial distribution of earthquake focal mechanisms in the Ryukyu-Kyushu arc reveal that change of stress field exists on the top 100 km area: it is down-dip extension in Okinawa trough; reversely, it is downdip compression in Ryukyu-Kyushu trench (Shiono et al. (1980); Zang et al. (1990); Kao and Chen (1991)). The slope of normal maximum compression is steeper than that of minimum compression of the strike in the top 40 km layer of the island arc (Christova, 2004). The unbalanced stress field also indicates the existence of active pull force in arc area. The Ryukyu arc marks the subduction of the Philippine Sea Plate beneath the Eurasian plate. The rate of the plate convergence in the southern part of the arc is around 70 mm/yr, more than that in the northern part of arc, 40 mm/yr (Seno et al., 1993).

The historical seismic data distribution of East Asia plate area (Fig. 2) depicts earthquakes mainly concentrated in the Ryukyu island arc, Taiwan, and the Manila trench. On the other hand, earthquakes rarely occur in the territorial of China sea, not to mention large earthquake series. This shows that subduction zones between the Philippines Sea plate and the Eurasian plate are stress concentration region of East Asian plate. The Okinawa trough and Manila trench are the largest seismogenic tsunami source that could seriously impact the Chinese eastern coastal area.

# 3 Probabilistic Forecast of Tsunami Hazards

## 3.1 Probabilistic Forecast of Tsunami and Seismic Hazards

Considering both seismic activities and tsunami numerical simulation, we have developed a new method called **Probabilistic Forecast of Tsunami Hazard (PFTH)** (Liu et al., 2007), which synthesizes the probabilities of potential seismic hazards and the probabilities of different heights of the waves along the coastline, that are obtained from numerical simulation of the waves excited by the earthquakes. Our probabilistic forecast of tsunami hazard (*PFTH*) is made up by the following three steps:

(1) *Probabilistic Forecast of Seismic Hazard **(PFSH)**.*
Probabilities and locations of earthquakes are estimated. The earthquake distribution is spatially heterogeneous. It is necessary to analyze the potential seismogenic tsunami sources in entire study region. We would know the potential seismic zone by analyzing the detail of the geological and geophysical background, the seismic activity, and the seismic solid simulation. The potential seismic assessment is the basis of forecasting of tsunami hazard. Here we get the Probabilistic Forecast of Seismic Hazard with statistics method by integrating the above geophysical background.

(2) *Tsunami Modeling.*
For each earthquake predicted, the hydrodynamical evolution of the waves reaching each coastal location is computed based on the results of the tsunami wave height predictions using the shallow-water equation. Here we only consider the risk evaluation on coastal area. Therefore, two processes are simulated: tsunami generation and propagation.

(3) *Probabilistic Forecast of Tsunami Hazards **(PFTH)**.*
After recording the wave highs information in tsunami simulation process, we sum up tsunami risks of all possible major earthquakes and provide the statistical risk distribution of different wave highs.

In contrast to the earlier work of Geist and Parsons (2006), which estimated tsunami probability only from earthquake magnitudes, we have determined the tsunami probability by using the wave height at each stations by numerically solving the shallow-water equations. Our method, based on numerical simulations of the actual equations, is different from empirical methods we mentioned in the introduction section and is valuable in areas where there are not many historical records.

In our method, PFSH is conducted by using four sources. The computation can be expressed by following Eq. (1):

$$P_{PFSH,i} = P_{e,i} \cdot P_{oc,i} \cdot P_{sh,i} \cdot P_{f,i} \tag{1}$$

$i$ is the sequential number of the forecasted earthquakes. $P_{e,i}$ is the major earthquake occurrence probability of an earthquake $i$ is estimated on the basis of the Gutenberg-Richter (GR) relationship (Gutenberg and Richter, 1949). $P_{oc,i}$ is the probability of these earthquakes occurring in the oceanic area, since only oceanic earthquake can cause tsunami. $P_{sh,i}$ is the probability of the oceanic earthquakes occurred in shallow depth ($\leq$ 10 km) in all of events. Shallow earthquake can generate big vertical displacement on sea floor, which could induce dangerous tsunami waves. $P_{f,i}$ is the probability of the rupture length as compare to the whole seismic zone. The properties of rupture have big contribution for tsunami wave generation. We simplify tsunami generation model with simple solid simulation. We assume the earthquake with same magnitude could happen evenly along the seismic zone. More detail of rupture computation is shown in Eq. (9).

As previously discussed, $P_{oc,i}$ and $P_{sh,i}$ can be easily induced by historical records. From Eq. (9), $P_{f,i}$ is fully decided by seismic magnitude. $P_{e,i}$ reflects the information of more seismic activities of different magnitudes. It is the core of this method. In this work, by estimating the probability of the major earthquake ($P_{e,i}$) with PFSH, we take into account the tectonic evolution, the GPS velocity field, and present-day geophysical stress field, that were analyzed above. At this point a few words are needed to explain our rationale based on the law of total probability in deducing the joint probability. According to the theory of plate tectonics, large-scale interplate earthquakes occur near the global subduction zone where the potential energy of elastic strains accumulated over tens to hundreds of years is released over a very short period of time. For predicting possibility of the earthquake occurrence, the most common assumption (Reiter, 1990) is that the frequency of seismic events follows the Gutenberg-Richter relationship. As a phenomenological tool based for estimating the probabilistic analysis of seismic hazard, the GR relationship has been widely applied for decades since its introduction in late 1940s. This relationship can be written as (Lomnitz, 1974):

$$log\, N = a - bM \quad (2)$$

where $N$ is the cumulative total number of earthquakes within a certain period of time for a given magnitude rang; $M$ is the magnitude of the earthquake in any linear or intensity scale or the log of seismic moment (Krinitzsky, 1993). The GR relationship also holds when $N$ is the earthquake number for particular regions or specific time intervals. Parameters $a$ and $b$ are empirically derived constants specific for each region. By determining the slope of a magnitude-frequency plot, we find that $b$ depends on the relative proportion of small, medium, and large shocks.

The seismic magnitude frequency data can also be well described, as one or more populations, each of which is normally distributed with respect to the magnitude. This holds true for large earthquakes, when it is sorted out by global subduction zone with the general USGS, NEIC catalog (Speidel and Mattson, 1997). For the PFSH method building, we first study the GR relationship along the global subduction zone. We have divided the global subduction boundaries into nine regions (Fig. 4). Here we have also employed NEIC database. The GR relationship computation of each region verifies the log-linear relationship between the magnitude and the number of events in the history (Fig. 5). Therefore, our prediction of the earthquake occurrence probability $P_{e,i}$ can be derived by GR relationship along Manila and Ryukyu subduction zones. It should be noted that the length and dislocation area limit the vertical displacement field of earthquakes. Thus the biggest magnitude of potential earthquakes come from local subducation zone must be taken into consideration. Two biggest tsunamogenic earthquakes in history occurred in Chile, $M_w = 9.5$ in May 1960, and in Sumatra, $M_w = 9.3$ in December 2004. It should be emphasized that we cannot assume the magnitude of potential earthquake without constraints of geophysical factor and the magnitude of largest local earthquake in history.

We use computational methods similar to *PFSH* to develop our probabilistic forecast of the tsunami hazard (*PFTH*). In our research, we do not consider the factors of wind and oceanic current. $P_{PFTH}(x, y, h)$ will be same as $P_{PFSH,i}$ in each tsunami case $i$, since the probability of generating certain maximum wave height $h$ is the same as probability of earthquakes that induce such tsunami. Therefore, the probabilistic risk of same tsunamic wave height is estimated by a combination of large earthquake occurrence probability and the numerical simulation results from tsunami wave propagation:

$$P_{PFTH}(x,y,h)=\sum_{i=1}^{m}P_{PFSH,i}\{\max(f(x,y),t)\} \qquad (3)$$

where $P_{PFTH}$ is the probability of a particular wave height ($h$) of tsunami in the position $x$, $y$ along the coast. $x$, $y$ are the latitude and longitude of the receivers. $P_{PFSH,i}$ is the probability of attaining a maximum wave height from each tsunamogenic earthquake. It is derived from **PFSH**, Eq. (1). $i$ is the index of the earthquakes. Here $m$ is the number of earthquakes. $f(x, y)$ is the wave height of tsunami. $h$ is $max$(f($x,y$), $t$). $t$ is the time of wave propagation. This case provides a method to calculate the cumulative or joint probability from the spatial variability in the probability pattern.

### 3.2 Linear and Non-linear Modeling Potential Tsunami Sources

In our PFTH method, tsunami simulation is employed in place of traditional statistical methods between wave height and seismic magnitude (Geist and Parsons, 2006). This is main characteristic of our tsunami forecasting method. The feature of tsunami waves is revealed by its fluid dynamical and physical mechanisms. Now most researchers have applied the shallow water equations in their tsunami modeling because of the long wavelength nature in tsunami propagation (Goto et al. (1997); Shokin et al. (1979); Pelinovsky et al. (2001)). In our research two kinds of tsunami models are used: linear and nonlinear shallow water equations in different ocean depths. The application of physically reasonable equations in different fields is very important to obtain accurate results on the shore. Two processes are simulated: tsunami generation and its subsequent propagation.

Seismogenic tsunami generation is a very complex dynamic problem. Certain factors affect tsunami sources, including the duration period of earthquake rupture, geometric shape of rupture, bottom topography near the epicenter of earthquake, seismic focal mechanism, and rock physical properties. Ward (1982) studied tsunamis as long-period, free oscillations of a self-gravitating earth, with an outer layer of water representing a constant depth ocean. The tsunami displacement field can be constructed by summing the normal modes of the spherical harmonics. Comer (1984) regarded the tsunami source excitation in the flat Earth as a point source.

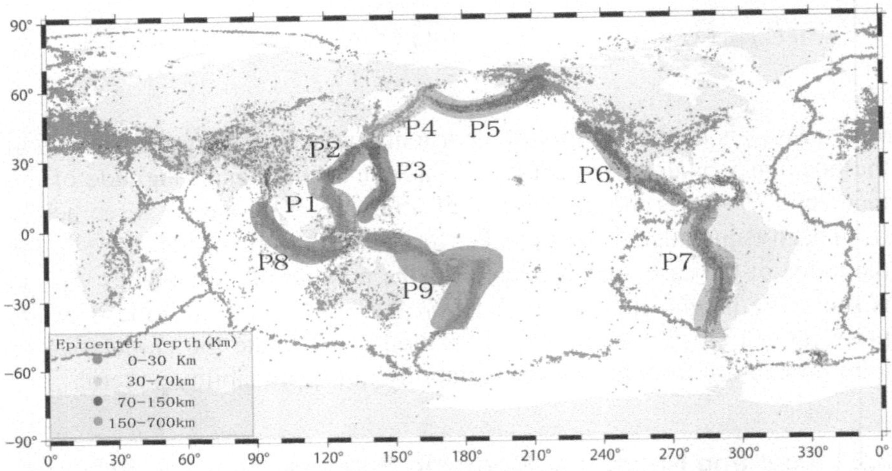


**Fig. 4** The global subductions and their GR relationships evaluated in this study. We divided global subduction zones into nine partitions. All of these zones followed the linear rule of GR relationship (Fig. 5). The relationship for Manila and Ryukyu subductions (P1 and P2) are found to follow the GR relationship separately, that are analyzed in our paper. Boundary subduction data are derived from Bird (2003). The historical earthquakes database comes from NEIC

He emphasized that source problem in the flat Earth differs substantially from the corresponding problem for the spherical Earth. Yamashita and Sato (1976), using the fully coupled ocean solid earth model, analyzed the influence of the parameters of seismic focal mechanism, such as dip angle, fault length, focal depth, and the rise time of the source time function of tsunami wave. They took wave form of tsunami as a long period gravity wave and Rayleigh wave.

We can employ the elastic dislocation theory in the numerical simulation of coseismic process as the part of tsunami generation, because the time for seismogenic tsunami generation is very short comparing to that of the wave propagation. The popular numerical method in the excitation of tsunami modeling is the elastic dislocation method. The initial condition of the linear shallow water equation is computed according to Okada's work on elastic deformation from a dislocation (Okada, 1985), which numerically predicts the water level changes due to earthquake faulting. Rectangular 2-D fault and half-space elastic model were adopted to represent major faults of the seismic origin for calculating the earthquake induced tsunamis. The historical seismic records are taken into account. In the seismic rupture models, source parameters (rupture length L, width W, and the average slip D) are derived from the theoretical and empirical relationships (Wells and Coppersmith, 1994) that have been widely applied. The fault planes

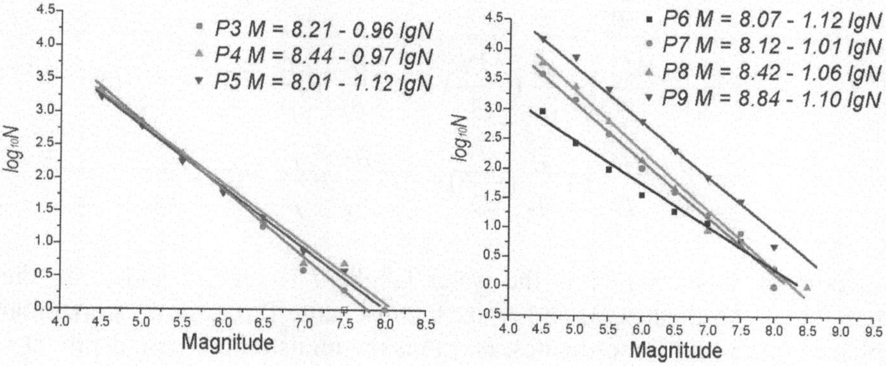


**Fig. 5** The GR relationships of global subduction zones. The partition numbers are marked in Fig. 4)

were chosen in accordance with the seismic tectonic situation. The fault dips and strikes from the composite fault plane solutions come from the average dip of the fault segments obtained from the Harvard catalog.

For tsunami propagation simulation, we have employed the linear tsunami propagation model Tunami-N1, which were developed in Tohoku University (Japan) and provided through the Tsunami Inundation Modeling Exchange (Time) program (Goto et al., 1997). And nonlinear model is developed under instruction of Prof. Imamura of Tohoku University. First, we take care of the simple model without the seabed bottom friction term. The following linear shallow water Eqs. (4) are employed.

$$\begin{aligned}
&\frac{\partial z}{\partial t}+\frac{\partial M}{\partial x}+\frac{\partial N}{\partial y}=0 \\
&\frac{\partial M}{\partial t}+gD\frac{\partial z}{\partial x}=0 \\
&\frac{\partial N}{\partial t}+gD\frac{\partial z}{\partial y}=0
\end{aligned} \tag{4}$$

Due to the existence of the shallow water regions, as a comparison, we also applied the non-linear shallow-water model. Here we include the effect of the friction coefficient on the wave height. The non-linear Eqs. (5) are given by:

$$\frac{\partial z}{\partial t}+\frac{\partial M}{\partial x}+\frac{\partial N}{\partial y}=0$$

$$\frac{\partial M}{\partial t}+\frac{\partial}{\partial x}(\frac{M^2}{D})+\frac{\partial}{\partial y}(\frac{MN}{D})+gD\frac{\partial z}{\partial x}+\frac{\tau_x}{\rho}=0 \quad (5)$$

$$\frac{\partial M}{\partial t}+\frac{\partial}{\partial x}(\frac{MN}{D})+\frac{\partial}{\partial y}(\frac{N^2}{D})+gD\frac{\partial z}{\partial y}+\frac{\tau_y}{\rho}=0$$

In both models where $z$ is the water height, $t$ is time, $x$ and $y$ are the horizontal coordinates, $M$ and $N$ are the discharge fluxes in the horizontal plane along $x$ and $y$ coordinates, $h(x, y)$ is the undisturbed basin depth, $D = h(x, y) + \eta$ is the total water depth, $\rho$ is density of water, $g$ is the gravity acceleration and $f$ is the bottom friction coefficient. $\tau_x$ and $\tau_y$ are the tangential shear stresses in either $x$ or $y$ direction. In the nonlinear model, the effect of friction on tsunami wave propagation is considered. The bottom friction is generally expressed as follows (Goto et al., 1997):

$$\frac{\tau_x}{\rho}=\frac{1}{2g}\frac{f}{D^2}M\sqrt{M^2+N^2}$$

$$\frac{\tau_y}{\rho}=\frac{1}{2g}\frac{f}{D^2}N\sqrt{M^2+N^2} \quad (6)$$

We will not get into the detailed discussion of function $f$ here. Rather, we will use the Manning roughness $n$, which is a familiar term to civil engineers. The friction coefficient f and Manning's roughness $n$ are related by $n=\sqrt{\frac{fD^{\frac{1}{3}}}{2g}}$. This relationship holds true when the value of total depth $D$ is small. Under this condition, $f$ becomes rather large and makes $n$ nearly a constant value. Thus, the bottom friction terms are expressed by:

$$\frac{\tau_x}{\rho}=\frac{n^2}{D^{\frac{7}{3}}}M\sqrt{M^2+N^2}$$

$$\frac{\tau_y}{\rho}=\frac{n^2}{D^{\frac{7}{3}}}N\sqrt{M^2+N^2} \quad (7)$$

Throughout this model, the expression of bottom friction in Eq. (7) is being used. *n* depends the condition of the bottom surface. We will make a detailed comparison of the linear and nonlinear models in different geographical conditions of China sea region and different Manning constants in the non-linear model in Sect. 4.3.

These tsunami codes ensure the numerical stability of linear and non-linear shallow water wave equations with centered spatial and leapfrog time

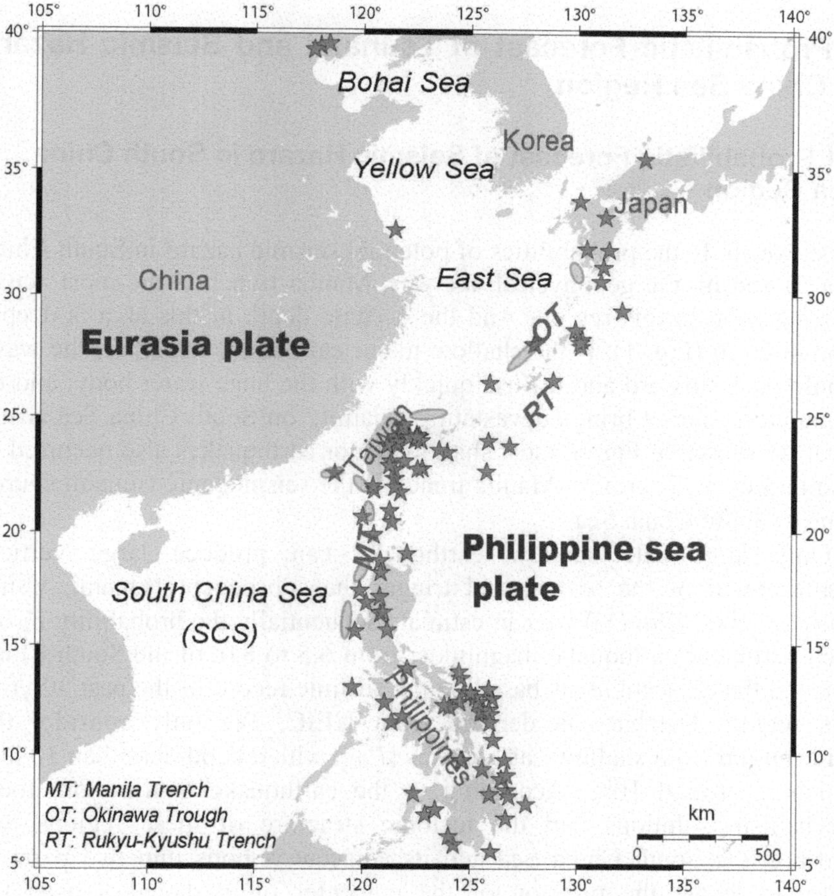


**Fig. 6** The Shallow Major Earthquake Distribution in China Sea and Adjacent Area. The magnitude is ≥6.0 and epicenter depth is ≤30 km. Data derived from Harvard CMT solutions (1976–2006). *Ellipse signs* are the potential tsunamogenic seismic sources in tsunami simulations

difference (Goto et al., 1997). The computational stability depends on the relationship between the time step and spatial grid-size. Furthermore, the computational stability is also constrained by the physical process in the models considered. Here, the numerical simulation in the models must satisfy CFL criterion, that is $\Delta s / \Delta t > \sqrt{gh}$, where $\Delta s$ is spatial grid size, $\Delta t$ is time step, $g$ is the acceleration of gravity, and $h$ is water depth. We use open boundary condition in these models which permits free outward passage of the wave at the open sea boundaries.

## 4 Probabilistic Forecast of Tsunami and Seismic Hazard in China Sea Region

### 4.1 Probabilistic Forecast of Seismic Hazard in South China Sea Region

First, we study the probabilities of potential seismic hazard in South China Sea. Based on the geophysical analysis, Manila trench is the most active seismic zone in this region. And the oceanic depth in this area is deeper than 4000 m (Fig. 1). If the shallow major earthquake occurred, the wave would push forward and diffuse quickly with the huge water body, and its cumulative energy brings devastating calamity on South China Sea costal area. As shown in Fig. 6, most shallow major earthquakes also occurred in Manila trench. Therefore Manila trench is the seismogenic tsunami source zone in South China Sea.

Only large-scale shallow earthquake can produce large vertical displacement of the seabed and trigger the subsequent tsunami. Using *PFSH* method (Eq. (1)) we can estimate sequentially the probability $P_{e,i}$ of each particular earthquake magnitude, from 6.5 to 8.0, of the South China Sea and the adjacent areas based on the seismic record of the past 30 yr in this region. Database is derived from NEIC. We only consider the contribution from shallow earthquakes ($P_{sh,i}$) with a depth less than 10 km, which is around 10%. According to the earthquake distribution, focal mechanism solutions and the tectonic structure of these regions, we partition the South China Sea and its adjacent regions into two parts in order to locate the position of the epicenter of earthquakes used for tsunami modeling. In our tsunami simulation, we set 5 potential tsunamogenic earthquakes, marked by ellipses in Fig. 6. The magnitudes and numbers of historical earthquakes (Fig. 2) in each part all satisfy the local statistical distribution of the GR relationship (Eq. (8)).

$$\begin{aligned} &\mathrm{M} = 6.29 + 0.89 \log \mathrm{N} \quad \text{Latitude}: (19^\circ - 23^\circ)\mathrm{N} \\ &\mathrm{M} = 7.11 + 0.98 \log \mathrm{N} \quad \text{Latitude}: (12^\circ - 19^\circ)\mathrm{N} \end{aligned} \tag{8}$$

Because we cannot accurately decide the epicenter for tsunami modeling, we must take account that every fault takes a portion of each seismic zone. This probability is expressed as $P_{f,i}$. The fault size or the surface rupture length is linearly related to earthquake magnitude distribution, as described in Eq. (9) (Wells and Coppersmith (1994); Bonilla et al. (1984)).

$$\begin{aligned} M_w &= 4.33 + 0.90 \times \log(LW) \quad \text{Reversefaults} \\ M_w &= 3.93 + 1.02 \times \log(LW) \quad \text{Normalfaults} \\ M_w &= 5.08 + 1.16 \times \log(L) \\ M_w &= 4.07 + 0.98 \times \log(RA) \end{aligned} \tag{9}$$

where $L$ is the rupture length. $RA$ is the area of the faulting, and $W$ is the width of rupture. The main rupture along the Manila trench is reverse faulting; on contrast, that along the Okinawa trough is normal faulting with the focal mechanism resolution of HCMT. To estimate the probabilities of potential seismic hazards in the eastern China Sea area, we take the same approach that we have described above.

## 4.2 Probabilistic Forecast of Seismic Hazard in Eastern China Sea Region

The eastern China sea area is divided into Bohai, Yellow Sea, and East Sea (Fig. 1). Only 9 earthquakes have been documented with magnitudes bigger than 6.0 in recent 1000 yr in Bohai Sea area (Gao and Min, 1994). Although there also have been some earthquakes with magnitude big than 7.0 in the last 30 yr in this region, the ruptures are mainly slide-slip strike (Huan, 1989). This kind of faulting can not generate the large vertical seabed displacement. On the other hand, as is shown by the visualized figures (Figs. 1, 8), huge waves cannot be produced in very shallow depths, only 20 m deep. For these reasons we can infer that destructive tsunami waves cannot be developed in this area. In other words, for tsunami waves generated from the Kyushu island region, the wave height along the coast would be very small because of strong bottom friction and energy dispersion. There also have been no historically destructive tsunamis found in Yellow Sea and East Sea. For these same reasons, this is due to upper shallow water layer and infrequent major earthquakes. But the Yellow sea and East Sea are vulnerable due to the existence of the productive seismic source zone along the Ryukyu-Kyushu arc (Fig. 1), which we previously elaborated in the geophysical background section.

The tsunamis in Ryukyu-Kyushu trench should not impact the Chinese coast because the Ryukyu arc serves as a blocker of the incoming waves.

Only the tsunamis coming from earthquakes in the Okinawa trough can possibly influence the Chinese coast. In fact, the tsunami sources recorded in ancient Chinese books that affect the eastern Chinese coast are mainly located on the northern and southern parts of Okinawa (Wang and Zhang, 2005). Although there also exists major shallow earthquakes in the oceans near eastern Taiwan island, the slope of shore is very steep, the wave is mainly reflected back (Grilli et al. (1997); Grilli and Svendsen (1990); Jensen et al. (2003)). In addition, the earthquake sources in this area are mostly generated from Ryukyu trench. There is not a large body of water in this region. Thus, it cannot induce large tsunami wave. Consequently, we only consider the potential seismiogenic tsunamis that are produced from Okinawa trough and can influence the eastern Chinese coast area.

Here we place three potential seismogenic tsunami sources that can seriously impact the eastern Chinese coast: south-western ocean of Kyushu island, Okinawa island area, and north-eastern Taiwan island ocean (the epllise areas in Fig. 6). From the focal mechanism results (Fig. 3), the main rupture of Okinawa trough is normal faulting. The GR relationship between magnitude and number of earthquake in this region is $M = 7.9717 - 0.98 \log_{10} N$. the probabilities of different magnitudes $P_{e,i}$, 7.0, 7.5, 8.0, and 8.5, respectably are 15%, 4.5%, 1.5%, and 0.48% respectively, which are derived by GR relationship. The probability of shallow earthquake is 11.37%. We also consider the probabilities of $P_{f,i}$ of different magnitudes by the same method as that applied to South China Sea region. The length of fault with specific magnitude can be obtained from Eq. (9).

## 4.3 Tsunami Numerical Simulation in China Sea Region

To estimate the near-field tsunami potential hazard in China Sea, two different types of shallow water equations are employed in different ocean parts with different natural geographical conditions. The linear shallow water equation (Eq. (4)) is applied to describe tsunami generation and subsequent wave propagation in the South China Sea regions, and the nonlinear model (Eq. (5)) for eastern China Sea area (Liu et al., 2008).

We need to find out what hydraulic theory should be applied to South China Sea and earthen China sea region. Therefore, the linear and nonlinear shallow water equations in describing tsunami generation and propagation are compared firstly both in South China Sea and eastern China sea regions. The South China Sea is an ideal setting for testing the linear and nonlinear properties of tsunami waves because of its great range in seafloor depth. The sea in this area varies from 7000 m to around 10 m deep. Since over three-fourth of this area is deeper than 500 m, the shallow water region is relatively narrow. Both linear and nonlinear models are simulated in South China Sea area. The bathymetry of the South China Sea was obtained from the Smith and Sandwell global seafloor topography (Etopo2) with grid resolution of near 3.7 km. The total number of grid points in the computational domain is 361, 201, or 601 × 601 points. The time step, $\Delta t$, in both models is selected to be 1.0 sec to satisfy the CFL temporal stability condition. In our simulation, since the bottom friction coefficient is larger than *0*, we have $D = h + \eta > 0$. $D$ is total water depth, $h$ is water depth, and $z$ is wave height. This means that shallow water wave equations can maintain computational stability only within a computational domain within the fluid computational domain (Wang, 1996). Since the wave height of tsunami wave is only a few meters in the propagation process, we have set the smallest computational depth as the order of ten meters along coast area in both linear and nonlinear models. In this way, the entire computation domain can satisfy this condition.

In linear shallow water theory we ignore the bottom friction (Eq. (4)). On contrary, for comparison between the linear and nonlinear models, we perform our nonlinear simulations under the condition of $n = 0.025$ recommended by F. Imamura, as the bottom friction coefficient in this application (Goto et al., 1997). The value $n = 0.025$ is suitable for the natural channels in good condition which is valid for the China Sea regions. We have visualized in Fig. 7 one set of simulated tsunami wave propagations with the linear and nonlinear model comparison. This hypothesized seismic tsunami occurs southwest to Philippines (14.5_*N*, 119.2_*E*), with a magnitude of 8.0. Tsunami occurring at this location lies the furthest to the coast of mainland China. Due to the longest propagation time and very strong wave energy of this tsunami, strong oscillations and reflection and interference characteristics of tsunami waves can be well observed near the islands. Wave diffraction is also observed among Taiwan, Philippines, and the small islands of south Philippines. In the simulation, tsunami waves are well absorbed along the open boundaries near Taiwan Island. No abnormal computational values have been observed. Notably, the wave front propagates forward steadily in our computation. Normally, the wave fronts in the shore do not disperse for numerical simulation reason in nonlinear model. The overall analysis from various locations indicates that the numerical simulations of both our linear and nonlinear models yield stable and reliable solutions.

In order to confirm the type of tsunami model which is needed to be applied to eastern China sea region, we also model the linear and nonlinear shallow water theory in this area. The situation of geography in eastern China sea region is extraordinarily different from that in South China Sea. The average depth in this area is less than 300 m, as illustrating in Fig. 1.

The bathymetry of the eastern China sea region was obtained from global seafloor topography (Etopo1) with grid resolution of near 1.8 km. The total number of grid points in the computational domain is 1, 442, 401, which is 1201 × 1201 points. The time steps in both models are also selected to be 1.0 sec to suit the temporal stability condition. The total wave propagation time is 12.0 h. Here we showed the comparison of tsunami propagation process in the China sea region with linear and nonlinear shallow water equations. We compare the wave characters of linear and nonlinear results for two regions. Figure 7 shows tsunami propagation in South China Sea area (see Movie 1 SCSlinear.mpg, available on accompanying DVD) and Fig. 8 shows that in the eastern China sea region (see Movie 2 Eastchinaseanonlin.mov, available on accompanying DVD). The waveforms are predicted to be the same in the deep ocean; on the other hand, they are different in the shallow ocean with the friction and convection terms not negligible. In South China Sea area, the wave appearances predicted by linear and nonlinear models are only different near the shore (Fig. 7). However, the wave height is small when the tsunami arrives at the shore with the energy dispersion, and our simulation does not include the run-up part. Thus, the wave heights of two models in the coastal area of South China Sea should not change much. Therefore, we can employ the linear model to predict tsunami hazard in this region. In contrast, in eastern China sea area the ocean depth of most part is extra shallow, the wave contours are very different in the entire propagation (Fig. 8). Hence, the nonlinear model must be applied for the eastern China sea region.

In order to validate further the wave propagation process in linear and nonlinear models, we placed receivers in different water depths both South China Sea and eastern China sea regions (Figs. 9, 10, 11). As illustrated in these figures, the features of first arriving waveforms in linear and nonlinear models are influenced by the topographical condition of entire computational domain. The first arriving waveforms of linear and nonlinear models are similar in South China Sea region. Because the average depth of South China is around 3000 m and water depth shallows 500 m is very narrow. The bottom friction influences less in nonlinear models in entire wave propagation process. This confirms that we can apply linear theory to a good accuracy for the South China Sea. On the other hand, the first arriving waveforms of linear and nonlinear models are very different both in arriving times and wave heights. Because most

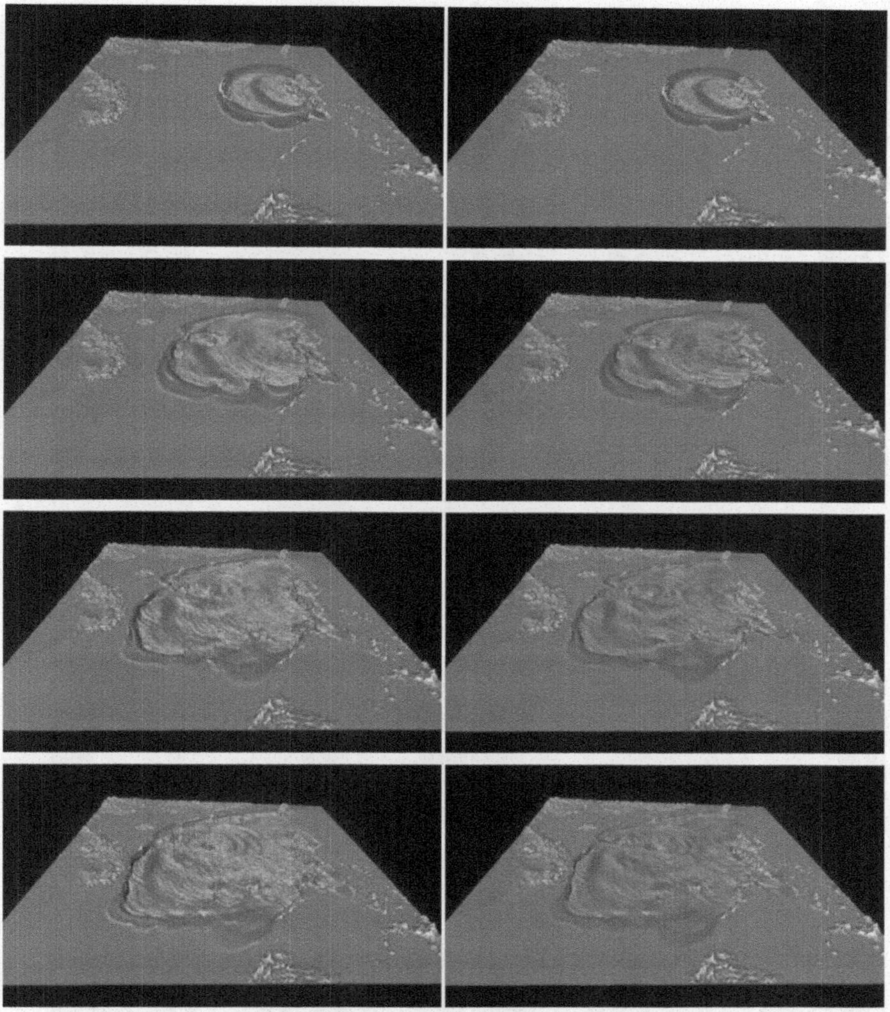

**Fig. 7** The tsunami propagation, comparing linear and nonlinear shallow water model in South China Sea (The *left* is from nonlinear model, the *right* is from linear model)

water depth of eastern Chinese area is less than 300 m, except that it is deeper than 1500 m in Okinawa trough. The sea-bottom friction and convection terms in nonlinear models practically act as main factor that make the first arriving wave very different in linear and nonlinear models with shallow water depth in eastern China sea region. The bottom friction makes the first arriving wave of nonlinear model lag that of linear model in shallow ocean area. The convection term induced by the wave height of first wave of nonlinear models is higher than that of linear models. At the

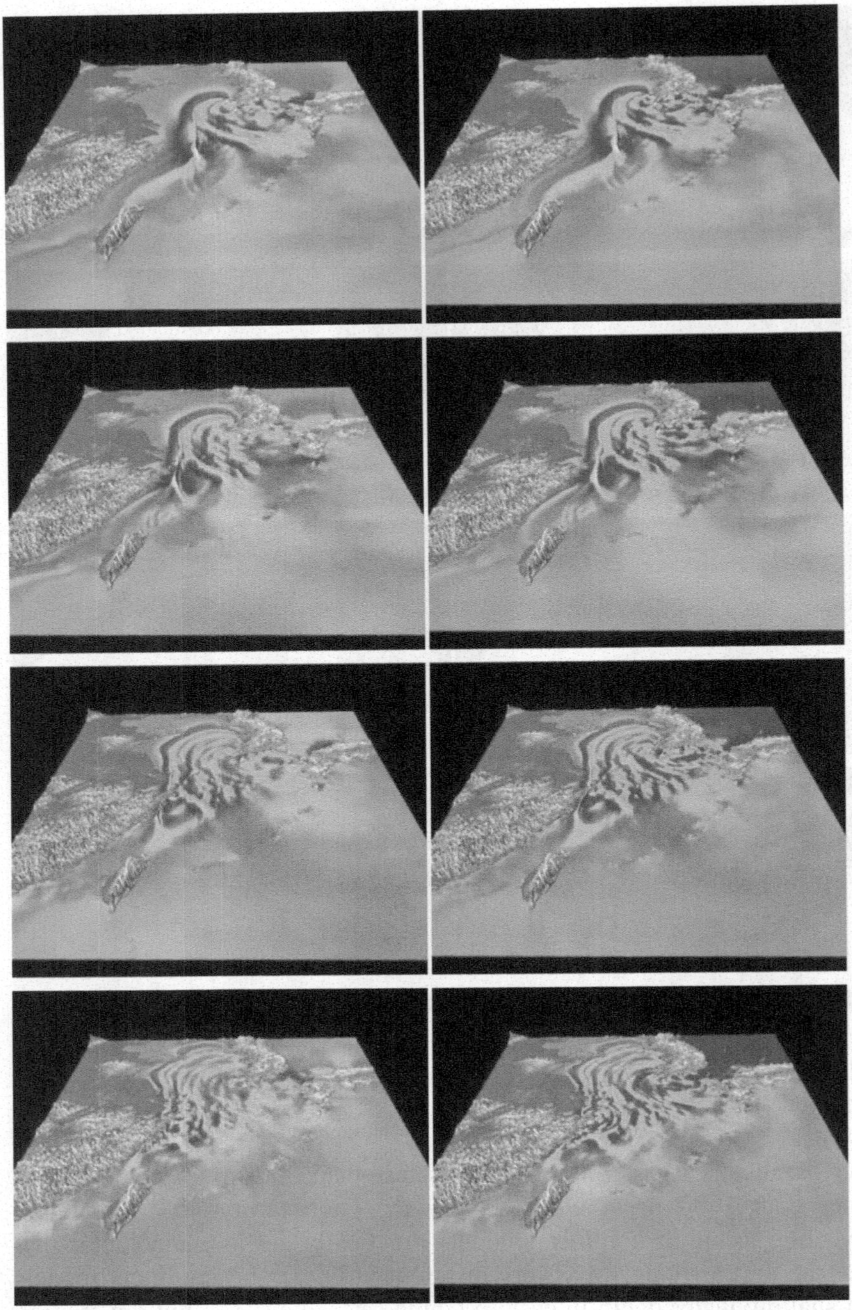

**Fig. 8** Visualization for comparison of linear and nonlinear shallow water model in eastern China sea region (The *left* is from nonlinear model, the *right* is from linear model)

same time, the waveforms of two models in deep ocean area reconfirm that the results of linear and nonlinear models are similar in South China Sea region. According to computational results of linear and nonlinear models in two areas, we must take care of nonlinear terms in shallow ocean computational area of tsunami simulation, such as eastern China Sea region.

From the figures, we also found there exist a critical region between 400 and 500 m depth (Liu et al., 2008). Above this depth, both linear and nonlinear models generate similar wave shapes and wave magnitudes. In other words, wave propagation can be modeled by linear theory with reasonable accuracy. Otherwise, the non-linear model is necessary. We also discovered the effect of friction values seriously influences on the prediction of tsunami wave heights in the case of the nonlinear modeling (Liu et al., 2008). In our study, the difference of calculation times based on two models is huge. The time of computing nonlinear model is around 5 times of that of linear model. Therefore, the suitable model for different areas is very important for issuing timely warning and accurate results in tsunami warning system.

Based on previous analysis, we can predict the potential tsunami hazard with the linear shallow water equation (Eq. (4)) in South China Sea with its natural bathymetric condition (Fig. 1). However, we have to employ the nonlinear model (Eq. (5)) in the eastern China sea region, for which the bottom frictional effect must be considered.

## 4.4 Probabilistic Forecast of Tsunami Hazard in China Sea Region

We forecast the probabilities of potential tsunami hazard along Chinese coast area in this century based on linear and nonlinear shallow water equation simulations in South China Sea and eastern China sea separably. Our $P_{PFTH}$ computation is based on the maximum wave height of all seismic tsunami together with the occurrence probability of each synthesized tsunami, the same as PFSH (Eq. (1)). Two tsunami wave height regions of [1.0 m, 2.0 m] and heights over 2.0 m are considered for our tsunami hazard prediction. Two meters in propagation could be enlarged a few times, up to ten times, based on local ocean topographical conditions after run-up process computation. Therefore, wave height over 2 m could cause economic hazard after run-up for China coastal area because continental altitude of major Chinese cities are only couple meters over the sea. In fact, the wave height of Chinese historical tsunami records is

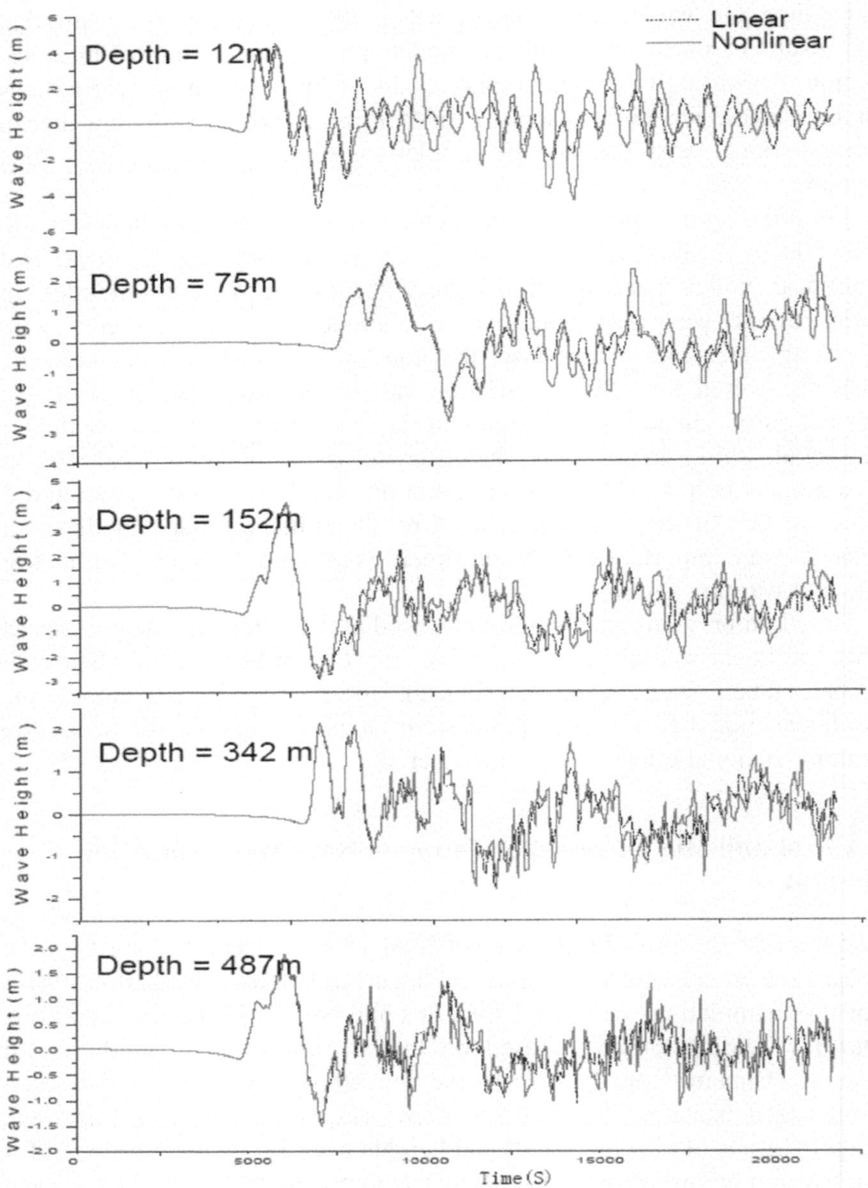


**Fig. 9** Comparison of water heights in time-series with linear and nonlinear models for various water depths in South China Sea region. The epicenter is located at in Manila trench (Adapted from Liu et al. (2008))

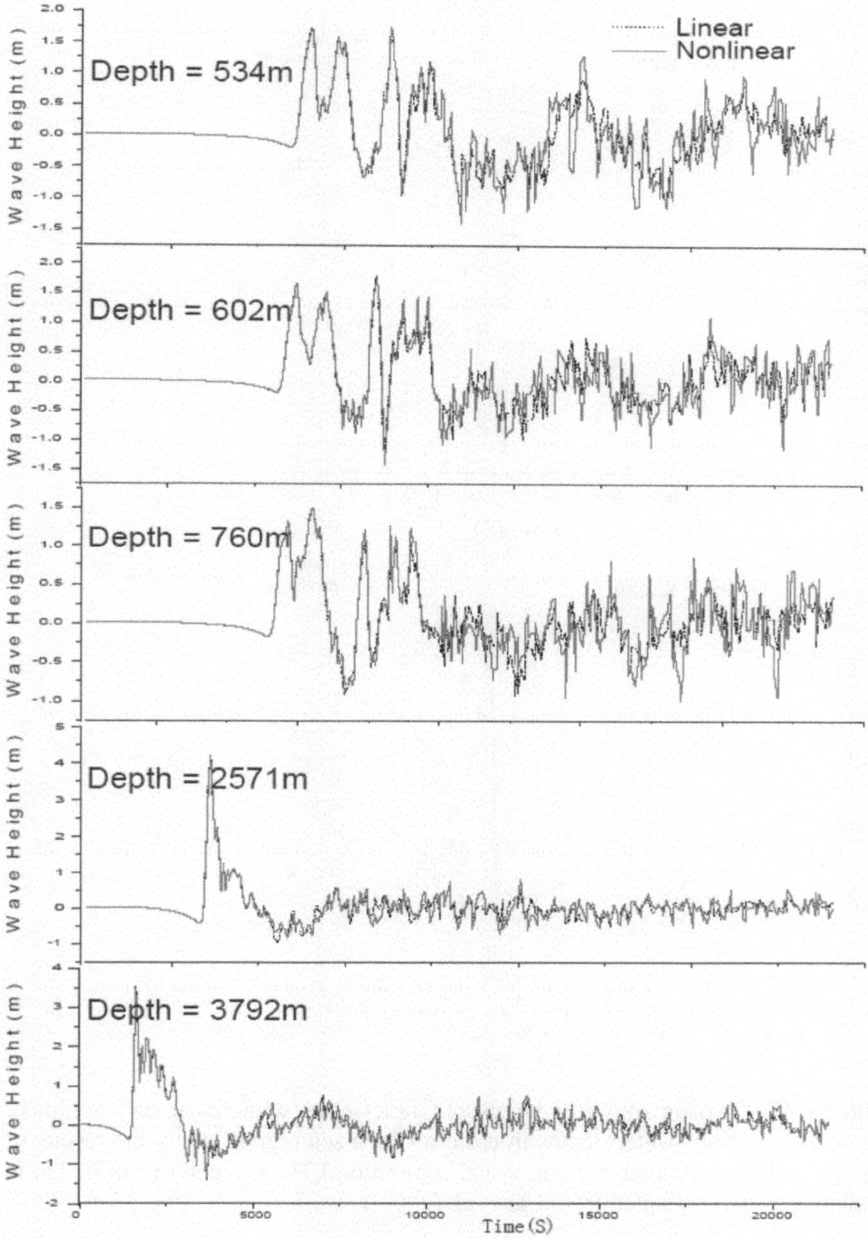


**Fig. 10** Comparison of water heights in time-series with linear and nonlinear models for various water depths in South China Sea region (Adapted from Liu et al. (2008))

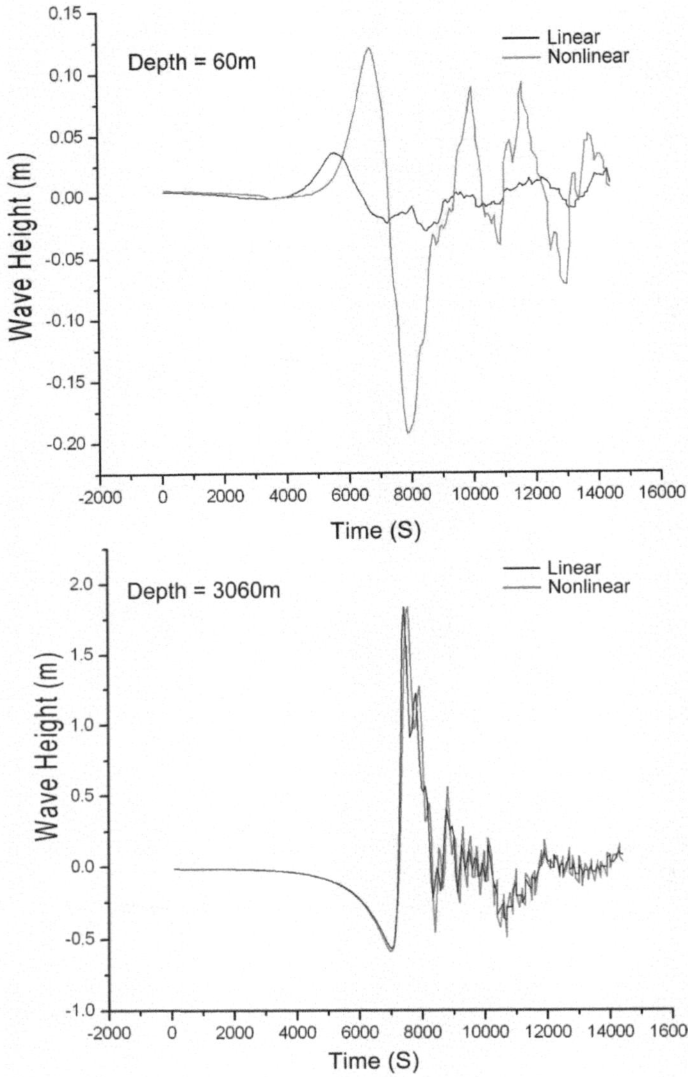


**Fig. 11** Comparison of water heights in time-series with linear and nonlinear models for various water depths in eastern China sea region. (Only the results of the first 4.1 h are plotted to avoid visual congestion.) The epicenter is located at in Okinawa trough (Adapted from Liu et al. (2008))

between half a meter to 7 m. Most of Chinese tsunamis are about 1–2 m. We consider both economic and scientific factors for wave scales for tsunami simulation. The purpose of this paper is to provide some suggestions for applications of realistic tsunami simulation and tsunami warning with linear and nonlinear shallow water equations. The critical wave height is chosen to be 2.0 m for illustrative purpose.

For local tsunamis hazard analysis, we examine a scenario in which the interpolate thrust along the Manila subduction zone and the north-western part faults of the South China Sea (Fig. 3), and Okinawa trough based on our geological and geophysical analysis in Sect. 2. To estimate precisely the probability of tsunami hazards around the China Sea, we have computed 13 seismic tsunami models with magnitudes ranging from 6.5 to 8.0 in five hypothetical epicenters (Fig. 3) in South China Sea; and 10 seismogenic tsunami nonlinear models for earthquake magnitudes ranging from 7.0 to 8.5 in three putative epicenters of Okinawa trough in eastern China sea region. Altogether 162 coastal receiver points are placed along the mainland coast of China Sea, Hainan, and Taiwan Island. The maximum absolute value of wave amplitudes are recorded down through out entire wave traveling time. According to $P_{PFTH}$ (Eq. (3)), we obtain the probabilities of tsunamis hazard of different maximum wave height by combining the total occurred probabilities from each tsunami sources.

Along the south-eastern coast of mainland and Southwestern Taiwan, in Fig. 12, we show the distribution of tsunami wave with greater than 2.0 m height hitting the coast. Shown in Fig. 13, this same place, together with southeastern Taiwan, also has a potential chance of being assaulted with tsunami waves with heights of 1.0–2.0 m. The cities of Shantou, Xiamen and Hong Kong are under direct impact from the tsunami earthquakes originated from the central basin of the South China Sea, the southwest and northwest of the Philippines (Fig. 12). The tsunami-risk probabilities of these three cities are high, around 10%. Tainan, Kaohsiung, and Nanwan are three major cities of Taiwan Island directly affected by the tsunami occurred in the western part of the central basin of the South China Sea and the north Manila trench. However, the historical seismic records show that the earthquake with a magnitude higher than 7.0 are rare in the oceanic area close to Taiwan, the probability of tsunami wave height higher than 2.0 m in this century (3.4%) in these regions are lower than that in the Hong Kong region (10%). The tsunami hazard probability along the Coast of the South China Sea is plotted in Figs. 12 and 13. As shown in these figures, from Shantou to Hong Kong, Macau, the southwest portion of Taiwan Island, the east Hainan Island are all in harm's way from tsunamis with wave heights more than 2.0 m tall.

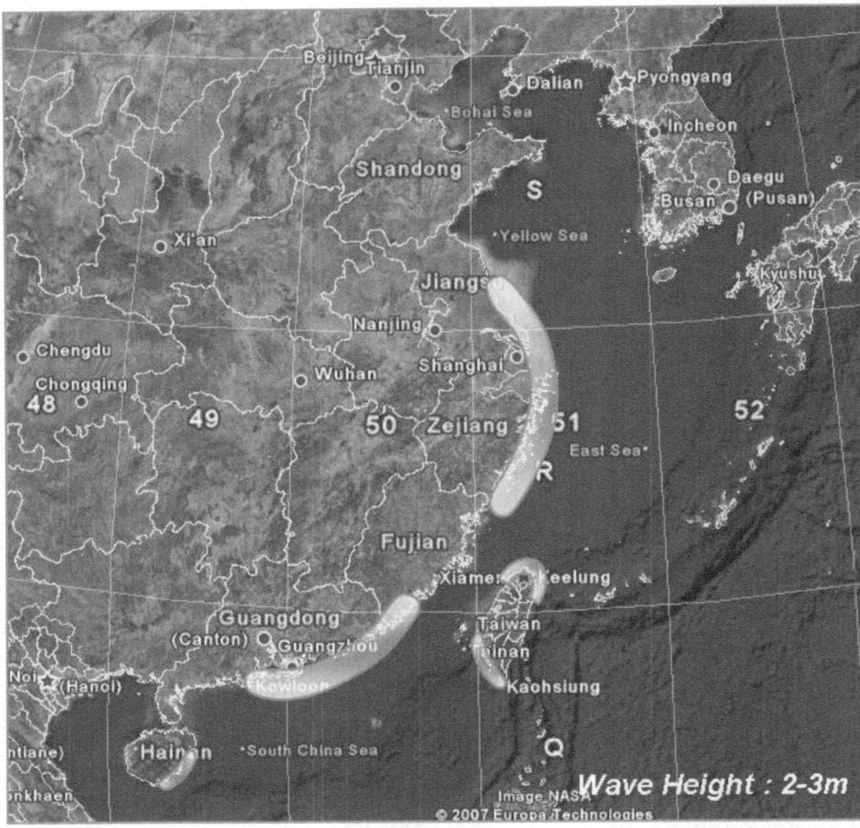


**Fig. 12** The spatial distribution of potential tsunami wave height impinging on the Chinese coast in this century. *Light grey zones* represent the case where the wave height is between 2.0 and 3.0 m.

The computed results of tsunami hazard probability with a 2-m wave height in this century for major cities along the coast of mainland China are shown (Fig. 12): 0.5% for Shanghai, 3.2% for Wenzhou, and 7.2% for Keelung. Additionally, the probability of tsunami wave height higher 1.0–2.0 m in Shanghai coastal area in this century is 7.2%, that with a smaller wave height 0.5–1.0 m is 13.2%. The Shanghai and Zhejiang (Fig. 12) coastal area and the delta region of the Yangzi River, are directly impacted by the tsunamis occurring in the southern Kyushu island, middle of Okinawa trough, and northern Taiwan island oceanic region, around 300 m depth, induces the lower probabilities of huge tsunami hazard in this area. In fact, the maximum historical documented tsunami heights in this area are around 1.0 m. However, Shanghai today is an extremely important

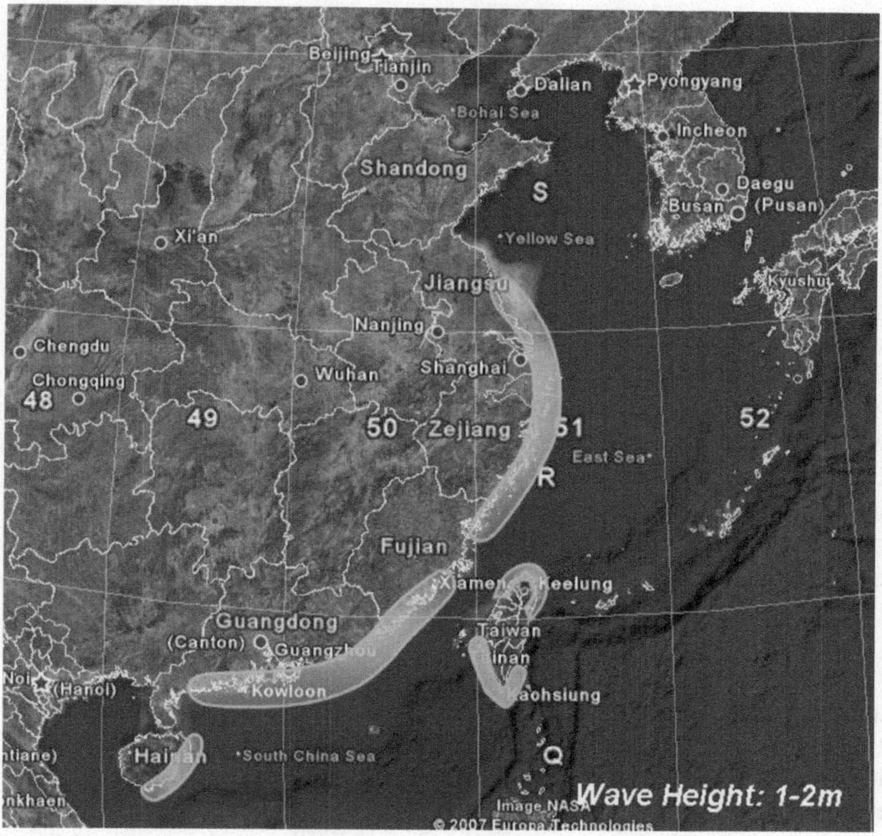


**Fig. 13** The spatial distribution of potential tsunami wave height impinging on the Chinese coast in this century. *Light grey zones* describe the wave height lies between 1.0 and 2.0 m

Chinese economic city, with a ultra lower terrain. Thus, disasters even from small waves would cause great economic losses. In historical records, Keelung is well-known for its tsunami related disaster. It is bordering Okinawa trough and Ryukyu-Kyushu trench. It is influenced by tsunamis originating in both of these regions. Due to the proximity to the tsunami sources, Keelung will be affected by destructive tsunami hazard because the waves would not be dissipated so easily. The tsunami wave also could quickly arrive at this place because of the Okinawa trough deep channel. Therefore, the potential probability of large destructive tsunami wave height to hit Keelung is greater than other cities along the eastern Chinese coast.

## 5 Discussions and Summary

In this section, we discuss first the validity of the linear model. We will also clarify the relationship of our simplified simulation resulting from a single earthquake and the probability of disaster occurrence in complex realistic scenarios involving lengthy rupture with long duration, as in the case of the Sumatran earthquake. In addition, we will discuss the effects of specific characteristics of various regions concerning the tsunami hazards.

Our simulations consider the result from the coseismic generation of the wave by a single shock. Thus a single large wave in the epicentral zone is selected as the initial input for the wave simulation. In reality, the rupture process of tsunamogenic earthquake is not so simple. The actual generated waves do not come from waves due to one single shock. During the earthquake occurrence, waves propagate outwards in the form of wavelets by continuous rupture dynamics. The constructive combination of the original waves and the following waves of a tsunami could pose even more dangerous situations. Based on these considerations, our modeling provides an estimate of the lower-bound of estimate of the tsunami hazard possibilities.

The geological and geophysical backgrounds of China sea and its adjacent areas are extremely complex. Much effort has been devoted to the research of this region. There are also an abundance of the literature of historical seismic records of this region. In Chinese historical tsunami record, the coastal areas of Shanghai, Jiangsu, Zhejiang, and Guangzhou were affected by large wave hazards. Especially the northern and southern tips of Taiwan Island experienced high frequency of tsunami hazard. The potential tsunamis in South China sea could be mainly generated from the subduction zone between Eurasian plate and Philippine sea plate. From the historical seismic distribution in western pacific region, the majority epicenters depths are between 60 and 70 km along the subduction zone (Seno and Eguchi, 1987 ). The largest earthquake happened in the Ryukyu island $M_s = 8.1$ (in 1911). A small portion of the earthquakes has shallow epicenters. Most of them took place in northern Manila trench, south-eastern and northern Okinawa trough. In addition, there exists a seismic gap between the north-western Taiwan Island and middle of Ryukyu Island along Okinawa trough. The Kirby report (Kirby et al., 2005), predicted that a potentially big earthquake, magnitude bigger than 8.0, could occur in the north-eastern Okinawa trough region. Although there is a small chance of tsunami hazard from local China sea region, there are

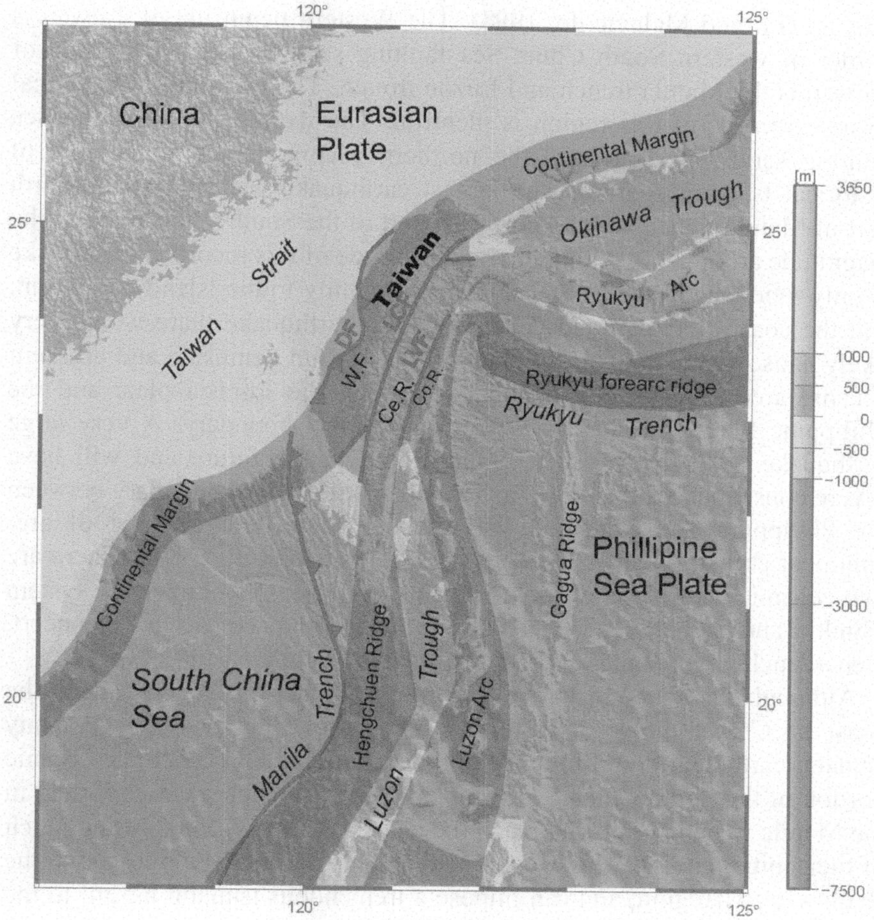


**Fig. 14** Taiwan tectonic map. DF stands for deformation front; LCS for Lishan-Chaochou suture; LVS for Longitudinal Valley suture; WF: Western Foothills; CeR for Central Range; CoR for Coastal Range; HP: Hengchun Peninsula; P is Outcrops of pillow lava along the western suture. Tectonic information is taken from Shyu et al. (Shyu et al., 2005)

much higher probabilities present from northern and southern subduction belts of the western Pacific.

The recent occurrence of the pair of Hengchun earthquakes has alarmed people of the danger from tsunamis coming from Hengchun peninsula, which may affect the Chinese coast. The Hengchun peninsula (Fig. 14) is located at the southern tip of the Taiwan mountain belt. Because of its strategic location near the present-day transition zone between collision and subduction, it is a key area to understand the orogenic evolution of

Taiwan (Lu and Malavieille, 1994). The Western peninsula of Taiwan is border of western South China Sea faulting series; eastern peninsula of Taiwan is the Manila trench and Luzon trough. The literature of historical seismic records of this region is plentiful. The western South China Sea faulting series, the same as the northern Manila trench, is a zone of historical tsunami sources. The largest earthquake occurred at the north part of Manila trench and the northeast part of the South China Sea has the magnitude around 7.0. Although the magnitude of this recorded earthquake is only moderately high, because of its proximity to the island of Taiwan, and the coast of Fujian and Guangdong, an earthquake there would very likely cause a tsunami catastrophe. The Hengchun peninsula and adjacent oceanic area are also the boundary between the Eurasia plate and the Philippine Plate with a belt of plate transitional boundary. A very large earthquake is likely to occur at the region in the future and will have severe consequences. In addition, Taiwan, located at the boundary between the Philippine Sea plate and the Eurasian plate, is a product of arc-continent collision (Chai, 1972; Biq, 1973; Bowin et al., 1978). There are two volcanic belts along Ryukyu-Kyushu arc system and Luzon arc system (Simkin and Siebert, 1994). If volcano and earthquake occur in concert, then a much larger tsunami disastrous scenario would ensue.

Although the southern part of the Manila trench is far away from the coast of China, the local historical records of this region have many tsunami earthquakes up to the magnitude of around 8.0. Since the oceanic portion of the South China Sea is mostly deep, tsunami wave generated in the Manila trench region can reach the coast of China without losing much of their initial energy. The wave kinetic energy can then be released in the shallow water region, and can impose a tremendous tsunami hazard to the coastal regions.

## 6 Conclusion

We have employed different shallow water equation models for evaluating the potential tsunami hazard in South China Sea and the eastern China sea regions by using realistic bathymetry conditions. Potential 2.0 m tsunami wave could hit Guangzhou, Fujian, Hainan, Zhejiang, Jiangsu, and Southwestern and northeastern tips of Taiwan Island. Shanghai only has a one twentieth times probability for facing 2 m wave of that for Hong Kong, since shallow water in eastern China region causes strong sea-bottom friction. Keelung almost has the same probability as that for Hong Kong. These probabilities of tsunami hazard are the same as that of big seismic hazard, which occurs once every 1000 yr. Chinese tsunami hazard is a potential threat with long recurrence time.

The computing time of nonlinear model is much longer than that of linear model. In addition, the computer requirement, such as the memory, used for nonlinear model is much higher than that for linear one. If we only simply estimate the travel time of the tsunami wave in local tsunami situations, linear model is enough, but the computing time is slightly shorter than real time. In far field tsunami simulation we could consider combining different models, different space steps, and different time steps. We can apply linear model to far field with a large spatial grid and time step when rough result required. On the other hand, we can apply the nonlinear model to the shore with small spatial grid and small time steps for obtaining the required accurate results.

The recent series of large earthquakes around the Sumatran region have raised further concerns. An outstanding question for the near future is the potential scenario for the next Sumatra mega event. Tsunami earthquakes are always generated along subduction zones. The Manila trench has already been identified as an extremely high risk zone, because of the Eurasian plate actively subducting eastward under the Luzon arc on the Philippine plate. Two other subduction zones of lesser risk are also in this neighborhood. Along the Ryukyu trench the Philippine Sea plate subducts northward beneath the Ryukyu Arc on the Eurasian plate, while along the North Sulawesi trench the Pacific-Philippine, Indo-Australian Plates and the Sunda Block collide together. These long subduction zones can also rupture and generate large tsunamis in the future that will have significant impacts on the countries in the South China Sea region, which includes not only China but also Vietnam, Cambodia, Thailand, Malaysia, Singapore, and Indonesia.

**Acknowledgements** We would like to thank Professor Fumihiko Imamura for his providing computational codes TUNAMI N1 and instructing nonlinear code programming, and his kind guidance on tsunami numerical method. We would also like to thank Professor How-Wei Chen at National Central University in Taiwan for his comments and suggestions. We would also like to thank Jessica Schmidt for her help to revise paper to print format. This research is supported by National Natural Science Foundation of China (NSFC-40574021, 40676039) and the EAR and ITR programs of the U.S. National Science Foundation.

## References

Annaka, T., Satake, K., Sakakiyama, T., Yanagisawa, K., Shuto, N., 2004. Logic-tree approach for probabilistic Tsunami hazard analysis and its applications to the Japanese coasts. Pure Applied Geophysics PAGEOPH 164 (2–3), 577–592.

Bao, L., 1991. Tsunami Disaster and Its Warning. Marine Press, Beijing, China.

Becker, M., Reinhart, E., Nordin, S. B., Angermann, D., Michel, G., Reigber, C., 2000. Improving the velocity field in South and South-East Asia: the third round of GEODYSSEA. Earth Planets and Space 52, 721–726.

Biq, C., 1973. Kinematic pattern of Taiwan as an example of actual continentarc collision. US-ROC Cooperative Science Program 25, 149–166.

Bird, P., 2003. An updated digital model of plate boundaries. Geochemistry Geophysics Geosystems 4 (3).

Bonilla, M., Mark, R., Lienkaemper, J., 1984. Statistical relations among earthquake magnitude, surface rupture, and surface fault displacement. Bulletin of the Seismological Society of America 69, 2003–2024.

Bowin, C., Lu, R., Lee, C., Schouten, H., 1978. Plate convergence and accretion in Taiwan-Luzon region. American Association of Petroleum Geologists Bulletin 62, 1643–1672.

Chai, B., 1972. Structure and tectonic evolution of Taiwan. American Journal of Science 272, 389–422.

Chen, Y., Chen, Q-F., Zhang, W., 2007. Tsunami disaster in china. Tsunami Disaster in China 16(2), 1–6.

Christova, C., 2004. Stress field in the RyukyuKyushu WadatiBenioff zone by inversion of earthquake focal mechanisms. Journal of Physics of the Earth 384 (1–4).

Comer, R. P., 1984. Tsunami generation: a comparison of traditional and normal mode approaches. Geophysical Journal International 77 (1), 29–41.

Cornell, C. A., 1968. Engineering seismic risk analysis. Bulletin of the Seismological Society of America 58 (5), 1583–1605.

Gao, H., Min, Q., 1994. The possibility of the earthquakegenic tsunami occurred in Bohai sea. Marine Forecasts 11 (1), 63–66.

Geist, E. L., Parsons, T., 2006. Probabilistic analysis of tsunami hazards. Natural Hazards 37, 277–314.

Goto, C., Ogawa, Y., Shuto, N., Imamura, N., 1997. Numerical method of tsunami simulation with the leap-frog scheme (IUGG/IOC Time Project. IOC Manual, UNESCO 35.

Grilli, S., Svendsen, I., Subramanya, R., 1997. Breaking criterion and characteristics for solitary waves on slopes. Journal of Waterway, Port, Coastal, and Ocean Engineering 123 (3), 102–112.

Grilli, S., Svendsen, I. A., 1990. Computation of nonlinear wave kinematics during propagation and run-up on a slope. Journal of Water Wave Kinematics. NATO ASI Series E. 78, 387–412.

Gutenberg, B., Richter, C. F., 1949. Seismicity of the Earth and Associated Phenomena. Princeton Univ. Press, Princeton.

Huan, W., 1989. Seismic activity in Bohai sea region. Journal of Seismological Research 12 (1), 1–9.

Jensen, A., Penersen, G. K., Wood, D. J., 2003. An experimental study of wave run-up at a steep beach. Journal of Fluid Mechanics 486, 161–188.

Kao, H., Chen, W. P., 1991. Earthquakes along the Ryukyu–Kyushu arc: strain segmentation, lateral compression, and thermomechanical state of the plate interface. Journal of Geophysical Research 96, 21443–21485.

Kirby, S., Geist, E., Lee, W. H., Scholl, D., Blakely, R., October 2005. Tsunami Source Characterization for Western Pacific Subduction Zones: A Perliminary Report. Report, USGS Tsunami Subduction Source Working Group.

Kreemer, C., Holt, W., 2001. A no-net-rotation model of present-day surface motions. Geophysical Research Letters 28, 4407–4410.

Kreemer, C., Holt, W., Haines, A., 2002. The global moment rate distribution within plate boundary zones, in Plate Boundary Zones. Vol. 30 of Geodynamics Series. Washington D.C.

Krinitzsky, E., 1993. Earthquake probability in engineering – Part 2: Earthquake recurrence and limitations of Gutenberg Richter b values for the engineering of critical structures. Engineering Geology 36, 1–52.

Li, Q., Xu, M., 1999. Tsunamis attacking Taiwan and its neighbouring areas. International Seismology 1, 7–11.

Li, S., 1981. Chinese Earthquakes. Seismic Press, Beijing, China.

Liu, Y., Santos, A., Wang, S. S. Y. L., H., Yuen, D., 2007. Tsunami hazards along Chinese coast from potential earthquakes in South China Sea. Physics of the Earth and Planetary Interiors 163, 233–245.

Liu, Y., Shi, Y., Yuen, D. A., Yuan, X., Sevre, E. O. D., Xing, H. L., 2008. comparison of linear and nonlinear shallow water equations applied to tsunami waves over the China Sea. Acta Geotechnica 1861–1125, DOI: 10.1007/S11440-008-0073-0.

Liu, Z., Yang, S., Chen, S., Liu, Y., et al., 1988. South China Sea Geology Tectonic and Continental Margin Extension (in Chinese). Science Press, Beijing.

Lomnitz, C., 1974. Global Tectonics and Earthquake Risk. Elsevier.

Lu, C., Malavieille, J., 1994. Oblique convergence, indentation and rotation tectonics in the Taiwan Mountain belt: insight from experimental modelling. Earth and Planetary Science Letters 121, 477–494.

McAdoo, B. G., Watts, P., 2004. Tsunami hazard from submarine landslides on the Oregon continental slope. Marine Geology 203 (3–4), 235–245.

Nakamura, M., 2004. Crustal deformation in the central and southern Ryukyu Arc estimated from GPS data. Earth Planetary Science Letters 217 (3–4), 389–398.

Okada, Y., 1985. Surface deformation due to shear and tensile faults in a half-space. Bulletin of the Seismological Society of America 75, 1135–1154.

Pelinovsky, E., Talipova, Kurkin, A., Kharif, C., 2001. nonlinear mechanism of tsunami wave generation by atmospheric disturbances. Natural Hazard and Earth Science 1, 243–250.

Polet, J., Kanamori, H., 2000. Shallow subduction zone earthquakes and their tsunamigenic potential. Geophysical Journal International 142 (3), 684–702.

Reiter, L., 1990. Earthquake Hazard Analysis: Issues and Insights. Columbia Univ. Press, New York.

Seno, T., Eguchi, T., 1987. Seismotectonics of the western Pacific region. Journal of Geophysical Research 98, 17941–17948.

Seno, T., Stein, S., Gripp, A., 1993. A model for the motion of the Philippine Sea plate consistent with NUVEL-1 and geological data. Journal of Geophysical Research 98, 17941–17948.

Shiono, K., Mikumo, T., Ishikawa, Y., 1980. Tectonics of the kyushuryukyu arc as evidenced from seismicity and focal mechanisms of shallow to intermediate-depth earthquakes. Journal of Physics of the Earth 28, 17–43.

Shokin, Y., Marchuk, A. G., and Chubarov, L. B., 1979. To the numerical simulation and propagation of tsunami according to the shallow water equations. Sixth International Conference on Numerical Methods in Fluid Dynamics, 487–491.

Shyu, J. B. H., Sieh, K., Chen, Y., Liu, C., 2005. The neotectonic architecture of Taiwan and its implications for future large earthquakes. JGR 110 (B08402), 503–515.

Simkin, T., Siebert, L., 1994. Volcanoes of the World. Geoscience Press, Tuscon, p. 349.

Speidel, D. H., Mattson, P. H., 1997. Problems for probabilistic seismic hazard analysis. Natural Hazards 16 (2–3).

Wang, F., Zhang, Z.-Q., 09 2005. Earthquake tsunami record in Chinese ancient books. Chinese Earthquakes 21 (3).

Wang, J., 1996. Global linear stability of the two-dimensional shallow-water equations: an application of the distributive theorem of roots for polynomials on the unit circle. Monthly Weather Review 124, 1301–1310.

Ward, S., 1982. On tsunami nucleation: an instantaneous modulated line source. Physics of the Earth and Planetary Interiors 27, 273–285.

Wells, D., Coppersmith, K., 1994. New empirical relationships among magnitude, rupture length, rupture area, and surface displacement. Bulletin of the Seismological Society of America 84, 974–1002.

Wong, F. L., Geist, E. L., Venturato, A. J., 2005. Probabilistic Tsunami Hazard Maps and GIS. 2005 ESRI International User Conference, San Diego, California.

Xu, T., 1981. Tsunamis Attacking Taiwan and its Neighbouring Areas. China Meteorological Press, Beijing.

Xu, Z., Zhong, S., 1997. A study on present-day tectonic stress in the Southern Yellow Sea and East China Sea region. Earth and Planetary Science Letters 40 (6).

Yamashita, T., Sato, R., 1976. Correlation of tsunami and sub-oceanic rayleigh wave amplitudes. Journal of Physics of the Earth 24, 397–416.

Yu, S.-B., Kuo, L.-C., Punongbayan, R. S., Ramos, E. G., 1999. GPS observation of crustal deformation in the Taiwan-Luzon region. Geophysical Research Letters 26, 923–926.

Yun, W. S., Huai, X. Z., Xiang, Y. Y., 1997. Inversion for the plate driving forces acting at the boundaries of China and its surroundings. Chinese Journal of Geophysics 40 (1), 17–25.

Zang, S., Ning, J., Xu, L., 1990. Distribution of earthquakes, figuration of the Benioff zone and stress state in the Ryukyu island arc. Acta Seismologica Sinica 3 (2), 137–148.

# Index

The manufacturer's authorised representative in the EU is Springer Nature Customer Service Centre GmbH, Europaplatz 3, 69115 Heidelberg, Germany. If you have any concerns regarding our products, please contact ProductSafety@springernature.com

Printed and bound by CPI Group (UK) Ltd, Croydon, CR0 4YY
25/09/2026
02222891-0010